The Neurobiological Basis
of Memory and Behavior

Hinrich Rahmann   Mathilde Rahmann

The Neurobiological Basis of Memory and Behavior

Translated by Steven J. Freeman

With 193 Illustrations

Springer-Verlag
New York Berlin Heidelberg London Paris
Tokyo Hong Kong Barcelona Budapest

Professor Dr. Hinrich Rahmann
Universität Stuttgart-Hohenheim
Zoologisches Institut
Stuttgart, Germany

Dr. Mathilde Rahmann
Universität Stuttgart-Hohenheim
Zoologisches Institut
Stuttgart, Germany

Translated by
Steven J. Freeman
Department of German and Russian
Drew University
Madison, NJ 07940, USA

Cover illustration: Model of memory formation through molecular facilitation in synapses by gangliosides. This figure appears on p. 259 of the text.

Library of Congress Cataloging-in-Publication Data
Rahmann, Hinrich, 1935–
   [Gedächtnis. English]
   The neurobiological basis of memory / Hinrich Rahmann, Mathilde
Rahmann.
     p.   cm.
   Translation of: Das Gedächtnis.
   Includes bibliographical references and index.
   ISBN 0-387-97545-4 (alk. paper)
   1. Memory. 2. Comparative neurobiology. I. Rahmann, Mathilde.
II. Title.
   [DNLM: 1. Brain—physiology. 2. Memory—physiology. WL 102
R147g]
   QP406.R34 1991
   153.1'2—dc20
   DNLM/DLC
   for Library of Congress   91-4997
                                           CIP

Printed on acid-free paper.

Original German edition: *Das Gedächtnis*, published by J. F. Bergmann Verlag München.

Production managed by Terry Kornak; manufacturing supervised by Robert Paella.
Typeset by Thomson Press India, New Delhi, India.
Printed and bound by Edwards Brothers, Inc., Ann Arbor, MI.
Printed in the United States of America.

9 8 7 6 5 4 3 2 1

ISBN 0-387-97545-4 Springer-Verlag New York Berlin Heidelberg
ISBN 3-540-97545-4 Springer-Verlag Berlin Heidelberg New York

Foreword to the English Edition

The understanding of the way in which the brain functions remains one of the great challenges of biology. In every regard, the brain is the last and greatest of the organ systems, that organ which has changed most in the process of evolution. This book provides an important perspective on the complicated functions of the brain, drawn as it is on the authors' wide experience as comparative neurobiologists.

Those characteristics which we consider to be uniquely human (memory, reasoning ability, emotions) are in fact not unique to man even though they are much more developed in those species with more elaborate central nervous systems. Most of the volume is not about memory, but rather about the building blocks from which memory is made possible. One cannot hope to understand the most complex functions of the nervous system without understanding the building blocks (neurons and glia), their intracellular structure, chemistry, and physiology and how they are organized and function in ensembles.

It is interesting and important that there are few major differences in the properties of single neurons from animals with very different levels of central nervous system development. The differences are rather in the number and complexities of their connections. This is what this book is all about.

By tracing the evolutionary development of the nervous system and detailing the complexities of its organization, the authors accomplish their goal of describing the neurobiological basis of memory. While there is not adequate knowledge to explain the mechanisms of memory, the integration of comparative information on the structure and function of neurons contained in this book does provide a basis for continued investigation of memory and other aspects of higher central nervous system functioning.

David O. Carpenter, M.D.
Dean of the School
  of Public Health
University at Albany,
  State University of
  New York
Albany, New York

Preface

Of all the areas of biological science, there is, perhaps, none that has experienced in recent decades so great an increase in findings as neurobiology, the discipline that concerns memory in all of its myriad aspects. The notion of exploring memory, that capacity to store and recall individual experience, has received attention increasingly in our society. Of course, animals can exhibit astounding powers of memory, but memory is of paramount importance to human beings due to the significant role it plays in the transmission of our cultural traditions. It is tradition, after all, that ensures the passing on of qualities established by lineage, a continuous link from generation to generation, between past and present. And it is tradition that inspires bodies of thought (knowledge and customs, for example) to be handed down by a multiplicity of information bearing devices (i.e., word, writing, picture, electronic data carriers).

The objective of this book is to inform the reader in one clear volume of the groundwork which has been established in memory research from the diverse disciplines of neurobiology. It is intended, primarily, for students of medicine, zoology, biology, psychology and psychiatry, but will certainly prove to be a valuable resource to others with a healthy interest in the area. Building upon the level of a college education with an emphasis in the biological sciences, or a comparable knowledge of biology, the book requires of the reader no specialized knowledge. Rather, it is the intent of the authors that the book serve to introduce the reader to recent scientific findings in the field.

Mankind has sought for ages to explain the phenomenon of memory. Present day science has brought us closer to this goal. Achieving it will be possible only through interdisciplinary research. Accordingly, the authors have attempted to view the present state of memory research from the perspectives of the various research disciplines which comprise neurobiology. Proceeding with an understanding of comparative anatomy, histology and cytology, the basics of neurophysiology, neurochemistry and behavioral research, the authors present the commonly debated hypotheses of memory. These areas of knowledge, enhanced by their own research of the last twenty five years, lead to the presentation of a functional model of memory explained in terms of molecular facilitation in the synapses. This model takes into consideration information from recent studies regarding the extraordinarily pronounced functional-morphological plasticity of the nerve endings (synapses) and the striking adaptive capacity of important metabolic processes of the synapses, in particular, of certain glycolipids, the gangliosides.

The authors are indebted to many colleagues at the Zoologisches Institut at the Universität Stuttgart-Hohenheim for their help in the preparation of sections of the manuscript. Special thanks go to Mrs. D. Freihöfer, Dr. E. Zimmermann, Mr. H. Beitinger, Dr. R. Hilbig, Dr. K.-H. Körtje, Dr. W. Probst and E. Ficker. The authors are especially grateful to H. Poeschel, graphic artist, Mrs. E. Herrmann and Mrs. Th. Predel for the meticulous attention given to the charts and photographic work presented in this work. Similarly, we are extremely grateful to Professor H. Rösner, Dr. H. Strebel, Mr. V. Seybold and Dr. B. Hedwig (Göttingen) for their efforts in preparing some of the photomicrographs.

Hinrich Rahmann
Mathilde Rahmann

Contents

1
The Cellular Basis of Memory

Without question, the nervous system of the more highly developed animals, especially of the vertebrates, represents the most complex level of organization of all living matter. In essence, it is comprised of only two different types of cells, nerve cells (neurons; Waldeyer, 1891) and glial cells (neuroglia). The neurons perform the specific, specialized activity of the nerve tissue, i.e., reception, processing, and transmission of information from cell to cell and, most importantly, the storage of information, thereby serving as the repository of memory content. The functions of the neuroglia, in contrast to those of the neurons, cannot be defined so readily. Indeed, the glial cells insulate, protect, and support the nerve cells from external, mechanical influences, but the neuroglia also perform metabolic tasks in the sense of metabolic symbiosis with the nerve cells. Certain impulse-conducting properties of the neuroglia as well as tasks of support and assistance in neuron differentiation must be considered, also.

The blood vessels of nonneuronal and nonglial components of the nervous system should be mentioned since they play a vital role in carrying out the overall function of this organ system. Likewise, the meninges are crucial, wrapping around the entire system as connecting tissue derivatives.

1.1 Nerve Cells (Neurons)

1.1.1 Neuron Theory

The elemental unit of impulse conduction in all nervous systems, whether vertebrate or invertebrate, is the nerve cell: the neuron inclusive of its processes, the nerve fibers. In spite of the great differences in the outer appearance of the individual types of neurons, the functional principle essentially remains the same. It is in this way that nerve cells differ from all other cells. Due to this unique status, and in light of research into the phenomenon of information storage within the nervous system, it is necessary first to examine the structure and function of the individual components of the central nervous system (Fig. 1.1).

Nerve cells are cell systems structured in an extremely polar manner. A cell body (perikaryon, soma) merely $40-60\,\mu m$ in size can have processes, i.e., nerve fibers, of extraordinary length, often extending 1 meter or more. Because of this, rendering a precise morphological description was once very difficult. Relative to the nature of nerve fibers and, in particular, their endings, it was not possible to determine by light microscope (LM; maximum enlargement with the most modern equipment was about 1,000 times) whether the fiber endings of the individual nerve cells crossed over one another in net-like fashion, lacking cell boundaries as neurocytium, or whether each cell was self-contained and merely lying in extremely close proximity to its neighboring cells. Both views were represented. Around the turn of the century, the neuroanatomists Held, Nissl, Bauer, and Stohr advocated the *"neurocytium/continuity theory."* Opposed to this theory were the anatomists Wilhelm His (1887) in Leipzig, August Forel (1888) in Zurich and, above all, Wilhelm Waldeyer (1891) in Berlin, all of whom represented the *"contiguity/contact theory."* Waldeyer and

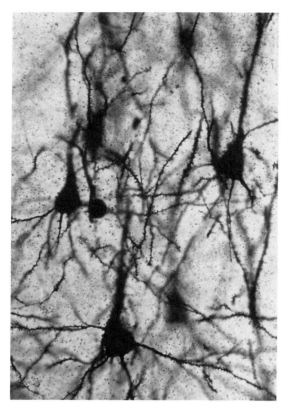

Fig. 1.1. Pyramidal cells from the cerebellum of a calf.

Santiago Ramon y Cajal (1888) in Barcelona applied to the nerve cell the "cell theory" based on the works of the botanist Matthias Schleiden (1838) and the zoologist Theodor Schwann (1839), according to which every cell is self-contained, and developed the so-called *neuron theory* or *neuron doctrine*. According to this theory, all nerve fibers are runners of nerve cells (ganglion cells, neurons) that were formed independently of one another from nerve-forming cells (neuroblasts) in special formation centers of the nervous system (matrix zones).

Around 1950, E. De Robertis, S. Palay, and G. Palade, among others, were able to show, with the help of electron microscopy (EM; capable of enlarging subject matter up to 100,000 times) that the fiber end regions of the nerve cells did not interweave, but that they only contacted each other there in *synapses*, morphologically and physiologically specially adapted nerve fiber end

formations. Meanwhile, it became clear from the standpoint of electrophysiology, as well, that neurons are discrete entities, i.e., separate from one another, and contact each other through synapses in which the transmission of impulses ensues from cell to cell in a specialized manner.

Thus, the most important recoginition of neuron theory is that a neuron, as the elemental unit of the nervous system, constitutes both an ontogenetic (developmental) and a physiological (i.e., trophic and functional) unit that is responsible for reception, processing, transmission, and, above all, storage of information in an organism.

1.1.2 Morphology of Nerve Cells

Nerve cells, especially of vertebrates, are quite diverse in shape. This is primarily a factor of the great variety in size of the *cell bodies* (*perikarya*) and the varied number and arrangement of the fiber-like cell processes. The latter can be classified into two categories:

1. *Dendrites*, tree-like ramifications, usually present in abundance, whose task it is to lead impulses to the cell;
2. *Neurites* (*axons*), of which only one extends from each cell and whose task it is to lead impulses from the cell.

The neurite begins to branch out only later down the course of its extension as it is transformed into *collaterals* (side branches) that terminate in the form of synaptic trees (*telodendria*) just as the main branch of other neurons terminate at cells of response organs such as muscle, gland, and sensory cells (Fig. 1.2).

Dendrites can be differentiated from neurites by the ribonucleoproteids, the so-called *Nissl* or tigroid strata, in their conically shaped base from which the nerve fibers originate as in other cell soma. These substances are absent in the base cone (axon hillock) of the neurite.

Based on the presence or absence of dendrites, nerve cells are characterized as dendritic or adendritic. The multipolar, multidendritic ganglion cells are the most commonly encountered of the dendritic neurons in the central nervous system (CNS). The most clearly defined cells of this category are the Deiters cells which are found in the spinal cord of mammals

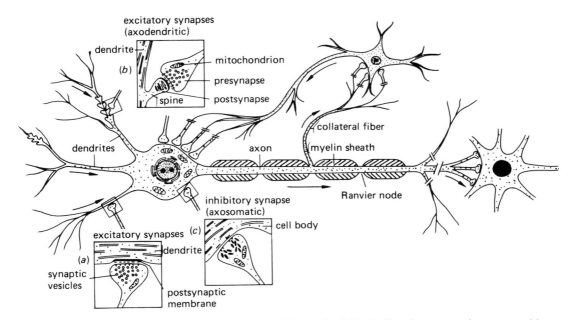

Fig. 1.2. Diagram of a multipolar ganglion cell with axodendritic (*a, b*) and axosomatic synapses (*c*).

(Fig. 1.3) and named after the German anatomist from Bonn who discovered them in 1865, Otto Deiters. They have numerous dendrites, or "protoplasmic processes," as Deiters called them, and one particularly long axon with collaterals that Deiters called an "axis cylinder." The axon is wrapped in a myelin sheath (myelin or medullary sheath, see Chap. 1.2.4) from the point where it leaves the CNS. In 1885, C. Golgi, from Pavia, succeeded in obtaining a dyed representation of neurons with the aid of a potassium bichromate recording and subsequent silver impregnation (Fig. 1.4). The *Golgi neurons* (Fig. 1.3a) named after him have only one short neurite which branches off right at the nerve cell body (perikaryon) into numerous collaterals that depart the CNS. *Mauthner neurons* (Fig. 1.5) occur in the hindbrain of fish and amphibian larvae. They, too, are multipolar. Numerous dendrites of other cells accumulate at the strong axon. The so-called *pyramidal cells* (Fig. 1.3c) of the cerebral cortex in mammals have an apical main process of the cell body which branches off into numerous dendrites. Other dendrites issue from the pyramidal mantle. The axon of the pyramidal cell emerges at the base of the cell (Figs. 1.3c and 1.1) and can

attain the extraordinary length of 1 meter or more. It leaves the cerebral cortex and enters the spinal cord via the *pyramidal pathway*, and proceeds through the spinal cord into the anterior horn cells situated in the lumbar part of the cord. The *Purkinje cells* of the cerebellum, named after the Czechoslovakian anatomist who discovered them in 1836, are bipolar dendritic neurons. One enormous, trellis-like, highly ramified dendrite originates at its apical pole, i.e., at the base opposite the neurite (Figs. 1.3d, 1.5a and b).

Neurons without axons and with only one branched process are found in the form of *amacrine cells* in the retina (Fig. 1.3e) and elsewhere. *Spinal ganglion cells* (Fig. 1.3g), *epibranchial ganglia*, and many neurons of the vegetative NS contain two processes. They derive ontogenetically from bipolar, primary sensory cells with a short apical process and one long neurite (Fig. 1.3f). As these cells shift from the exterior into the interior of the body, there emerges a neuron with two opposing processes. This kind of nerve cell can change in structure from the state of being bipolar to one that is pseudo-unipolar in nature (Fig. 1.3h).

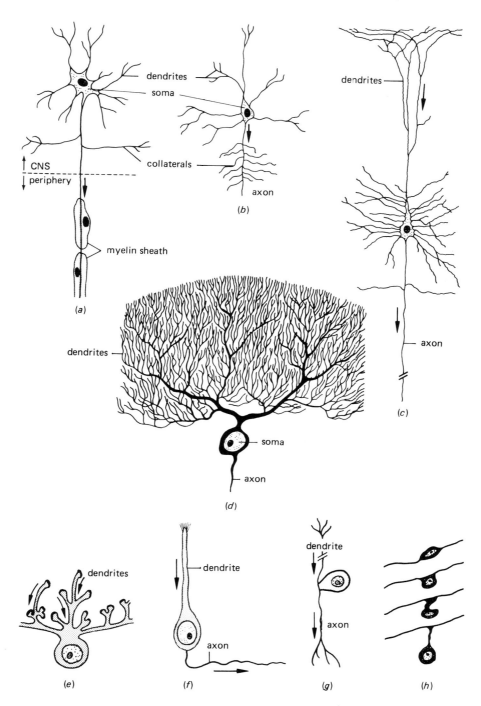

Fig. 1.3. Diagram of various types of nerve cells: (*a*) Deiters type; (*b*) Golgi type; (*c*) pyramidal cell from the neocortex; (*d*) Purkinje cell from the cerebellum; (*e*) cell without axon (amacrine cell from the retina); (*f*) bipolar cell from the olfactory epithelium; (*g*) metamorphosis of a bipolar nerve cell into a pseudo-unipolar nerve cell (spinal ganglion cell, *h*).

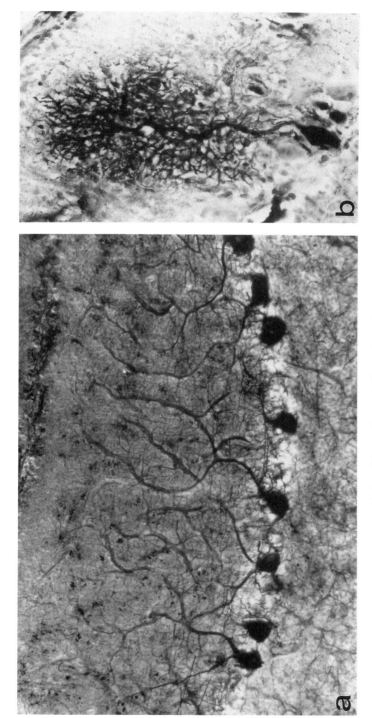

Fig. 1.4. Purkinje cell from the cerebellum of a guinea pig: (*a*) overview, (*b*) single cell.

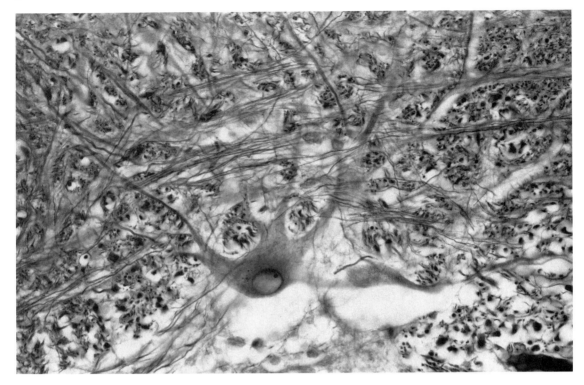

Fig. 1.5. Mauthner neuron from the CNS of a fish.

1.1.3 Fine Structure of Nerve Cells

Despite the diverse morphology evidenced by
nerve cells, their fine structure is quite uniform.
The following section addresses the fine organization of the cell body including the fibers that
originate in them. Due to the distinctive functional characteristics that they exhibit, the nerve
fiber end formations, together with their points of
contact to other cells, the synapses, will be
discussed separately.

1.1.3.1 Nerve Cell Membrane (Neurilemma)

As are all other cells of the body, the nerve cell,
including its fibers, is encased in a plasma
membrane that is 7.5 nm (75 Å) thick. An electron
microscope reveals its structure to be of three
layers, two of which are electron dense and one
of which is not. Contemporary views hold that
the neuroplasmic membrane is essentially an
arrangement of lipids oriented in a contrasting
manner as well as completely or partially de-

posited (intrinsic or extrinsic) proteins that, to
varying degrees, can be interconnected.

Lipid and protein components are found,
relative to the "*fluid-mosaic-membrane model*," in
a predominately liquid (fluid) state (see Fig. 7.5).
Lateral movements of the lipids as well as the
proteins within the membrane provide for the
constant shifting and repositioning of substances. This occurs in concert with the structure of
the underlying neuroplasm. Ultimately, it is
dependent upon the fact that the various connections constituting the membrane are not
equally distributed in it, but rather form a pattern
of "*microdomains*" of various substances. This
pattern of membrane composition varies from
area to area within the nerve cell (perikaryon,
dendrite, axon, synapse). To be sure, this has an
effect on the considerable functional characteristics of the membrane activity of a nerve cell.

Details of the molecular structure of the neuronal membrane and its role in impulse conduction and transmission are discussed in
Chap. 7.1.1.

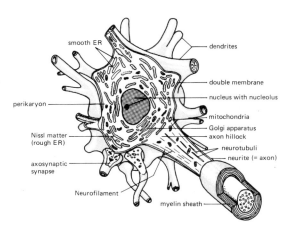

smooth ER

dendrites

double membrane

nucleus with nucleolus

perikaryon

mitochondria

Golgi apparatus

axon hillock

Nissl matter
(rough ER)

neurotubuli

neurite (= axon)

axosynaptic
synapse

Neurofilament

myelin sheath

Fig. 1.6. Diagrammatic overview of the ultra-structural organization of a nerve cell.

1.1.3.2 Nerve Cell Body (Soma, Perikaryon)

All of the cytoplasmic components of a nerve cell vital for life are found in the neuroplasm of the cell body, the perikaryon (Fig. 1.6): nucleus, Golgi apparatus, smooth and rough endoplasmic reticulum (ER), ribosomes, mitochondria, lysosomes, neurotubuli, and neurofilaments. *Centrioles* are not present; they are unnecessary because neurons lose their capability to undergo mitotic division in the postnatal stage. This assures that the most important fundamental metabolic processes take place only centrally in the nerve cell body. Damage to the cell body of the neuron, therefore, leads to the destruction of the nerve cell whereas damage to the nerve fibers can be reversible since, for lack of rough ER, no independent protein synthesis occurs in them.

The *nucleus* is particularly conspicuous in the perikaryon due to its loose, electron-dense matter [*chromatin*, containing deoxyribonucleic acid (DNA), and *nucleolus*] (Fig. 1.7). The size of the nucleus varies considerably with the type of neuron (3 to ca. 10 μm in diameter). It is independent of any functional state of the cell and is, thus, not subject to a strict nucleus–plasma relation. Compared to the nuclei of normal neurons and glial cells, doubled amounts of DNA have been measured in the nucleus of the Purkinje cells of the cerebellum and the pyramidal cells of the hippocampus. In this regard, the question remains as to whether a tetraploid

chromosome set or giant chromosomes (polytene chromosomes) lie at issue. Several nuclei per cell have been observed in the case of some neurons.

The loosely arranged euchromatin and heterochromatin within the nucleus derive from the *chromosomes*, which consist of *deoxyribonucleic acid* and proteins and are the genetic basis for the regulation of metabolic functions (cell differentiation and synthesis of proteins and other building blocks of cells). They ensure this through interactions, sufficiently known in molecular biology, between the chromosomal DNA in the nucleus and the *messenger ribonucleic acid (m-RNA)* synthesized there by means of *transcription processes*. The mRNA passes through pores in the nucleus into the neuroplasm and moves on to the ribosomes with their *ribosomal RNA (r-RNA)*, where they synthesize neuronal proteins (*translation*) with the help of *transfer RNA (t-RNA)* and amino acids, which occur freely in the neuroplasm.

The membrane of the nucleus is especially prominent when viewed under the electron microscope. Each of the two single membranes, 7 nm in thickness, cross into one another at regular intervals and, in doing so, they form *nucleus pores* ca. 70 nm in size and as many as 3,000 per nucleus. These pores are closed by a membrane (diaphragm) 5 to 10 nm in thickness with a swelling in the middle. The RNA synthesized by the chromosomal DNA passes through the pores of the nucleus out of the karyolymph into the neuroplasm. The nucleus membrane branches off in other locations and passes smoothly into the *endoplasmic reticulum* in the neuroplasm (Fig. 1.8).

The *nucleolus* within the nucleus (occasionally there are two) does not exceed one-fifth of the total size of the nucleus. It consists of one part fibrous r-RNA (pars fibrosa) and one part pars granulosa proteins with special enzyme functions [phosphatases, dehydrogenases, adenosinetriphosphatases (ATPases)]. Proteins that are synthesized in the neuroplasm and progress via the nucleus pores to the karyolymph come into contact in the nucleus with the r-RNA of the nucleolus. Beyond its significance as an indicator of the metabolic activity of a nerve cell, the nucleolus is also especially important in sex determination. An appendage was discovered on

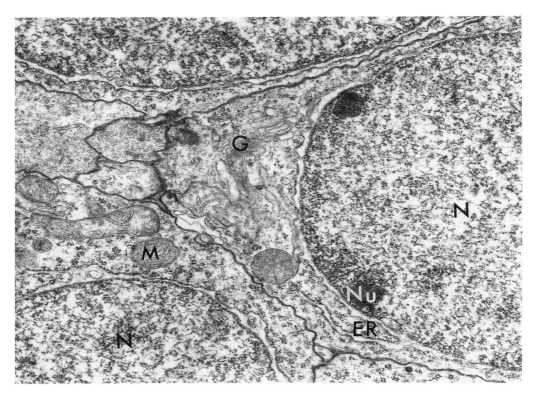

Fig. 1.7. Electron micrograph of three contiguous nerve cell bodies from the optic tectum of the cichlid fish *Sarotherodon sp.* N: nucleus; M: mitochondrion; G: Golgi apparatus; ER: rough endoplasmic reticulum; Nu: nucleolus.

the nucleolus of female cat neurons (*Barr's appendage* or *body*, Fig. 1.9) that is not present in male cats. The Barr appendage is attached only to the X chromosome. In the meantime, it has been detected in all mammals and serves in the diagnosis of genetic abnormalities.

After the r-RNA passes from the nucleus through the nucleus pores, the smallest cell organelles, the *ribosomes*, develop in the neuroplasm (Fig. 1.10). They are 12 to 25 nm in size and fall into two groups. One group, the polysomes, comprises ribosomes that, individually or in rosette form, consist of 5 to 7 separate particles and occur freely in the neuroplasm. The other ribosome group is stored with its subunits in a highly ramified inner membrane, the *endoplasmic reticulum* (*ER*). The so-called

rough ER, which contains ribosomes, gives the nerve cell body a "tigroid" appearance due to its great capacity to take on coloring with histochemical dyes (gallocyanine, toluidine, cresyl violet, etc.) which is why it is also characterized as a *tigroid substance* (*Nissl substance*). The occurrence of this Nissl substance is limited to the perikaryon and origination points of the dendrites at the cell body (see Fig. 1.7). Since the rough ER is associated with the protein synthesis of the nerve cell and is absent in the base cone (axon hillock) of the neurites as well as in all other sections of the nerve fiber distal to the cell, it is impossible for protein synthesis to take place in these sections. This is actually the reason for the phenomenon of *neuronal transport* (see Chap. 9.1), which aids in the continuous trans-

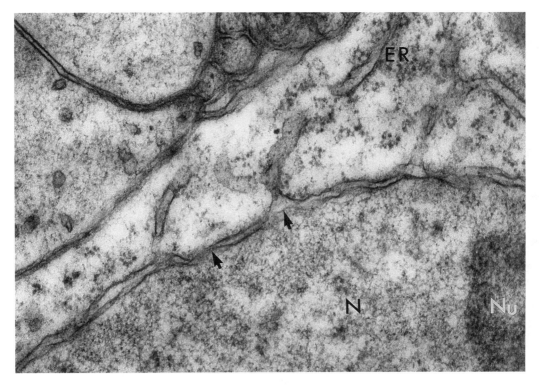

Fig. 1.8. Electron micrograph of a section of a nerve cell body of the optic tectum of a trout. Transition of the membrane of the nucleus (↓) to the rough ER (see Fig. 1.7 for remaining component labels).

port, proximodistally, of the new products of synthesis formed in the cell body to supply the nerve fiber end formations (synapses). This is occasionally compared to the ergastoplasm of the glandular cells due to the great amount of protein synthesized by the rough ER in the perikaryon. Independent of the functional state of the neuron, i.e., of the intensity of protein synthesis, the histological phenotype of the nerve cell is quite variable. Thus, the RNA content of a neuron changes, for example, in the course of ontogenesis during various functional developmental phases or even due to pathological influences whereby the Nissl substance can be dissolved completely (chromatolysis).

Smooth ER, to which the ribosomes do not attach, is found in the perikaryon as well as in the distal nerve fiber regions, including the synapses.

The smooth ER is attributed functional importance in regard to lipid synthesis and neuronal transport.

The so-called *Golgi apparatus* must be regarded as a special formation of the smooth ER. It consists mostly of various *dictyosomes* (coherent formations of lamella, vacuole systems, and vesicles) that are arranged around the nucleus in belt-like fashion (Fig. 1.10). These membrane formations, measuring 20 to 60 nm in diameter, lie, more or less, arching astride one another with a *cis*- and a *trans*- side. Protein- and lipid-laden vesicles from the rough and smooth ER fuse on the *cis*- side with the lamella where the vesicles can be filled with sugar molecules, phosphate or sulfate groups. In like manner, the vesicles are attached to the *trans*- side of the Golgi apparatus, and depending upon content, either migrate to

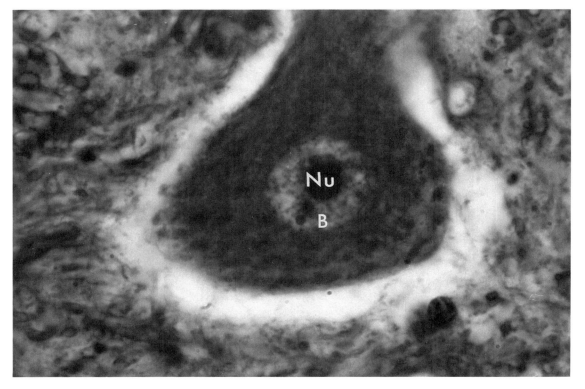

Fig. 1.9. Light micrograph of a Barr body (B), i.e., of a body attached to the X chromosome of a structure at the nucleolus (Nu) of a female cat neuron.

the plasma membrane or are filled with products of secretion, or even migrate to the lysosomes.

Lysosomes of the residual type occur in the perikaryon of aging neurons as so-called *age pigments* or *lipofuscins*. They are thick bodies ca. 350 to 600 nm in size, homogeneous in structure and with a restrictive membrane. They absorb products of catabolism that are metabolic in nature or alien. They are referred to as *primary lysosomes* as long as they do not come into contact with the decomposing substratum. If, however, they have absorbed degrading substratum, then they are called *secondary lysosomes*.

Further, a good many neurons, for example, cells of the *substantia nigra* in the mesencephalon, or *ferric compounds* (cells of the nucleus ruber), contain melanins. *Neurosecretion granulae* are formed by specific cells, e.g., in the hypothalamus, and, assisted by anterograde neuronal transport, travel from the cell body into the nerve endings

from which they move to neurohemal organs or into the bloodstream by processes of exocytosis (see Chap. 3.2.2.4.1).

Mitochondria are found in nerve cells, just as in all other cells of the body. These are organelles of varying size, usually between 1 and 2 μm long or round. They are composed of a wraparound double membrane (outer membrane) with a ruffled inner membrane (cristae mitochondriales) or tubuli (tubuli mitochondriales) that project into the matrix. The mictochondria of the nervous system are, as a rule, of the crista type. They occur in the perikaryon (as many as 100,000) as well as in the fibers and synapses that they reach by means of slow neuronal transport. Recent EM research indicates that the outer mitochondrial membrane can be in contact with the smooth ER. This outer membrane is water and ion permeable, but the inner membrane is relatively impermeable and, accordingly, special transport mechanisms are necessary to overcome it. The essen-

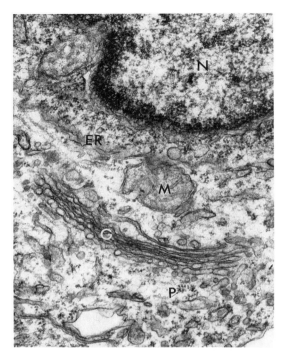

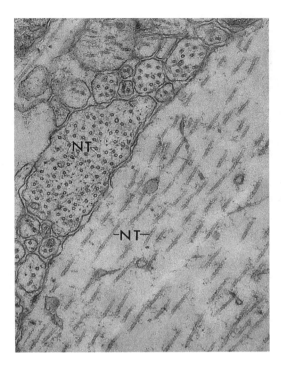

Fig. 1.10. Electron micrograph of the perikaryon of a nerve cell from the optic tectum of a cichlid fish (*Sarotherodon sp.*).

Fig. 1.11. Electron micrograph of neurotubules (NT) in longitudinal and cross sections from unmyelinated nerve fibers of the optic tectum of a trout.

tial functions of the mitochondria serve cell respiration. It should be regarded from the perspective of the endosymbiont hypothesis that mitochondria contain their own ribosomes and DNA.

In addition to the organelles previously discussed, the neuroplasm contains still other structured, thread-like elements that can be distinguished by diameter size: neurotubuli (20 to 30 nm), neurofilaments (10 nm), and microfilaments (5 nm).

Neurotubuli (Fig. 1.11) are morphologically similar to the microtubuli of other animal and plant cells as well as to the mitosis spindle apparatus. They are made of *tubulin*, a protein with a molecular weight of 120,000 that consists of two subcomponents and that has an extraordinarily high bonding capacity for *colchicine*. This alkaloid of meadow saffron (*Colchicum autumnale*) inhibits the function of the

tubuli by disassociating the tubulin into its two components. The neurotubuli perform important functions in the rapid neuronal transport (see Chap. 9.1.2) of such compounds that are synthesized in the perikaryon and then pass (anterograde transport) into the nerve fiber endings, and of such substances that are absorbed in the nerve fiber end formation of the synaptic membrane by endocytosis and supplied back to the perikaryon (retrograde transport).

The *neurofilaments* (Fig. 1.12) of vertebrate neurons consists of protein filaments that are about 10 nm in thickness and of indeterminate length. The neurofilaments are tubular in shape, as the neurotubuli are; however, their protein belongs to the category of globulins and has a molecular weight of 80,000. They are found primarily in nerve and glial fibers and, indeed, in far greater number than the tubuli. They change in shape and in number with increasing age and with the occurrence of certain dysfunctional

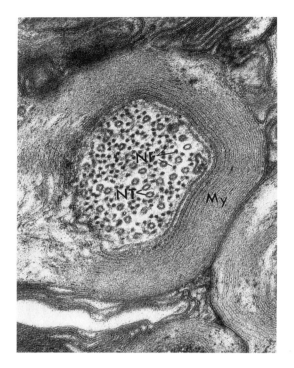

Fig. 1.12. Myelin sheath of a myelinated axon of the optic tectum of a goldfish. Note the cross sections of neurofilaments (NF) and neurotubules (NT) in the axoplasm. My: myelin sheath.

and filaments) in accordance with cell activity at any given instant.

1.1.3.3 Nerve Fibers (Dendrites, Neurite/Axon)

Various fibers extend from the soma of a nerve cell, numerous *dendrites*, and a single *neurite*, or *axon*.

A distinction is made between two different types of nerve fiber based on the direction of the impulse flow, or the transmission of information dependent upon it. There are

1. *Effector* (having an effect upon other cells), *efferent centrifugal* (leading away from the center) nerve fibers that
   (a) As motor efferences, innervate the skeletal musculature,
   (b) As vegetative-motor efferences innervate the heart and the smooth musculature of viscera, and
   (c) As vegetative-secretive efferences innervate glands, or
   (d) Regulate the sensitivity of peripheral receptors;
2. *Sensory* (having the ability to sense), *afferent centripetal* (leading to the center) nerve fibers that conduct sensations to the CNS from the sensory organs and, indeed,
   (a) As visceral afferences from the viscera, and
   (b) As somatic afferences from the musculature, the skin, and the sensory organs.

Further distinctions in nerve fibers can be made in terms of function and morphology in particular. Two categories emerge:

- *Unmyelinated fibers*, which are older phylogenetically and have a slower rate of impulse conductivity than
- *Myelinated fibers*, which are wrapped in Schwann cells (myelin sheath, see Fig. 1.23) and are capable of conducting impulses at a much greater speed due to the insulating effect afforded by the myelin sheath.

Dendrites

With regard to fine structure, the dendritic processes of a neuron do not differ from the soma region except that they lack rough ER in the distal section and, accordingly, are incapable of

conditions such as Alzheimer's disease. It is possible that the neurofilaments fulfill a function similar to that of the tubuli in neuronal transport. Clearly, they may add support to the longitudinal orientation of the nerve fibers.

The *microfilaments*, with a diameter of only 5 nm, are the smallest fibrillary structures of the neuroplasm. They consist of neuronal actin very similar to that of muscle cells and are particularly abundant in growing nerve cell processes.

All of the previously described organelles are found in the neuroplasm where they freely float about, albeit not as in a clear liquid. Rather, they move about oriented toward a fine mesh net of neuroplasmic fibers, the so-called *microtrabecula system*. This type of *cytoskeleton* of the nerve cell constitutes, so to speak, a "cell musculature" that redistributes and rearranges the cell organelles and the elements of the cell framework (tubuli synthesizing their own proteins. The most essential job of the dendrites may well be to increase the receptive area of a nerve cell. As a rule, numerous dendritic processes several hundred micrometers in length extend from a cell body of 40 to 60 μm in diameter. It bas been calculated, in the case of Purkinje cells of the cerebellum, that these processes increase the receptive area of the soma 90-fold, enabling some 100,000 synapses of other cells to contact this type of branching system. In addition to their sensory, afferent function as the structure for receiving impulses, the dendrites perform the task of integrating the impulses that are received in the various terminal regions. It remains unknown if and with which part of their original partial impulse, a single dendritic ending is involved in the origination of a change in potential to be conducted in the area of the axon hillock. The peripheral sensitive, or afferent, fibers of the *spinal ganglia cells*, i.e., leading from the periphery to the spinal cord, are unique in their dendritic form. These strikingly long, specialized dendrites perform tasks of conduction. They receive impulses at their receptive end, for example, as mechanoreceptors, and conduct these impulses past the soma of their own ganglion cell directly to a motor neuron located in the spinal cord.

Neurite (Axon)

In like manner to the dendrites, the neurite (axon), a component occurring as a solitary element, has the same fine-structural composition as the soma of a nerve cell. Rough ER is absent in the axoplasm, too; thus, supplying the axon and its endings with new products of synthesis can be ensured only with the help of axonal transport (see Chap. 9.1). Axons can attain the extraordinary length of 1 meter or more (pyramidal pathways in the spinal cord, or motor neurons of extremity musculature, for example). Extending from the axon are axon collaterals (see Chap. 5), which are highly ramified (as is the main neurite) and can come into contact with many other neurons. Impulses in the form of action potentials are formed at and conducted away from (see Chap. 6) the axon hillock, the base cone of the neurite at the cell soma, whose distinct lack of rough ER and

other organelles is its only clear distinguishing characteristic vis-a-vis the exiting dendrites. The main function of the axon is to facilitate effector-efferent conduction of these electrical potentials to the presynaptic ending and to transmit information with the help of chemical transmitter molecules to the receiving cell.

The chemical transmitter (*Dale's principle*) is the same for the presynaptic endings of an axon and for all of its collaterals.

1.1.3.4 Fine Structure of Synapses

In 1897, the English physiologist Charles Sherrington termed the nerve fiber end formations of axons as well as dendrites *"synapses."* Cells contact other cells (other neurons, glandular, sensory, or muscle cells) through synapses in the sense that a morphological and physiological connection is made. Because of the significant role played by synapses in the formation of memory, it is essential that a detailed discussion of their morphology, as well as their essential functional aspects, be undertaken at this juncture. There are three types of synapses:

1. *Effector synapses* innervate glandular or muscle cells with their axon terminals.
2. *Receptor synapses* facilitate sensory innervation, for example of muscle spindles or the tactile bodies in the skin.
3. *Interneuronal synapses* establish contact between nerve cells, indeed in the most diverse types, in which an impulse is conducted via an axon to the presynapse located at the end of the fiber and from their is carried to the postsynapse of the "receiving" nerve cell:
   - *Axosomatic synapses* connect the ends of the axon with a postsynapse that is located directly at the cell body of a "receiving" or receptor cell (see Fig. 1.2c);
   - *Axodendritic synapses* are those whose axon endings lead to a dendrite whose postsynaptic shape can stretch around the axon, in part in the form of a so-called spine (see Fig. 1.2b);
   - *Axoaxonal synapses* form the contact between the presynapse and the neurite of a neighboring cell;
   - *Interaxonal synapses* produce synaptic axon swellings in a neurite. This type is

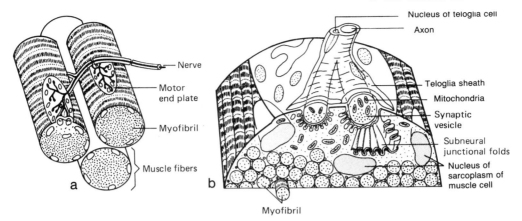

Fig. 1.13. Diagram of light (*a*) and electron micrographs (*b*) of the organization of the motor end plates. KS: nucleus of the sarcoplasm of the muscle cell. (From Rohen, 1971.)

found especially in the gray matter of the CNS.

Only very rarely does a single type of synapse occur alone; usually, synapses are organized in patterns as in the form of numerous axodendritic synapses occurring serially one behind the other. *Reciprocal synapses*, in which two fiber terminals form reciprocal synaptic contact zones, are also common. This type of contact pattern is termed *synaptic glomeruli* and is characterized by the

nerve fibers, with many synaptic points in a very small space, being surrounded by glia structures.

In contrast to the interneuronal synapse, the so-called *effector synapses* do not innervate other nerve cells, but rather innervate smooth and transversely striped muscle cells and gland cells. Here, the so-called *motor end plates* (see Figs. 1.13 and 1.14) have been examined particularly closely. These are extreme ramifications of peripheral nerve fibers into many individual fibers; termed *end trees* or *telodendria*, each branching nerve

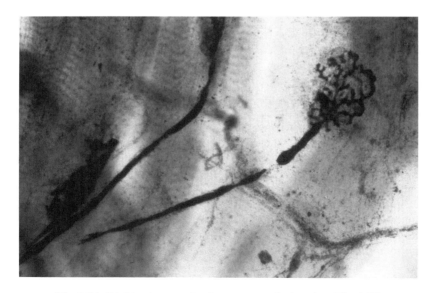

Fig. 1.14. Light micrograph of a motor end plate (see Fig. 1.13).

fiber leads to an end plate on a muscle fiber. Within this fiber is the sarcolemma, the wrap of the muscle cell that is greatly enlarged by junctional folds, the palisade-like subneuronal folds situated parallel to and beneath the end area of the telodendria.

So-called electrical synapses are distinguished from chemical synapses by significant differences in the morphological makeup of the synaptic contact areas and, especially, by principal differences of functional mode in the transmission of impulses.

In the case of *electrical synapses*, the nerve endings of two neurons lie with their membranes very close together, such that they are separated by a gap of only 2 to 4 nm. An electrically coded impulse signal can be conducted across this gap in either direction since the membrane has a reduced electrical resistance through the incorporation of transcellular gap junctions (Fig. 1.15a). These gap junctions consist of six protein subunits. They penetrate, in tubular form, the double lipid layer of each of the two plasma membranes involved and thereby ensure an intercellular exchange of lower molecular substances as well as electrical currents (Fig. 1.16).

Electrical synapses are found primarily in the invertebrate nervous system. Only rarely have they been discussed in regard to lower vertebrates, for example, fish. In mammals, they may represent the exception as chemical synapses predominate.

In contrast to electrical synapses, the conduction of impulses in *chemical synapses* takes place only in one direction (unidirectionally). This polarity can be traced back to (a) differences in the composition of the nerve cell membranes involved, (b) the fact that the presynapse is equipped with *synaptic vesicles* that contain chemical *transmitters*, and (c) a gap 20 to 30 nm wider (Fig. 1.15b) than in electrical synapses. This large synaptic gap is, optically, not empty, but filled with a fine, filamentary *molecular fuzz* that is electron dense. In vertebrates, this is probably composed of the oligosaccharide side chains of certain glycoproteins and glycolipids (gangliosides) containing neuraminic acid. Due to its strong affinity for Ca^{2+} ions, this intersynaptic molecular interlacing obviously

plays an important role in the process of functional *facilitation* between nerve cells as the basis for memory formation (see Chap. 11). In chemical synapses, the transmission of electrical excitatory impulses across the gap takes place by means of chemical transmitters (see Chap. 7).

A closer analysis of the synapses (French: boutons terminaux), which are 1 to 2 μm in size, was possible only after 1950 with the invention of electron microscopy. The most prominent structures are the *synaptic vesicles* in the presynapse. In addition to enzymes, they contain ions and various mediators, especially *transmitter substances*. Of these, more than 40 different compounds have been discovered. They can be categorized into two groups: inhibitory and excitatory substances (see Chap. 7). Excitatory transmitters effect the depolarization of a postsynaptic membrane. They thereby enable an electrical impulse to be conducted to the next nerve cell. Acetylcholine, noradrenalin, and glutaminic acid are examples of excitatory transmitter compounds. Inhibitory transmitters such as γ-aminobutyric acid and glycine, on the other hand, bring about a *hyperpolarization* of the postsynaptic membrane (an increase of the membrane potential above the normal resting value). The membrane potential is thereby removed from a critical potential (at which stimulus conduction can occur) and the further conduction of an impulse is inhibited. One distinguishes among cholinergic, adrenergic, serotoninergic, glutaminergic synapses, etc., depending upon the type of transmitter substance to which the synapse is reacting.

In contrast to the relatively uniform nature of electrical synapses (gap junctions), the comparative morphology of chemical synapses reveals diversity among the different types of synapses (Fig. 1.17). The fundamental type is represented by a *simple synapse*, which has specially formed, swollen pre- and postsynaptic membrane areas, as well as with a distinct synaptic cleft of defined width (*A*). All other more or less specialized *synapses* might well derive from this one, including the commonly occurring "spine" synapse (*B*) and the "subjunctional" synapse with its additional particles (*C*). The subsynaptic swelling can be replaced by so-called subsurface cisterns (*D*). In the case of the neuromuscular synapse, the

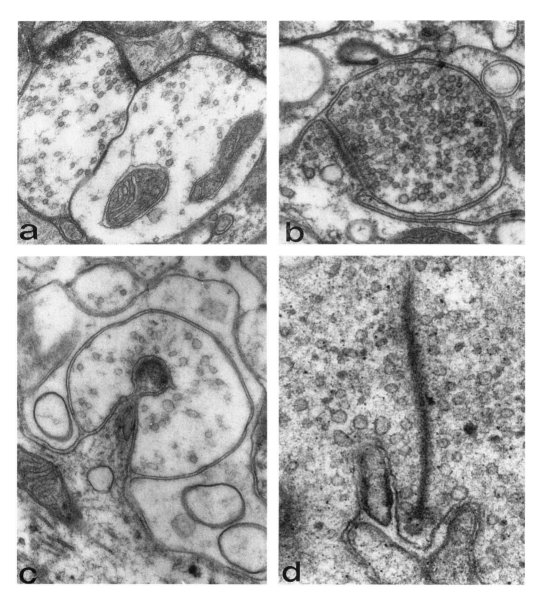

Fig. 1.15. Electron micrograph of (*a*) an electrical synapse (gap junction); (*b*) a typical chemical synapse; (*c*) a spine synapse from the optic tectum of a carp; and (*d*) a "ribbon" synapse from the retina of a goldfish.

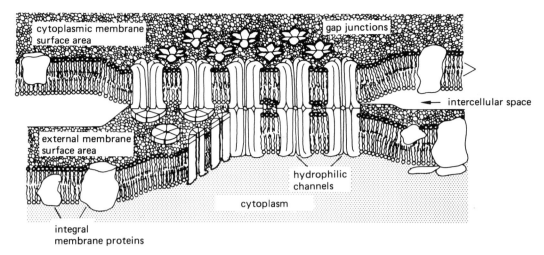

Fig. 1.16. Three-dimensional diagrammatic reconstruction of a gap junction (electrical synapse). Composition of the channel walls from six protein subunits that bridge the gap between the double lipid layer of each membrane.

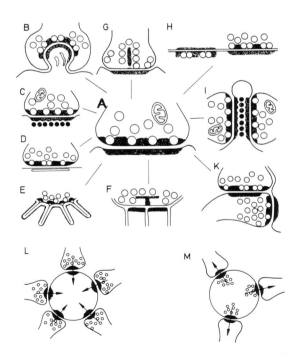

Fig. 1.17. Diagrammatic overview of various types of synapses: (*A*) basic type; (*B*) spine synapse; (*c*) subjunctional synapse; (*D*) subsurface cistern synapse; (*E*) neuromuscular synapse; (*F*) "T-conductor" synapse from insects; (*G*) ribbon synapse; (*H*) reciprocal synapse; (*I*) crest synapse; (*K*) serial synapse; (*L*) convergence synapse; (*M*) divergence synapse.

nerve ending and the muscle cell are separated by a basal membrane (*E*). In insects, "T"-shaped conductor reinforcing structures can be found in the presynapse opposite the intercellular space of postsynaptic elements (*F*). So-called synaptic ribbons appear in the synapses of retinal rod cells (*G*, Fig. 1.17D). Further, synapse areas that interact reciprocally can be found between two cells (*H*); then there are crest synapses (*J*) or synapses firing serially (*K*). Additionally, several synapses can coverage upon one receptor cell (*L*), or, conversely, impulses can be transmitted from a single nerve ending divergently to various connector cells (*M*). These diverse modes of synaptic "pitching and catching" might be of special significance in the emergence of the various neuronal circuits (see Chapter 5). Viewed through the electron microscope, the presynaptic vesicles appear most clearly as additional elements in the fine structure of chemical synapses, and diverse membrane shapes are revealed in the area of actual synaptic contact, i.e., in the active zone.

Synaptic vesicles can differ widely in shape, size, and content depending upon the type of synapse. In the majority of synapses, there are found "*small, clear vesicles*" measuring ca. 40 to 60 nm in diameter that, as transmitters, contain acetylcholine and amino acids. similar small but "*flattened vesicles*" containing glycine or γ-aminobutyric acid (GABA) can found in other synapses. "*Coated vesicles*," comparatively small but coated with a glycoconjugates, are found occasionally in very close proximity to the presynaptic membrane. These are discussed in connection with the renewal of the building blocks of the synaptic vesicle membrane. The "*dense core vesicles*" are nontransparent to the electron microscope and vary in size from 50 to 150 nm in diameter. They contain catecholamines (i.e., noradrenaline), which function as a transmitter, or polypeptides, which function as neurohormones.

Lysosomes, which contain hydrolytic enzymes and store metabolites, are even larger vesicular synaptic structures (200 to 400 nm in diameter) the contents of which are nontransparent to the electron microscope.

A membrane peculiarity, the systematically arranged "*dense projections*," is conspicuous on the presynaptic side. Shaped like presynaptic vesicular lattice, they are hillocks of electron-dense matter (film-like proteins). The synaptic vesicles employ the space in the gap-like areas between them to come into direct contact with the inside of the synaptic membrane (Fig. 1.18). Accordingly, upon the event of a stimulation impulse and subsequent to the presynaptic membrane having opened, the transmitter substance can be released directly into the synaptic cleft (see Chap. 7).

Compared to the "dense projections," *post-synaptic densities*, located opposite the presynaptic lattice, are even more strongly apparent and usually run parallel to the contact zone. Receptor molecules for the specific recognition of transmitter substances are localized here as are the enzymes that catabolize the transmitter. In some synapses, additional "*subjunctional bodies*" can be observed beneath the postsynaptic densities. The significance of these bodies is not yet known. In addition to these synapse-specific organelles, or membrane structures, there are located in the nerve endings organelles of the remaining neuroplasm such as mitochondria, which serve the production of chemical energy. Of course, most metabolic processes occur in the perikaryon of the nerve cell. Nonetheless, among other processes, a resynthesis of transmitters and their reintroduction (recycling) into the synaptic vesicle ensues in the synapse. In addition to the mitochondria, structures of the *smooth endoplasmic reticulum* (*ER*) are found frequently in the synaptic endings. The function attributed to these structures is to transport substances synthesized in the perikaryon to the periphery. Beyond this function, the ER components could serve the neoformation of the membranes of vesicles and/or the plasma membrane.

Neurofilaments and *neurotubuli*, the latter facilitating neuronal transport (see Chap. 9), are located in the synaptic endings, which are strengthened, or stabilized, additionally by the microstructures of the cytoskeleton.

Various, indeed, often quite elaborate methods have been developed for the *identification of neuronal synapses* (pre- or postsynapses, etc.). In one such method, it is possible to reconstruct the synaptic juncture based on an analysis of serial sections. Additionally, the presynapse can be accentuated vis-à-vis the postsynapse with a dye by using a specific antibody developed for a neurotransmitter substance. Intracellular injections of contrast-enhancing molecules (procion yellow, Lucifer yellow, horseradish peroxidase) in conjuction with the phenomenon of neuronal transport (see Chap. 9) facilitate locating nerve fiber endings by means of both light and electron microscopy. The dyed image of individual nerve cells within the ventral ganglion chain of grasshoppers is an especially effective example of this (Fig. 1.19). A further potential means of identifi-

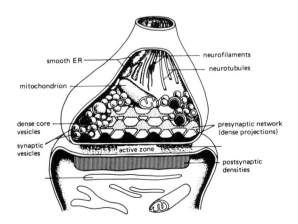

Fig. 1.18. Diagrammatic representation of the pre- and postsynaptic membrane differences and the intracellular organelles present in a chemical synapse.

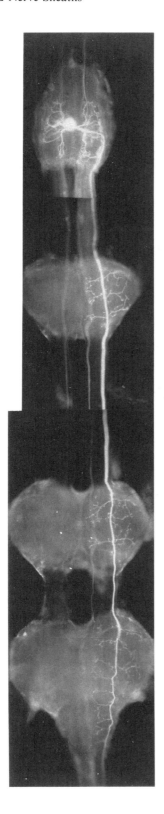

cation is offered by the degenerative changes following the lesion of defined nerve fibers. In this circumstance, one takes advantage of the fact that degenerating fibers develop a strong affinity for silver salts and, thus, can be dyed accordingly (Nauta method).

1.2 Glial Cells and Nerve Sheaths

The nerve cells of the central nervous system are wrapped, more or less completely, by neuroglial cells or, in the case of the peripheral nervous system, by Schwann cells. Neuroglial cells (glia in the CNS) were first observed in 1856 by Rudolf Virchow.

In fish, one glial cell is allotted to about eight neurons. In the CNS of mammals, in contrast, one neuron is allotted to up to ten glial cells. Generally, glial cells fulfill the functions of protection, support, insulation, and nourishment, the last function in the sense of metabolic symbiosis. Until recently, the view was overprized that the glia plays an essential role in the formation of the so-called *blood–brain barrier*, that controls which substances pass from the bloodstream to the nerve cell and which do not. Unlike neurons, glial cells retain the capacity to divide over their entire life span. They often occupy the place of neurons that are lost to injury (lesions). A state of degenerated glial growth can occur during periods of increased neoformation of cells (*proliferation*). Due to their extremely minimal bioelectric activity, the influence that glial cells have upon information processing is negligible. It is for this reason that the main types of glial cells are treated here only briefly (Fig. 1.20).

Fig. 1.19. Representation of nerve pathways in the ventral ganglion of the field grasshopper *Omocestus viridulus* enhanced with an intracellular injection of the fluorescent dye Lucifer yellow. The marked neuron travels from the hypopharynx ganglion to the abdominal ganglia and is active rhythmically during stridulation. (Photograph: B. Hedwig, Zoologisches Institut of the Universität Göttingen.)

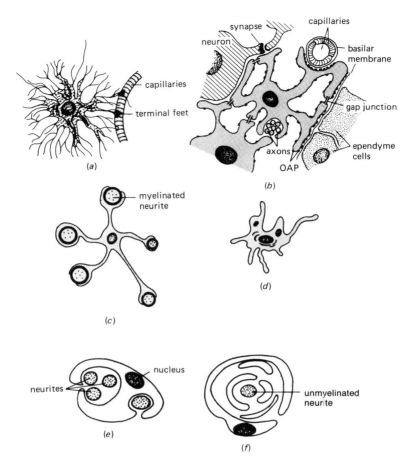

Fig. 1.20. Types of glial cells in the central nervous system of vertebrates $(a-f)$; (b) OAP (orthogonal arrays of particles) and lamina of the basilar membrane are ascribed functional involvement in the regeneration of nerve fibers: (a) protoplasmic astrocyte; (b) astrocyte functions; (c) oligodendroglial cell; (d) mesoglial cell; (e) unmyelinated glial cell; (f) unmyelinated loose glial folds.

1.2.1 Macro- or Astroglia

Macro- or astroglia consist of large, star-shaped cells with processes of varying length. Astrocytes of the white substance of the brain, rich in nerve fibers, contain dense fibril bundles in the cell body and in the processes, hence the alternate designation *fibrous astrocytes*. In contrast to these are the so-called *protoplasmic astrocytes* of the gray substance, which are poorer in filaments and richer in glycogen (Fig. 1.20a).

Both types of astrocytes enwrap the blood capillaries that course through the CNS and they envelop practically every neuronal membrane surface as well as the areas of the perikaryon, the dendrites, and the axon, including their contact points. Further, they send out widened cell runners (*gliapods*) to the cells of the soft menin (*pia mater*), which lies in close proximity to the brain, the spinal cord, and monolayered glial cell lining of the *ependyma* (Fig. 1.20b). The following functions are ascribed to the astroglia: mechanical support of the nerve cells, repair of certain damaged nerve tissue based on its increased proliferation activity (*glial growth*), insulation and bundling of nerve fibers, orientation assistance in the outgrowth of neurons, involvement in the metabolic process of nerve cells, particularly with regard to the regulation of the ion-metabolite balance in the sense of metabolic symbiosis. Most recently, the astroglia has been found to be especially important in the *regeneration* of nerve fibers (see Chap. 9.4) due to the so-called orthogonal areas of particles (OAP; see Fig. 1.20b) located in its membrane.

1.2.2 Oligodendroglia

The oligodendroglia, with only a few thin ramifications, contains on the ultrastructural level only few filaments and glycogen grains, but considerably more microtubuli. Consequently, it is difficult to distinguish its fibers from nerve fibers, although they never form synaptic contact zones. Oligodendroglial cells form the myelin

sheaths around the axons in the CNS and the Schwann myelin barriers in the peripheral nervous system (Fig. 1.21). Additionally, they are in a kind of metabolic symbiosis with the neurons that they envelop. In the CNS, one oligodendrocyte is capable of surrounding three to five nerve fibers simultaneously with one myelin barrier (Fig. 1.20c).

1.2.3 Meso- or Microglia

Also called *Hortega cells* (Fig. 1.20d), the first name most likely derives from the mesodermal connecting tissue that migrates jointly with the blood vessels into the CNS. Microglial cells are extraordinarily polymorphous. Following liquefaction of their extremely thin, spiny processes,

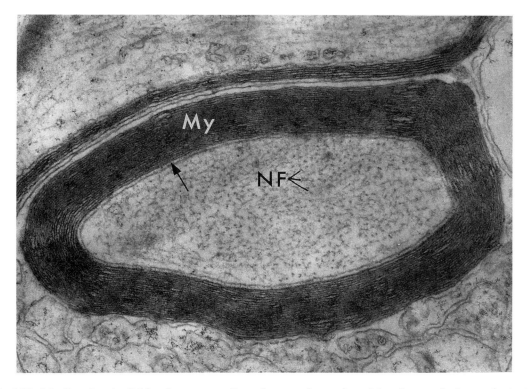

Fig. 1.21. Myelin sheath (My) of an unmyelinated axon from the vision layer of the optic tectum of a goldfish. Cross sections of neurofilaments (NF) can be seen in the axoplasm. The axon is bounded by the axolemma (↓).

they recast themselves into phagocytic wandering cells, removing accumulated tissue debris when nerve tissue is damaged. Accordingly, they are especially important in fending off infections of the CNS.

1.2.4 Neural Sheaths

The neuroglia, in particular the oligodendroglia, plays the important role of *neural sheath* in insulating the longer nerve fibers that perform connective functions between organs lying distant from neuronal centers of control. As such, the task of this glia is not simply to protect the nerve fiber, but to enable a more rational conduction of

impulse (*saltatory transmission of impulse*). In the simplest sense, neural or myelin sheaths are formed when an individual glial or Schwann cell loosely envelops a single axon or group of axons. This occurs with thin fibers in both invertebrates and vertebrates (Fig. 1.2e, f). The region where the Schwann cell meets the axon is called the *mesaxon*. The Schwann cells envelop short sections (0.5 to 3 mm in length) of an axon. Gaps emerge between two Schwann cells, the so-called *Ranvier nodes*. This is where the myelin sheath of one cell ends and the next one begins (Fig. 1.22). The myelinated section of the axon between two knots is called the *internodium*. The strong electrical insulating effect of the *myelin sheath* is the basis for *saltatory impulse conduction* by which an electrically coded impulse signal jumps from one node to the next (see Chap. 6.2.7).

Composition and development of the myelin sheath during ontogenesis were clarified with the help of the electron microscope. During *myelinization*, oligodendrocyte cells accumulate

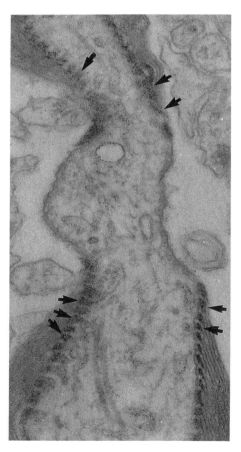

Fig. 1.22. Ranvier node with myelin sheath (↓) from the optic tectum of a trout (see Fig. 1.21).

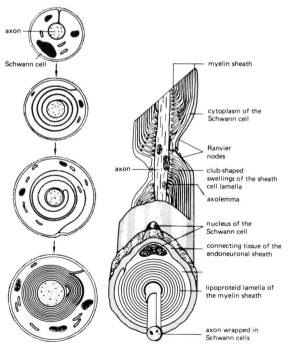

axon

Schwann cell

myelin sheath

cytoplasm of the Schwann cell

Ranvier nodes

club-shaped swellings of the sheath cell lamella

axolemma

axon

nucleus of the Schwann cell

connecting tissue of the endoneuronal sheath

lipoproteid lamella of the myelin sheath

axon wrapped in Schwann cells

Fig. 1.23. Diagram of the composition of a myelinated axon (*right*) and the development of the myelin sheath of a myelinated nerve fiber (*left*). (Reprinted with permission from Rohen, 1971) (see Fig. 1.22.)

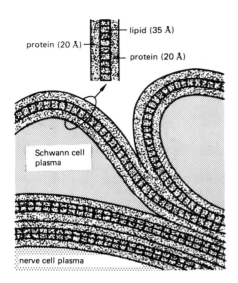

protein (20 Å) — lipid (35 Å)
— protein (20 Å)

Schwann cell plasma

nerve cell plasma

Fig. 1.24. Diagram of the lamella composition of the myelin sheath (see Fig. 1.23).

at the outgrowing nerve fiber and partially attach themselves to it. Owing to the normal outwardly spiraling path characteristic of the nerve fiber when growing, the membrane of the Schwann cell is rolled and pressed outward and its cytoplasm is

pressed to the periphery (Fig. 1.23). Schwann cell rolls mesh into one another dovetail fashion at the Ranvier nodes. The insulating effect of the myelin lamella is interrupted here to facilitate processes of substance exchange as well as to boost the electrical stimulus impulses. Unlike the membranes of other cells (liver cells, for example, the dry weight of which amounts to about 60% proteins and 40% lipids), the membrane of the myelin envelope is composed of about 25% proteins and 75% lipids, of which glycolipids in particular, such as cerebroside, as well as phospholipids and cholesterol are present in the greatest quantities. The so-called basic protein, glycoproteins and lipoproteins, are characteristic myelin proteins. Due to the large portion of lipids in myelin sheaths, the electron microscope represents these as a chiaroscuro of membranic deposits with a regular periodic distance of about 20 nm. This corresponds to the thickness of two cells lying one upon the another (Fig. 1.24).

Nerve sheaths organized in this way are found only in vertebrates. They are responsible for the fact that impulse conduction occurs so much more rapidly over great distances and without loss of intensity than is the case for invertebrates (see Chap. 6.2).

2

Development of the Nervous System in Vertebrates

Having introduced the essential building blocks of the nervous system in the previous chapter and having discussed their morphology, a closer look will be directed in this chapter toward the development and differentiation of neuronal systems. In particular, the cell-specific and molecular fundamentals of neuronal differentiation, i.e., the morphogenesis of larger neuronal assemblies in the central nervous system in vertebrates, will be examined in greater detail. An understanding of the processes upon which development is based leads to an understanding of the way in which the nervous system functions. These processes also constitute the basis for higher associative brain functions, in particular, the formation of the more complex functions of learning and memory.

The ability of the nerve cells to conduct impulses and, thus, to transmit information from one cell to the next is predicated upon the precise wiring capability of cells leading to specific synaptic activity. Of particular interest in this regard is the discussion of the requirements and possible cellular mechanisms that underlie the formation of neuronal structures as well as the depiction of the normal course of development of the nerve cell. Furthermore, the question lingers as to the basis of neuronal circuitry in general, i.e., how the projecting nerve fibers actually reach their target cells. This matter relates to the fundamental problem of *fiber routing* or *alignment*. Further questions concern those mechanisms by which neurons of various origins differentiate among numerous types of target cells so that functionally efficient synaptic connections result. This

latter question concerns the subject of *neuronal specificity*.

In an attempt to respond to the questions posed above, this chapter treats morphogenetic aspects of the processes of induction and differentiation in nerve cells in the formation of larger neuronal networks, including assisting structures. A few important cellular and molecular aspects of neuronal differentiation are then discussed in somewhat greater detail.

2.1 Morphogenetic Aspects of the Formation of Neuronal Structure

The central nervous system (CNS) of vertebrates (spinal cord and brain) originates during embryonal development from the neural plate, which is located on the back of the embryo. (In contrast to this, the corresponding structure for a central nervous system in invertebrates is on the ventral side.) The *neural plate* rolls inward in the course of its development and ultimately forms a neural tube around a *ventricle* filled with *cerebrospinal fluid*. The ventricle runs the entire length of the tube. The brain ventricle widens and wall sections thicken at the head area. From these occurrences emerge the three original main sections of the brain, i.e., the prosencephalon, mesencephalon, and metencephalon. The neural tube and the brain are surrounded by protective sleeves (meninges) and bone formations (backbone and skull cap).

The progressive development and maturation

of the brain, considered both embryonically and phylogenetically, is founded on processes of growth and differentiation on a cellular level.

There are eight separate steps, or stages:

1. Induction of the neural plate;
2. Multiplication of cells of specific regions;
3. Migration of cells from the place of their origin to the places where they ultimately remain;
4. Formation of anatomically identifiable cell groups;
5. Maturation (differentiation) of individual nerve cells;
6. Formation of bonds between the nerve cells;
7. Selective demise of some nerve cells;
8. Restructuring of some initially formed groups while remaining groups become stabilized.

2.1.1 Induction of Neural Plate and Neural Crest

The nervous system is formed very early in human embryonic development. The amniotic and yolk sac cavities are formed from the cell complex of the *embryoblast* about six days after fertilization of the egg cell and implantation of the germ in the uterine mucosa. Then, at seven to eight days, both germ layers, the ectoderm and endoderm, take shape from the cell complex. The chordomesoderm plate then moves between these two, representing the first organ structure for future organs to be located within the organism. Induction effects, not yet fully understood, originate from this mesoderm, and cause a portion of the ectoderm substance located above it to become nerve tissue and to form a neural plate (Fig. 2.1).

Two low molecular weight proteins clearly play an important role in the *induction* of these local processes and do so through their concentration ratio. One of these proteins appears to be responsible for the fact that nerve tissue is formed at all; the other seems to bring about regional differences in the formation of the neural structure through varied concentration in the separate sections of the neural plate. The substances in question have not, as yet, been isolated.

A chronological component also plays a significant role in induction. The chronological sequence is firmly established by which the neural plate is divided into separate areas of induction. First, the areas for the prosencephalon are induced, then those for the mesencephalon and metencephalon, and finally the areas for the spinal cord.

Neural Plate

Regarding their later development, the individual sections of the neural plate are determined chemically during their formation. Through this induction, a field forms at the anterior end of the neural plate from which the prosencephalon and the nerve cells of the eye originate. Damage to this field at a very early stage is balanced out by increased cell growth. Further development follows in a normal manner. If, however, the same field sustains damage of the same magnitude at a somewhat later point in its development, increased cell growth no longer occurs and the result is lasting damage to the prosencephalon and/or retina. Thus, development is passed on even at this early stage of the first structuring of a single prosencephalic field in the region of the neural plate.

In the course of further development, the prosencephalic field subdivides into different fields—a separate prosencephalic field and an occipital region. Then there occurs in the prosencephalic field the separation of individual, special cell groups whose development can be followed to very specific areas of the differentiated prosencephalon.

The number of cells of the neural plate is relatively small, about 125,000 according to research on amphibians. From about the 18th day, however, the neural plate, through increased cell division, begins to form folds on both sides of the ever-deepening neural groove and, in this way, unfolds. The *neural folds* increase in size until they ultimately come into contact and fuse with one another. In this way, between the 21st and 31st day, they form the full length of the *neural tube*, starting from the center, which is, at first, open at both ends. These openings disappear with the inward-rolling growth of the neural folds and, at this point, the separation of the neural tube from the ectoderm is complete. Occasionally, malformations occur in human

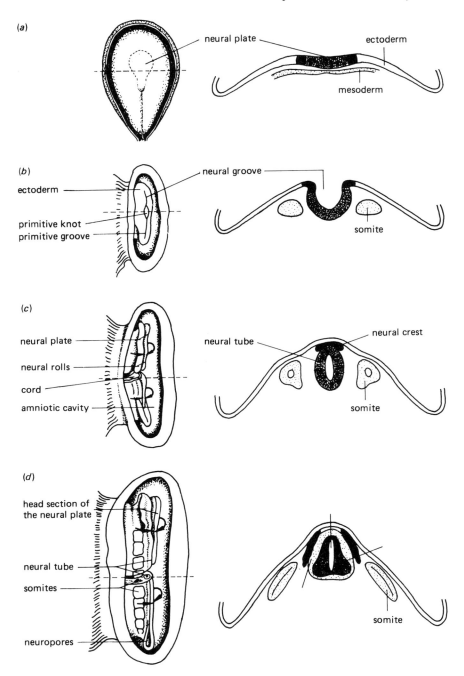

Fig. 2.1. Formation of the nervous system from the ectoderm of a human embryo in the course of the 3rd and 4th weeks after fertilization. Cross section of the embryo is represented on the *right*: (*a*) neural plate phase; (*b*) formation of the neural groove; (*c*) neural tube phase; (*d*) formation of the spinal cord.

beings, for example, when development in the neural tube does not run its full course, as in children with openings on the lower end of the spinal column (*spina bifida*) or with openings in the caudal region (*myeloschisis*, myelocele).

Neural Crest

Even during the formative process of the neural tube, a special cell formation becomes visible at the boundary between the integumentary ectoderm and the neural plate (Fig. 2.2). It becomes detached from the ectoderm shortly prior to the merging of the neural rolls, but then remains connected to the upper edge of the folds. Thus, during closure of the neural tube, this cell substance, brought from both sides, also becomes connected to one unified body, the *neural crest*, which undergoes its own development subsequent to the closure of the neural tube, in particular, by way of peripheral growth, migration of cells to other parts of the body and diverse manners of differentiation (Fig. 2.3). The following cells derive from the neural crest of the spinal cord:

- Neuroblasts of the *spinal ganglia* and of the peripheral ganglia,
- Sympathicoblasts of the *sympathetic trunk of the sympathetic system* and of the peripheral *vegetative ganglia*,
- *Parasympathetic ganglia,*
- *Pigment cells* of the adernal medulla (pheochromoblasts),
- All glio- and spongioblasts of all of the *peripheral glia*,
- *Melanoblasts* for the pigment cells of the trunk and the extremities.

2.1.2 Multiplication of Nerve Cells

The rate of cell division increases dramatically as soon as the neural tube is closed. The formerly thin wall of the neural tube becomes a thick epithelium in which the cell bodies can be found away from the inner wall. All cells undergo division prior to differentiation of typical tissue complexes and do so in a very unusual way: cell division is possible (ventricular mitoses, see Fig. 2.2c) only on the inner wall of the neural tube

and not the outer wall. Thus, as soon as a cell has divided, it develops a process by which the cell body stays in contact with the inner wall. The process then grows outward and, in doing so, pushes the cell ever further from the inner wall (Fig. 2.4). Meanwhile, DNA is synthesized in the cell body for the next step in cell division. In addition, the process involutes back to the inner wall and, thus, the cell body is drawn back to the inner wall. Apparently, the necessary mitosis apparatus can be formed only here. Subsequent to cell division, the push-out/draw-back event is repeated as often as is required for each separate section of brain and spinal cord.

After the relevant number of division cycles, the future nerve cells lose their ability to synthesize DNA and, thus, to divide. They are by no means functionally capable nerve cells in this stage. Indeed, they remain quite immature and must undergo further, special steps in development. In addition to the future nerve cells, glioblasts, i.e., the future glial cells of the brain, are also formed in the multiplication of primary neuroblasts (Fig. 2.5). These retain the ability to divide throughout their life span and now function in extremely close conjunction with the nerve cells as astrocytes or oligodendrocytes, protecting them and insulating them against electrical current (see Chap. 6.2.7).

The neural tube does not retain a typically tubular form in the region of the head; rather, *brain vesicles* are created very early from successive expansions and contractions of the neural tube: first, the prosencephalon vesicle subdivides further into the actual telencephalon vesicle and diencephalon vesicle, followed by the rhombencephalon vesicle. Finally, the mesencephalon vesicle is included first in the rhombencephalon and then it emerges secondarily (Fig. 2.6).

Cell multiplication in the wall of the neural tube ultimately leads to (a) various modes of distribution of nerve cells later on, (b) migration during the formation of neural networks, and (c) concentrated cell growth and migration to cell repositories. The rate of cell multiplication is, however, not the same in all places. In some areas, there are terrifically high rates of cell multiplication that would lead to a particular thickening of the brain enclosure. In other areas, multiplication is retarded. This results later in the

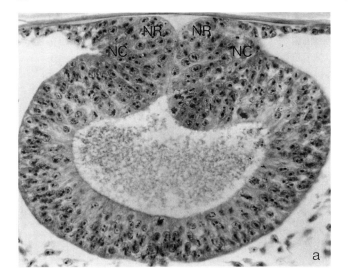

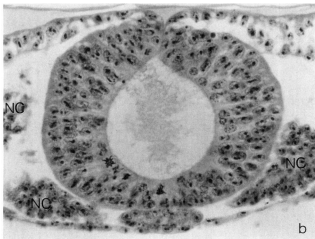

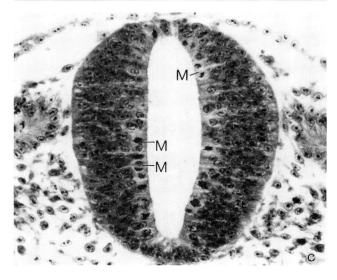

Fig. 2.2. Development of the neural tube and neural crest of the chick: (*a*) deepening of the neural groove to the neural tube, leveling of the neural rolls (NR), and emergence of the neural crest (NC); (*b*) migration of the cells of the neural crest; (*c*) periventricular mitoses (M) of the first neuroblasts.

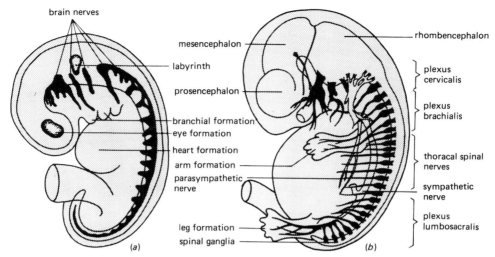

Fig. 2.3. Development of the spinal nerves from neural crest matter (*black*) in a 5-week-old (*a*) and an 8-week-old (*b*) human embryo.

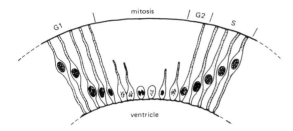

Fig. 2.4. Cycle of cell division in nerve cells and their migration in the wall of the neural tube between the germinal zone and the outer meninges. G and S: different stages of mitosis in the neuroblast.

complicated relief of individual brain structures.

A particular characteristic that marks the development of the neural tube in the brain region is the absence of universal cell multiplication in some regions of its wall. At this point in development, there is an epithelium here of only one layer. Later, migrating vessels of a tangled weaving, the *plexus choroidei*, develop at these locations which then are connected very closely to the lumen of the ventricles (see Fig. 3.5.4).

Three factors determine the maximal number of brain cells in each region of the brain:

- The duration of the cell division phase, which can last from a few days to several weeks,

- The duration of the individual cell division cycle, which elapses ever more slowly the more advanced the development, from only a few hours in the young embryo to a duration of five days in later stages,
- The number of cells, from which the development of a specific region of the brain primarily emerges.

In a number of cases, multiplication in the original ventricular layer is insufficient for the formation of particular brain structures (see Fig. 2.7). In addition, a *subventricular layer* can form here in which, contrary to the norm, future nerve cells, having already migrated, continue to divide. This layer is well defined primarily in the prosencephalon and occurs in the metencephalon, especially in the cerebellum. Additional billions of particularly small nerve cells are produced here, in this subventricular level, in the few weeks of activity.

This *neurogenesis*, in the sense of *mitotic* multiplication of nerve cells, ceases early in the higher vertebrates: in mice, at about three weeks after birth, in humans, up to two years old. Only in lower vertebrates (fish and amphibians) does the mantle zone continue to form neuroblasts that can be differentiated as neurons in the adult stage. The great regeneration capacity of the CNS in primitive vertebrates is based on this.

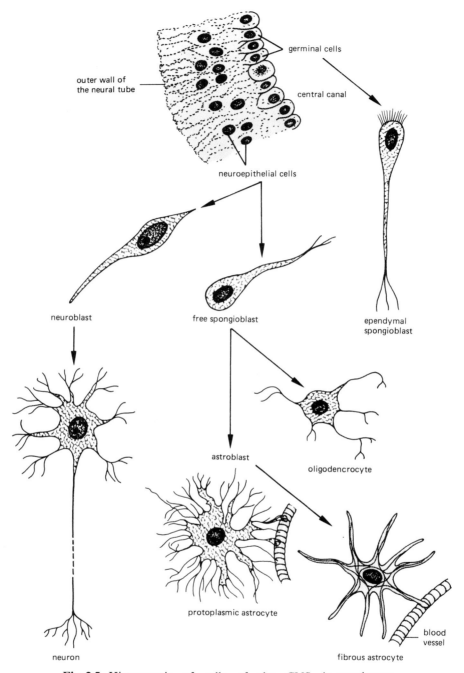

Fig. 2.5. Histogenesis of cells of the CNS in vertebrates.

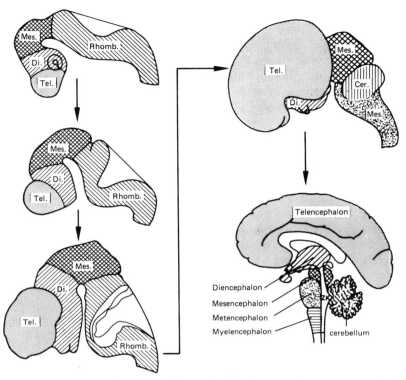

Fig. 2.6. Ontogenetic development of the CNS in higher vertebrates through differentiation of the spinal tube in the prosencephalon (tel- and diencephalon), mesencephalon, and metencephalon (rhombencephalon with met- and myelencephalon).

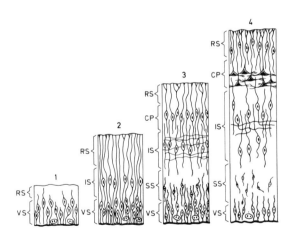

Fig. 2.7. Ontogenetic formation of cell groups (stratification) in the CNS in higher vertebrates. VL: ventricular layer; BL: boundary layer; IL: intermediary layer; SL: subventricular layer; CP: cortical plate; 1: neuronal tube phase; 2: cell division and migration phase; 3: formation of the cortical plate in the prosencephalon; 4: differentiation of a layered cortex.

Governance of cell multiplication is not yet fully clarified, but, apparently, it is predetermined precisely. The chronological component seems to play an essential role in controlling cell multiplication, for it is the particular moment in time that a cell is formed during embryonic development that is critical for its further development. A cell departing the cycle of division, i.e., a young nerve cell that presumably will arrive at a very specific location, is differentiated very specifically, and it appears to be established at this point with which cells' subsequent connections will occur.

There may well be quite natural explanations for the processes of cell-to-cell recognition. A key to its understanding probably lies with the gangliosides which, indeed, are found throughout the body but are particularly abundant in the brain. They have come, of late, under increasingly intense scientific examination (see Chap. 8.2).

2.1.3 Migration of Nerve Cells

After the cycles of cell division have concluded, their numbers for each area of the brain appearing to have been preestablished, the cells migrate from the periventricular matrix layer to an intermediary cell layer (Fig. 2.7).

This layer consists of young, as yet undifferentiated nerve cells (*neuroblasts*), which have lost their ability to divide, and undifferentiated glial cells (*glioblasts*), which never lose their ability to undergo division.

The neuroblasts, from which especially large nerve cells and/or nerve cells with extremely long processes develop, emerge relatively early on during ontogenesis. Neuroblasts for smaller nerve cells originate later. It appears that the first functionally capable glial cells emerge simultaneously with the formation of the first neuroblasts. This is essential for further differentiation and organization of neuronal structures inasmuch as the first glial cells constitute a supportive structure for the migrating nerve cells. These first, specialized glial cells function at least as long as it takes for the nerve cells to complete their migration and, most likely, until their connections to the first neural networks have been established.

It has been shown in numerous cell culture experiments (first by Harrison in 1910), that cell migration occurs by ameboid movement. It is also known, however, that a fully functional nervous system never develops from cells multiplying under cell culture conditions. The cells that are created in a culture medium are simpler than, and are structured differently from corresponding cells in tissue under natural conditions. It is most likely, in this regard, that the interaction with glial cells plays a critical role. Disorders during the formation of glial cells, or their premature demise, lead to errant routing and concomitant incorrect final positions, or result in the incapacity of certain nerve cells to migrate at all. The cells that are routed incorrectly do not establish contact with others and later degenerate.

How does an "ameboid" movement occur? What are its causes? As early as 1911, Ramon y Cajal discovered prominences at the ends of nerve fibers, the so-called *growth cones*. Electron-microscopic examinations showed uniquely small, thin prominences as the processes of larger *filopodia*. The cell attaches itself to the surrounding structure by means of these burr-like projections. The growth cones and their processes contain an entire network of *microfilaments* that are capable of contraction brought on by the protein *actin*. Hence, the basis for movement by which the entire cell is propelled.

The growth cones also contain mitochondria, microtubules, vesicles, and ribosomes. The cones are places of highly energetic processes of metabolism. New membranes and other particles are being formed here constantly and, in this context, intensive transport of material takes place from the cell body through axons and dendrites (see Chap. 9). The growth cones of axons and dendrites apparently have similar properties; by and large they are more simply built than the ultimately differentiated synaptic endings of the functionally active nerve cell. The growth cones of the glial cells, too, are structured similarly and function according to the same principle.

Growth of the nerve cell processes can be affected by many chemical substances, particularly by those that react with the membrane (see Sect. 2.2).

It has been shown electrophysiologically that the growth and function of the growth cones and, thus, the active "amebic" movement of the cell, are dependent upon the presence of Ca^{2+}. Ca^{2+} is associated with contractility of the filaments. It influences the shape of the growth cone and other structures as well as the manner of movement. The functional model of the ganglioside offers a sensible hypothesis for the explanation of these findings. This model is more closely examined in connection with the function of synapses (see Chap. 8.2).

2.1.4 Formation of Identifiable Groups

Normally, cell migration proceeds from the ventricular layer to the intermediary layer. In the prosencephalon (telencephalon), however, some of the cells transmigrate through this layer and accumulate in a third layer, the cortical plate (Fig. 2.7) from which the cerebral cortex develops. Furthermore, a portion of the cells does not reach the intermediary level, but rather gathers beneath it forming an interlayer, the *subventricular layer*, lying between the ventricular and intermediary layers. These cells, moreover, are capable of dividing and generating great quantities of nerve cells, especially in the cerebrum and cerebellum, which then migrate to the end positions. The massive basal ganglia of the cerebrum's hemispheres take form in the *cerebrum* from these cells, and a portion of the small cells of the cerebral cortex originates from here.

An exception to the rule, cells still capable of dividing migrate in the metencephalon from the subventricular layer and, in a second migration phase, reach the region beneath the surface of the cerebellum, where they then form a new layer through cell division. In the brief weeks of their activity, billions of granular cells and interneurons of the cerebellum are formed here.

In the final stage of migration, the ventricular layer remains as a thin lining of the ventricle and differentiation gradually occurs in the more distant layers.

The basic shape of the *spinal cord*, clearly recognizable in the brain as an elementary structure, develops as a result of dissimilar rates of multiplication in the matrix zone whereby more cell matter is generated in the lateral sections and, through appropriate migratory movements, is distributed in a cross-sectional pattern reminiscent of a butterfly. Thus, there emerges the characteristic subdivision into relatively thin upper and lower plates, consisting primarily of fibers, and the lateral base plate and lamina alaris consisting of matter rich in cell bodies (see Fig. 3.12, Chap. 3.2.2.1).

2.1.5 Differentiation of Nerve Cells

Differentiation of nerve cells goes hand in hand with their overall development, from their origin as neuroblasts to their active engagement as nerve cells.

Neuroblasts accumulate in cell aggregates, based on the molcecular surface of their membrane, after having reached their final position. The neuroblasts are, at this point, so specifically equipped that they correctly reassemble even upon experimentally induced dispersal. The specificity of molecular recognition even applies to the alignment of the individual cell, as is evidenced by the particularly large pyramidal cells in the prosencephalon. These cells are always in the same orientation: dendrites directed upwardly and perpendicular to the brain surface, axons directed downwardly.

Both the formation of the fibers and the utilization of the cell and its fibers with regard to its specific function play significantly into the differentiation of nerve cells. Thus, the mode of impulse reception and conduction is precisely determined for each cell as well as whether, and with which transmitter, synaptic impulse potentials are triggered, or which transmitter is produced by the cell under which conditions. Intensive study is still necessary in the area of nerve cell differentiation in order to afford us a better understanding of the process.

On the other hand, far more progress has been made in researching the more morphologically traceable steps both in individual cell differentiation and in the various areas of the brain and spinal cord, including their chronological sequence of development. *Multipolar nerve cells*, with their numerous dendrites for impulse reception and their singular axon for impulse conduction, result, in most cases, from the maturing nerve cell as it develops numerous processes. As a rule, these processes begin to form after the cell has migrated to its point of destination. The manner in which specific fiber differentiation is triggered is, as yet, unknown. Cell culture experiments reveal that no external structure is required, that merely a solid base is sufficient to induce formation. However, functional associations have yet to be created in cultures. Thus, very specialized structures of conduction must be present that also provide for the correct linkage of the fibers. This was also required for the migrating cell. The general structure of dendrites, in particular, in brain cells is largely uniform under both natural and artificial conditions leading one to conclude that a genetically established blueprint exists. The ultimate form of the complete cell, on the other hand, is dependent upon the number of fibers from other cells and where contact is established. In the final analysis, that is what determines the exact position of the individual cell within its cell group. The manner in which contact is established from cell to cell is discussed in greater detail in Chap. 9.2.

Differentiation of neurons, characterized, in particular, by the outward growth of the nerve fibers, leads to the *mantle zone* of the neural tube, which is filled with the perikarya of neurons and which can be depicted readily histologically. The *marginal zone* (*zona marginalis*), consisting of nerve fibers, derives from this mantle zone. Since many nerve fibers of higher vertebrates are myelinated with oligodendroglial cells (see Chap. 1.2), this kind of fiber layer stands out clearly from the gray cell body layer (*substantia grisea*) as white matter (*substantia alba*). As the neural tube distends in the anterior body section into brain ventricles that are filled with cerebrospinal fluid, there occurs a differentiated formation of the mantle and marginal zones in the form of variously layered brain structures. This will be discussed more closely under functional morphology in Chap. 3.

2.1.6 Elimination of Surplus Matter

Surplus cells are eliminated during ontogenetic differentiation of nerve cells. It can be concluded from this that considerably more neuroblasts are created originally than actually are needed later on. This would relate to the general rule of nature whereby, to offer a margin of security, more matter is produced than is necessary, especially when that matter provides for essential tasks. Reproductive cells, for example, are produced in great surplus.

It is apparently the case with nerve cells that, during differentiation, all cells that did not establish contact with other neurons or response organs are eliminated. Before this backdrop, the requirement discussed in another connection (for an adequate system of conduction during early development in humans) takes on great significace in assuring the best possible innervation early on (see Chap. 9.2). The actual utilization of original nerve matter and the ultimate size of a nervous structure do not depend upon the original neuronal matter available, but rather upon the size of the structures to be innervated and the demands placed upon them. In this regard, it has been shown experimentally that increasing the normal demand, for

example, brought on by the implantation of an additional leg structure in amphibians, leads to a corresponding spinal cord structure larger than normal. On the other hand, the brain structure becomes smaller when the area to be innervated is reduced in size artificially (*inactivity atrophy*). In this regard, extensive research of individual structures of the brain has been carried out, but studies of the entire brain are still incomplete.

However, it has been observed thus far that the elimination of unused cells can result in *shrinkage rates* of between 15 and 85% of the original mass.

In addition to the phenomenon of surplus neurons, there is also the massive outgrowth of nerve fibers that occurs during differentiation. To date, there have been only a few studies on this subject, such as those that have shown that the muscle cells of rats are connected at first to several synapses during their adolescent stage. This number is reduced at a later point in time. Ultimately, a muscle cell in an adult rat connects to the nerve cell by only one synapse. The rate of this neuromuscular *elimination of synapses* is dependent upon the overall activity of the system. Thus, on the one hand, a *muscle paralysis* brought on by treatment with pre- or post-synaptic neurotoxins reduces the extent of synapse elimination. On the other hand, the rate of elimination is shown to be accelerated by chronic stimulation of neonatal nerves or muscles. The conclusion that may be drawn from this is that the synapse elimination process is a measure of the specific actuation of few individual, but functional, synapses with a simultaneous reduction in many "pioneer" synapses, those of early ontogenetic existence.

It is possible that a change can take place in the convergence and divergence of synaptic connections (see Chap. 5.1) during early ontogenesis in vertebrates. Thus, considerably more preganglionic fibers are introduced into the ganglion cells of autonomic nerves than is the case in the mature stage, and preganglionic fibers innervate considerably more postsynaptic cells. At birth, numerous climbing fibers advance upon the large *Purkinje cells* in the cerebellum of the rat. In the mature state, however, each Purkinje cell receives information from only one individual climbing fiber.

Findings of this sort, about the function-dependent, early ontogenetic elimination of synapses, might be of critical importance not only with regard to the actuation of neuromusculature systems, but also in regard to interneuronal wiring during, for example, the formative phase and probably life-long in the formation of new neuronal assemblies that, indeed, might be the basis for the everlasting capacity to form memory.

2.2 Cellular and Molecular Aspects of Neuronal Differentiation

In connection with the developmental description of neuronal elements in the course of both neurogenesis and during the formation of structured nerve cell groups into a functional whole, attention will be directed more closely, at this juncture, to the still largely unanswered questions pertaining to neuronal differentiation, specifically to the manner in which the processes of the nerve cells actually reach their target cells, i.e., the sensory, glandular, and muscle cells.

It is known that, at first, only individual cell processes, the so-called *pioneer fibers*, grow from the nerve cell body outward to the periphery and that these pioneer fibers serve as elements that guide the fibers that later emerge (Fig. 2.8). In the

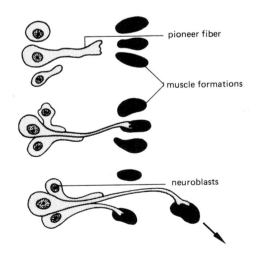

pioneer fiber

muscle formations

neuroblasts

Fig. 2.8. Growth of "pioneer" fibers from neuroblasts during early ontogenetic differentiation.

early ontogenetic stage, the distances between the nerve cells taking form and the structures to be innervated are, indeed, small. Nonetheless, the distances increase as the embryo grows in size and, as a result, establishing contact can become ever more difficult a task.

There are various experiments to explain the many processes of fiber guidance, or alignment, and neuronal specificity.

2.2.1 Nerve Fiber Growth Through Neurobiotaxis

The *neurobiotaxis hypothesis* holds that there are structures already present in the tissue that serve as directional pathways for the growing neurons. Weiss was able to show as early as 1943 that growing nerve fibers orient themselves mechanically even to the most ordinary structures that might be present as guide markers, such as scratches in the tissue culture dish. Furthermore, the manner and direction of the impulses affect the development of the ultimate nerve pathways.

Observations of axon growth in the extremities of the young chick offer clues to this kind of fiber alignment and pathway selection. Growing nerve fibers from the anterior horn roots of the spinal cord locate their effector musculature even if the muscles to be innervated (for example, the musculus sartorius opposite the musculus ischioflexorius) were interchanged. If, on the other hand, the nerve structures in the corresponding segments of the spinal cord are interchanged, then faulty innervations occur attended by corresponding malformations.

One concludes from this that nerve fiber growth is preprogrammed more or less rigidly and that a strong directional influence is exerted on that growth by the tissue to be innervated. Additionally, the effects of adjacent nerve cells on one another are of great importance, as has been demonstrated in research performed on the nervous system of *leeches*: large *Retzius cells* contained in a culture medium will form electrical synapses among themselves, something they will not do in the presence of other neurons. Furthermore, the Retzius cells will form chemical synapses with so-called P cells, but not when a third cell is present. The dynamics of this behavior are, as yet, not fully understood.

2.2.2 Nerve Fiber Growth Through Galvanotropism

The galvanotropism hypothesis of nerve fiber growth is based on the fact that differing *electrical fields* are formed during the growth of varying fiber populations and these fields influence the direction in which the fibers grow (Ariens Kappers). Given the extraordinary heterogeneity in the chemical composition of nerve cells, particularly of their membranes and their functional interaction with charged particles, one should not dismiss this hypothesis out of hand, but consider it very carefully until one understands the significance of changes in electrical field strength as it relates to neuronal processes. It is already known that completely endogenous electrical current on the scale of 1 to 10 $\mu A/cm^2$ exist in various adult and embryonic tissues. Furthermore, it has been shown in single neurons from the neural tube of clawed frogs in tissue culture that their growth is directed toward the cathode when they are exposed to an electrical field of $7\,mV/cm^2$.

2.2.3 Nerve Fiber Growth Through Chemoaffinity

According to the *hypothesis of chemoaffinity/neurotropism*, from the work of Ramon y Cajal, the organs that are to be innervated give off chemical "attractants" that stimulate the nerve fibers to grow in a specific direction. Presumably, nerve cells and their fibers are equipped to perform processes of identification. Additionally, growing nerve fibers, especially in the case of synapse formation, control very specific chemical affinities with regard to the receiving cells to be innervated.

Because of the particular minuteness of the structures in question, it is difficult to furnish proof of the correctness of such a *chemoaffinity hypothesis*. Thus, attempts have been made at solving the mystery of topographic classification by employing neurosurgical methods. However, experiments that seem to hold greater prospects for success have been carried out in which antibodies were prepared for specific antigens, in this case from the neurons of leeches. Some of these antibodies recognize particular populations of neurons or even single cells. This might represent the right path for future experimentation in substantiating the chemoaffinity hypothesis of neuronal specificity.

2.2.3.1 Nerve Growth Factor

The *nerve growth factor* (*NGF*), discovered in 1954 by Levi-Montalcini, Cohen, and Hamburger, offered conclusive proof of the existence of *chemotropism* in nerve fiber growth. It centered on an extract of a skin tumor from a mouse, which is readily obtainable nowadays from the salivary gland of the mouse and from snake venom. Administering this substance to cells of neural crest derivatives (sensory cells and sympathetic ganglia) in a cell culture had both a *neurotrophic effect*—the life span of the cell is increased—and a *neuronotropic*, or *neuritogenic* effect—axon growth is directed to the point where the NGF was applied. After administering the NGF antiserum in newborn rats, there is a complete disappearance of the sympathetic nervous system.

The NGF molecule, similar in composition to insulin, has a molecular weight of 130,000 and exists in several forms: the largest subunit, based on the sedimentary coefficient is the so-called 7S-NGF; the most active form, however, is beta-NGF, a polypeptide consisting of 118 amino acids that is loosely related to 7S-NGF. The NGF bonds specifically to receptor molecules in the nerve fiber endings. With the help of receptor-supported *endocytosis* (the taking up of substances through involutions of the cell membrane in the region where the substance is to be absorbed), it reaches the synapse and, from there, it is carried to the cell body by means of *retrograde transport* (see Chap. 9.1.3). A change in the rate of the synthesis of neuroplasm compounds is regulated from the cell body and is accompanied by accelerated fiber growth (Fig. 2.9). It is in this context that one speaks of the NGF–receptor interaction being dependent upon the Ca^{2+} concentration and cyclic nucleotides, or that the NGF brings influence to bear primarily upon the activity of the K^+-Na^+ pump.

Only the fibers of sympathetic neurons grow in the presence of the NGF in the tissue culture. The

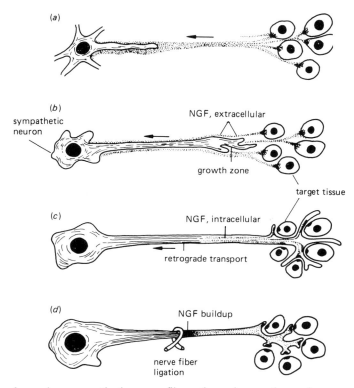

Fig. 2.9. Attraction of growing sympathetic nerve fibers through secretions of nerve growth factor (NGF) from cells of a target organ (*a, b*), reception of the NGF by the nerve terminals through endocytosis (*c*), and retrograde transport represented by fiber tying (*d*).

factor has no effect on the cell bodies. One concludes from this that the NGF has an effect only on the nerve fiber terminals, i.e., the growth cones, and that it is only here that the appropriate receptors are found.

2.2.3.2 Other Neuritogenic Substances

The NGF is a substance that has demonstrated a clear-cut, i.e., specific, neuronotropic effect on the nerve fiber growth of neural crest cell derivatives. It is reasonable to assume that there exists an array of compounds, each of which has a similar effect on other nerve cell fibers. Numerous other substances have a neuritogenic effect similar to the NGF: nerve fibers grow in the direction of the source of each substance after *dibutyryl-cAMP, cGMP, phosphodiesterase inhibitors,* and *calcium ions* are administered in the presence of a specific carrier, *ionophore "A 23187."* Intensified, di-

rectional growth was introduced each time by an increase in the formation of *filopods*, i.e., small, foot-like growths of protoplasm, and then by changes in the membrane in the region of the growth cone.

Important clues about chemoaffinity result from experiments in which the growth of ciliary ganglia cells in chicks was accelerated greatly when a culture medium containing *laminin* was used to target the neurons [laminin is an essential factor of the basal lamina (see Fig. 1.20b) that is produced by many cells in vivo.]

Additionally, *steroidal hormones* might also have a neuronotropic as well as neuritogenic effect.

Experiments on zebra finches show that *steroidal hormones (testosterone)* administered supplementally to male birds not in their reproductive period and to female birds, as well, increase the number, the size, and the dendritic branchings of neurons in the vocalization centers of the brain resulting in the birds singing out of season. Similarly, administering supplemental *ecdysone*, a steroidal hormone, to the silk moth, a butterfly, inhibits neuronal death.

Special attention should be given, as well, to the glycosphingolipids, which contain sialic acid and accumulate in the outer neuronal membrane. *Gangliosides* (see Chap. 8.2) merit particularly close attention in this regard. They clearly have both neurotrophic and neuritogenic effects on *primary neurons* in vitro and undifferentiated *neuroblastoma cells*. In tissue culture cells of varying origin, ganglioside mixtures and individual gangliosides, added exogenously—

depending upon the type of cell—not only increase the longevity of the cells, often in extremely low concentrations (*neurotrophic effect*), but also greatly affect the growth of nerve fibers (*neuritogenic effect*, Fig. 2.10). Presumably, the gangliosides modulate the activity of *protein kinases*, which are bound to the membrane. It remains to be discovered to what extent these ganglioside effects, which promote growth in nerve fibers, are molecule or cell specific.

Numerous other compounds, such as Ca^{2+} ions in conjunction with an ionophore (ion carrier molecule), trigger similar effects. Since sodium salts of the gangliosides are always administered in ganglioside application studies,

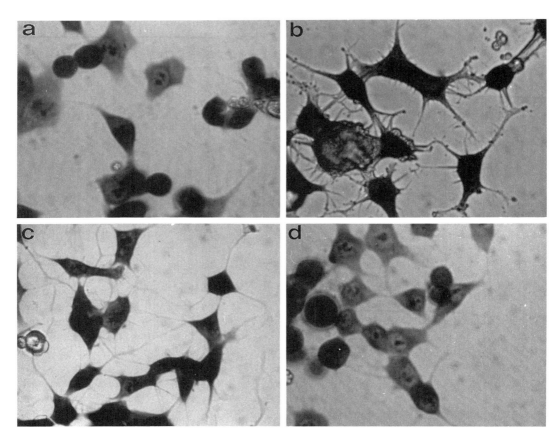

Fig. 2.10. The effect of externally administered gangliosides on the growth of in vitro neuroblastoma cells: (*a*) control cells; (*b*) growth-promoting effect (neuritogenic effect) following the application of a ganglioside mixture (40 μg/ml) from a bovine brain; (*c*) weaker effect following a dose double that of (*b*); (*d*) inhibitory effect of a highly polar ganglioside mixture from the brain of a pigeon.

and since the gangliosides show a greater affinity for Ca^{2+} ions, it would be possible to trace the neuritogenic effects to calcium activity. It is essential that further studies be carried out in this area, particularly in regard to in vitro findings, because, in the meantime, exogenously administered gangliosides are used in the treatment of *neuropathies*, even though they are unable to overcome the barrier systems (blood–brain barrier) that exist vis-à-vis the nerve cells.

2.2.3.3 Cell Adhesion Molecules

In addition to those substances that have a neurotrophic and/or neuronotropic (neuritogenic) effect on nerve fiber growth, compounds have been identified that play a significant role in *cell-to-cell recognition* and reciprocal cell adhesion. These "*cell adhesion molecules*" (*CAM*) are glycoproteins with an unusually high sialic acid content and a complex ramified pattern of sialic acid residue. Their job might be to attach individual neurons to one another and to regulate the exchange of matter between cells. A third topic of consideration is whether and to what extent they might be involved in the formation of *electrical lines of force* that surround the neurons.

The *nerve cell adhesion molecule (NCAM)* has been recognized to be an integral *sialoglycoprotein* of the membrane that consists of three subunits of varying molecular size. Another neuronal CAM is the so-called *NgCAM* which is limited to the neuroglia. Other CAMs, such as the liver CAM (*LCAM*), are not found in the nervous system. However, both CAMs, the N- and the LCAM, are produced at first by non-neuronal cells during early embryonic development. Present research is seeking to determine if NCAMs are responsible for *cell-to-cell attachment* in the region of the synapses. NCAM antibodies thwart synapse formation between growing optic nerve fibers and the visual centers in the brain. There is still considerable uncertainty as to the *mechanism of cell adhesion*. In general, the attaching of a cell to a surface is dependent upon two opposing forces: the tendency of surfaces to attract one another and their tendency to repel one another. Adhesion is ultimately the result of an equal balance between these two tendencies. The cells could be held together by two mechanisms: the relatively weak molecular attraction of the van der Waals' forces and specific bonding reactions similar in nature to interactions between antigens and antibodies, or between an enzyme and a substrate, or even between a ligand and its receptor. The conceivable mechanisms of bonding in regard to cell-to-cell adhesion on the molecular level could be

- A receptor of one cell is connected to the receptor of a neighboring cell by way of a bivalent ligand (Fig. 2.11a),
- A receptor of one cell connects with a ligand from another cell (Fig. 2.11b),

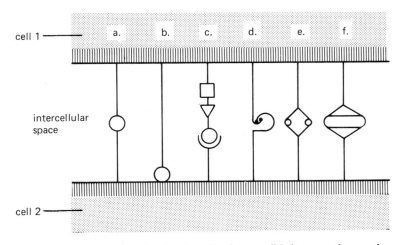

Fig. 2.11. Diagram of the various modes of molecular adhesion possible between the membranes of two cells.

- An enzyme-substrate interaction exists between two cells (Fig. 2.11c),
- A bond exists between two cells and an identical receptor (Fig. 2.11d),
- Based on the similarity of their sugar side chains, the glycoconjugates of the membrane are held together with the help of two divalent plant lectins (for example, concanavalin A, "wheat germ agglutinin") (Fig. 2.11e),
- Like the above item but assisted by a multivalent lectin (Fig. 2.11f).

The following possibilities come under discussion with regard to the triggering, or emergence, of these kinds of intercellular bonds:

- *Brown's molecular movement* of components of the intercellular substance could separate both cell surfaces from one another under certain conditions, or bring them together.
- *Steric effects* derived from macromolecules such as collagen, which are secreted into the intercellular space, inhibit, more or less, a close cell-to-cell bond.
- *Hydrodynamic forces* resulting from the intercellular medium (especially in the case of cells that are still growing) enable the cells to separate from one another.
- Finally, *electrostatic forces* from carriers of a negative charge in the membrane surface could regulate cell-to-cell contact. In vertebrates, glycoproteins containing sialic (neuraminic) acid and, especially, glycolipids (gangliosides) draw particular attention. Effects due to their negative charge might influence reciprocal cell recognition and cell adhesion in the adjacent area, for example, in synaptic cell-to-cell contact. The negative charge of the ganglioside might also have an effect on the formation of electrical fields which are especially important in the synaptic region. A reversible, weak bond of cations, mostly from calcium, seems to have a considerable effect on gangliosides, which are oriented toward the surface (see Fig. 8.16 and Chap. 8.2).

2.2.3.4 Gangliosides as Marker Substances of Functional Neuronal Differentiation

The accumulation of *sialoglycolipids*, especially *gangliosides*, in the outer nerve cell membrane of vertebrates received particular emphasis both in the discussion of the manner in which molecules of neurotrophic and/or neuronotropic (neuritogenic) nature affect neuronal differentiation (see Sect. 2.2.3) as well as in the discussion of cell adhesion molecules (Sect. 2.2.3.3). In addition to the extraordinary number of ways in which gangliosides influence the sphere of neuronal events (see Fig. 2.12), they are also credited with being the marker substances in neuronal differentiation. Ultimately, that means that they also play a part in the modulation of memory processes (see Chap. 11.2.4).

Tremendous increases in the concentration and polarity of brain gangliosides occur (Fig. 2.12a) during early ontogeny, particularly during critical and progressive phases of development (development of the eyes; birth or hatching, formation of the earliest reflex reactions; transition to free swimming in fish; formation of visual acuity: Fig. 2.12d). These can be correlated to increases in activity of the corresponding enzyme systems, i.e., on the one hand, to anabolic *sialyltransferases* which are involved in the composition of gangliosides, and, on the other hand, to catabolic neuraminidases which regulate the breaking down of neuraminic acids (Fig. 2.12c). The increase in the amount of gangliosides does not occur uniformly for the individual ganglioside fractions. Rather, conspicuous shifts can be seen in the combination of fractions that are less polar as opposed to highly polar, i.e., containing many neuraminic acids (Fig. 2.12b). Generally, *mono-* and *disialogangliosides*, which are less polar in nature, are synthesized first, i.e., beginning approximately with the transition from the gastrulation phase to the neurulation phase. The formation of more polar *polysialogangliosides* follows this stage and, although these remain for life in lower, cold-blooded vertebrates, especially fish (Fig. 2.12b) and amphibians, these polar fractions are reduced in warm-blooded birds and mammals in the course of further development in favor of less polar ganglioside molecules. This results in a preponderance of di- and trisialogangliosides in the brain of these animals (Fig. 2.13).

The main stages in the synthesis of gangliosides in vertebrates can be correlated in detail with successive morphogenetic steps of differentiation. The main synthesis of each specific gan-

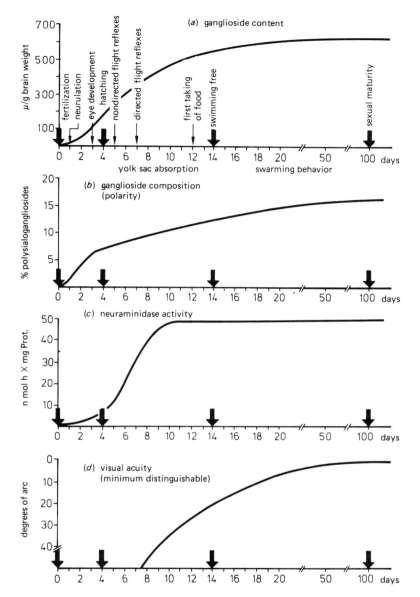

Fig. 2.12. Correlation of the degree of morphogenetic differentiation during the early ontological development of the cichlid fish *Sarotherodon sp.* and sialoglyco-conjugate metabolism of the CNS. Development profiles of the ganglioside concentration (*a*), composi-tion (△ polarity; (*b*), and increase in neuraminidase activity (*c*); parallel to this is a significant increase in visual acuity (minimum distinguishable = capacity to make visual distinctions; (*d*).

glioside at established points in time during neuronal differentiation means that each specific stage in development must be regarded in terms of a functional connection to the particular biochemical properties of each synthesized ganglion. Thus, specific gangliosides outwardly

mark important phases of neurogenesis (Fig. 2.14). Science still seeks to gain insight into the phenomenon of differentiation by illuminating the molecular–biological role that gangliosides play in it.

The correlation between gongliosides and

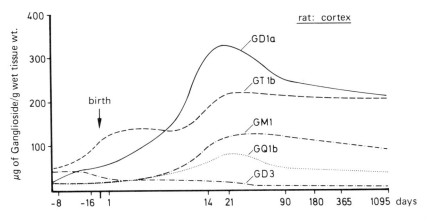

Fig. 2.13. Change in the composition of gangliosides in the rat cortex during the period from 18 days prior to birth to senescence (3 years).

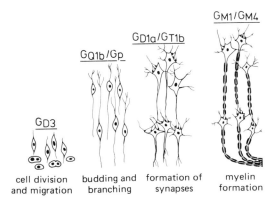

Fig. 2.14. The "marker" significance of gangliosides for neurogenesis: shifts in the main biosynthesis of individual gangliosides with differing amounts of negatively charged sialic acids ($\triangle$ polarity) characterize important stages in morphogenic differentiation. See Chapter 7.3 for the nomenclature of gangliosides.

neurogenesis can be seen in the following phases:

- The *phase of cell division and migration*, notably in newly formed neurons, is characterized by extremely intensive biosynthesis of the relatively simple disialoganglioside GD_3.
- The *phase of fiber growth, sprouting*, and *neuron ramification* is associated with the formation of highly polar tetra-(GQ_{1b}), penta-(GP_{1c}), and more highly sialylated gangliosides;
- The *phase of the growth spurt*, so-called, of neurons is associated with the *formation of*

synapses; it is signaled by an increase in the synthesis of the disialoganglioside GD_{1a} in mammals and birds, and by the trisialoganglioside GT_{1b} in fish;

- The *phase of myelination*, which concludes neurogenesis, is characterized by an increase in the formation of two relatively simple monosialogangliosides (GM_1 and GM_4);
- During the *senescence phase* in mammals, the ganglioside pattern finally is altered in individual regions of the brain, for example, in the cortex, once more in favor of polar fractions and is brought about by a strong breakdown of the GD_{1a} ganglioside.

A few of these marker phases can be shown very effectively histochemically in their corresponding brain structures with the help of *monoclonal antibodies* sensitive to particular gangliosides. Thus, for example, it can be demonstrated by cross sections of the spinal cord of a five-day-old chick embryo (Fig. 2.15) that a monoclonal antibody (AbR24) that specifically bonds to the disialoganglioside GD_3 reacts immunohistochemically only in the region of the periventricular neuroblast cell body which is still able to divide. Another antibody (Q_{211}), on the other hand, only reacts specifically to the nerve fibers already growing at this moment in time and not at all to the cell body regions. Electron microscopic cytochemistry, using the monoclonal antibody Q_{211} in application with extremely fine

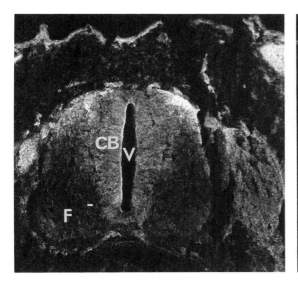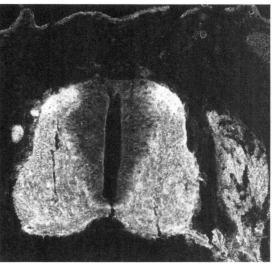

Fig. 2.15. Immunohistochemical representation of gangliosides in the spinal cord of five-day-old chick embryos. *Left*: labeled proliferating neuroblast cell bodies (CB) facilitated by the monoclonal antibody AbR24 which is sensitive to the GD$_3$ ganglioside; *right*: fibers (F) labeled with the Q$_{211}$ antibody which is sensitive to polysialogangliosides. V: ventricle. (Photograph: H. Rösner.)

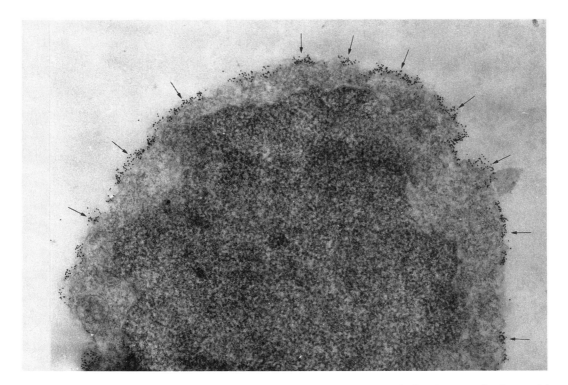

Fig. 2.16. Immunocytochemical (electron microscopic) representation of the Q$_{211}$ antibody sensitive to polysialogangliosides on the outside of the membrane of retina ganglia cells of the chick. Immuno-gold is used here. (Photograph: V. Seybold.)

gold particles, reveals that the highly polar gangliosides cluster almost exclusively on the outside of the membrane of growing nerve fibers (Fig. 2.16).

Presently, intensive research is being carried out on the development of other ganglioside antibodies for further identification and direct localization of additional marker gangliosides in nerve tissue. This seems increasingly important as recent work has shown that the ganglioside composition of nerve tissue (as evidenced in pathogenic changes such as *neuroblastoma* and *glioma brain tumors*) is radically altered in the direction of a very primitive, less polar configuration. In the future, the development of defined monoclonal ganglioside antibodies could be profoundly important in the treatment of tumors.

3
Functional Morphology of the Nervous System in Vertebrates

The preceding chapter dealt with cellular, molecular, and morphogenetic considerations as they apply to the formation of neuronal structures. Several molecular aspects of neuronal differentiation were addressed, as well. This background will serve as a foundation for the discussion of functional morphology. Functional morphology in vertebrates is of particular interest to us in that it serves as the basis for our own higher associative brain and memory performance. A brief discussion of the invertebrate nervous system is presented in Chapt. 4.

Cellular processes and differentiation in important regions of the CNS during development will be examined more closely. Changes in the course of development in the formation of regions of the brain, which have led to considerable structural variations of a relatively simple structure, can be traced to the extreme diversity of manner in which the mass development of nerve cells and their modes of associating ensues. Prior to treating the individual neuronal categories of function, a few general comments should be made regarding the basic structure and the interrelation of the central and vegetative nervous systems.

3.1 Basic Structure of the Nervous System in Vertebrates

The simplest arrangement of neurons in vertebrates is that of a diffuse web as exemplified in the intramural nervous system which courses through the walls of hollow organs such as the intestines (Fig. 3.1) and which is very similar to the nervous system of the lowest invertebrates such as polypi and jellyfish. In other respects, vertebrate nervous systems are arranged more rigidly, i.e., on the one hand, as a *central nervous system* with an associated *peripheral nervous system* and, on the other hand, as *autonomic* or *vegetative nervous systems* which function mostly in an independent manner (Fig. 3.1):

- Generally speaking, the *central nervous system* (*CNS*) consists of the brain and the spinal cord; the brain is subdivided into three main regions: the *telencephalon*, the *mesencephalon*, and the *metencephalon* including the *medulla spinalis*. In the CNS, which derives ontogenetically from a neural tube (see Chap. 2.1), there develop *tractus* (nerve fiber groupings) or *fasciculi* (conduction bundles) which are mostly free of connecting tissue. The cell bodies are distributed in a rather diffuse manner within the so-called gray matter. However, they frequently aggregate into *nuclei* or they form a layered arrangement, a *cortex*, which can be quite varied in degree of differentiation.
- The *peripheral nervous system* (*PNS*) does not represent, in a real sense, an autonomous organ; rather, it consists of the nerve fibers (*axons, dendrites*) that connect the CNS with the organs of the body, or with the concentrations of neurons (*ganglia*) that lie outside the CNS. The PNS has nerve fiber bundles that, more or less, are well wrapped in an insulating material (*myelin sheath*) and connec-

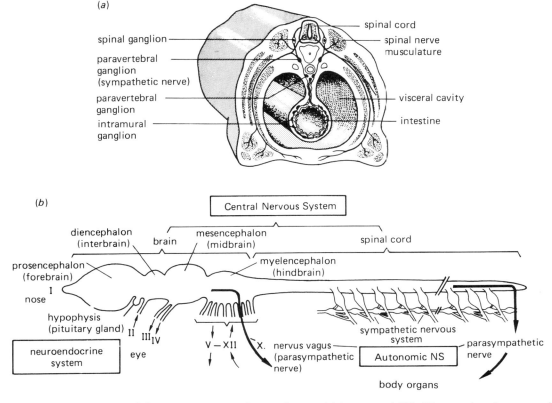

Fig. 3.1. Basic structure of the nervous system in vertebrates: (a) intramural NS; (b) central and autonomic nervous system.

tive sleeves of tissue (*endo-, peri-, epineurium*). They lead away from the CNS to the periphery as efferent or centrifugal bundles of fibers, or they lead to the CNS as afferent or centripetal bundles of fibers. In addition to these nerve fiber bundles, concentrations of nerve cell bodies, *peripheral ganglia*, belong to the PNS. They accumulate in many places in the body as well as in primary receptor organs.

• The *vegetative*, or *autonomous nervous system* represents a special anatomical and functional entity that, to a great extent, regulates many body functions independently, i.e., it does so without conscious control. The two constituent, opposing elements of the vegetative NS, i.e., the sympathetic and parasympathetic components, originate in different regions of the CNS. The original nuclei of the *sympathetic system* lie laterally in the region of the lateral

horn of the thoracal lumbar spinal cord, much like a string of pearls, as a double cord (boundary cord) and are connected to the spinal cord by means of delicate connecting strings (*rami communicantes*). The fibers of the *parasympathetic system*, on the other hand, course through the pathways of the cranial nerves, in particular, of the nervus vagus and within the sacral region of the spinal cord.

In addition to these morphological categories, a functional division also underlies the nervous system as a whole: the *animal*, or *somatic* nervous system, also known as the oikotropic portion of the NS, regulates the manner in which the organism articulates itself consciously into the environment. It allows for the sensory perceptions (light perception, hearing, taste, smell, sense of rotation, touch, and perception of water pressure, temperature, and gravity) as well as for innervation of the muscles of the body trunk and the extremities.

The *autonomic, visceral,* or *vegetative nervous system,* on the other hand, as the *idiotropic portion of the NS,* regulates the metabolic processes of the organism. It is responsible for the unconscious regulation of general sensitivity and for regulating motor innervation of the skin and intestines.

In addition to the systems of neuronal regulation that have been presented here in a cursory manner for the sake of overview, there is also the *neuroendocrine system,* a significant component of the NS, which controls various organs' functions by regulating neurosecretions, or neuropeptides entering the bloodstream and adjacent nerve structures. The *hypophysis,* or *pituitary body,* is the control center of the neuroendocrine system. Neurosecretory nerve pathways from the hypothalamus of the diencephalon terminate here (*neurohypophysis*), and it manufactures its own hormones to control peripheral hormone glands (*adenohypophysis*). The regulation of this hormone production, in turn, is subject to control by *releasing* or *inhibiting factors* that are created in the hypothalamus and are carried into the anterior portion of the hypophysis by way of a special portal vein system. In addition to the "classic" neurohormones of the hypothalamus, neuroendocrine production of numerous other substances has been discovered recently, particularly in CNS structures situated periventricularly (see Sect. 3.5.2) and in the limbic system (see Sect. 3.2). These compounds serve hormonal self-regulation in the CNS, lending great importance to the cerebrospinal fluid (liquor cerebrospinalis) in the brain ventricles and to neuromodulation (see Chap. 8), i.e., the modulation of functions specific to the conduction of impulses.

3.2 The Central Nervous System

3.2.1 Phylogenetic Aspects

The central nervous system in vertebrates, listed in order of ontogenetic differentiation, consists of the spinal cord (*medulla spinalis*) and its expansion in the head area, the brain (en-

cephalon, cerebrum). In the course of its 500 million years of phylogeny (Fig. 3.2), the vertebrate brain, with its pros- or *telencephalon* (forebrain), *diencephalon* (interbrain), *mesencephalon* (midbrain), and *rhombencephalon* (hindbrain), (consisting of the *metencephalon* and *medulla oblongata*), underwent a far greater increase in size than did its body.

Hardly any other organ has changed in its form so fundamentally as the brain, and with this change came changes in its functions. *Relative brain weight,* or the ratio of brain weight to body weight, offers a vivid illustration of this: in the carp, it amounts to 0.12%, in the house cat, 0.8%, in the human, 2.2%.

The absolute brain weight of approximately 200 vertebrate species is shown in relationship to their body weight in Fig. 3.3. Using a double-logarithmic representation, allometric growth ratios emerged among which three aspects are of particular significance:

1. The projection of brain weight lines for animals of the same family is constantly around 2:3. That means that physically larger animals within a family group have relatively smaller brains. Since, with increasing body size, the surface area increases in the second dimension and body weight increases in the third dimension, a CNS of relatively low volume is sufficient for supplying nerve service to the periphery of the body.

2. It can be inferred from the distribution of values in Fig. 3.3 that the rise of a curve within a family group (for example, within primates or bony fish) is parallel and, in contrast, that the starting position of the groups of values varies greatly. The b value in the allometry formula $Y = b \cdot x^{\alpha}$, which represents this, is variable for the different vertebrate groups. In this formula of growth relationships, b is the value of Y if x equals 1, as is the case in Fig. 3.3. Accordingly, the primates have the greatest b value, whereas the bony fish and reptiles have the lowest.

3. Building upon these findings, Jerison (1973) developed the so-called *encephalization quotient* (*EQ*) which attempts to illustrate whether, and to what extent, there are individual species within family groups whose

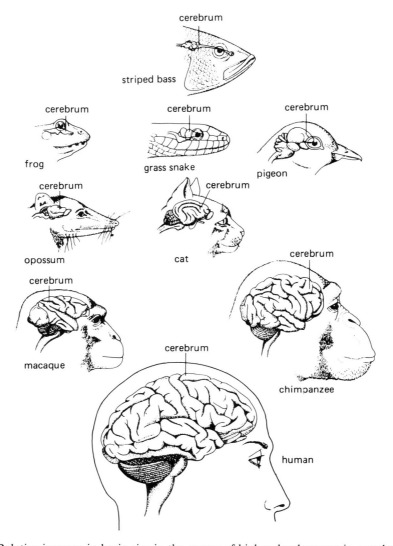

Fig. 3.2. Relative increase in brain size in the course of higher development in vertebrates.

degree of encephalization deviates greatly in either direction from their average. Thus, if one were to set the EQ of mammals at 1, then the EQ of primates (excluding humans) amounts to 2.1. At 7.6, the *EQ of the Homosapiens* lies far above the average for mammals. That means that a specific selection apparently occurred with regard to brain size, and this took place independently of brain size. A plausible, or even causal explanation for this has yet to be advanced. In this context, it should be noted with due regard that the degree of encephalization in *dolphins* is as great as it is in Homosapiens.

As evidenced by longitudinal representations, a fundamental, increasing phylogenetic differentiation of the brain can be shown for all areas that were brought to a comparable scale (Fig. 3.4). However, the higher brain centers connected to the spinal cord above the medulla oblongata (Fig. 3.5.1–7) have undergone the greatest changes, in particular, the structures that emerged secondarily which developed from a

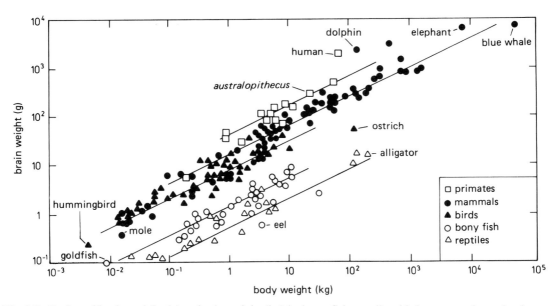

Fig. 3.3. Ratios of brain weight (g) to body weight (kg) in bony fish, reptiles, birds, mammals, and primates.

dorsal differentiation of the gray matter located on the surface. The change in form of several sectors is obviously correlated with the close association of these structures with the most important sensory organs (eye, nose, inner ear).

These phylogenetic trends in vertebrate brain development are recapitulated briefly in mammals during the course of ontogenesis (*Haeckel's rule*): as in all vertebrates, the spinal cord tube (see Chap. 2), originally stretched longitudinally, undergoes expansion of its anterior portion during early ontogenesis to become three primary brain vesicles (pros-, mes-, and rhombencephalon) whose different growth rates result in a branched or irregular brain structure. Two lateral vesicles, the two hemispheres of the future telencephalon, evert from the prosencephalon. Eventually, they will constrict the rearward region of the prosencephalon, and in so doing create the diencephalon. Both optic nerves, which induce formation of the eyes subsequent to their growth and eventual contact with the outer germ layer which forms the skin, the ectoderm of the integument, become tied off from the diencephalon. Additionally, the neurohypophysis (pituitary gland), which is connected to a peduncle (infundibulum), emerges ventrally and positions itself together with the adenohypophysis

which is segmented from the palatine vault. Dorsally, the epiphysis and, in lower vertebrates, the parietal organs form from the diencephalon. In reptiles, the parietal organs can lead to the formation of a third, unpaired median eye.

To better portray the structural and functional facets of the CNS, a few *figures* should be presented *regarding the human brain*: the brain weight of a central European adult averages about 1,360 g; among men, it ranges in weight from 1,180 g to 1,680 g (average = 1,375 g), and among women, the range is from 850 g to 1,280 g (average = 1,245 g). However, the figures present quite a different picture when correlated with body weight: on average, the relative brain weight of women is greater than that of men (Blinkov and Glezer, 1968). Likewise, the relationship of brain weight to physical type in humans shows the following: men and women of the pyknic type have the highest brain weight, men of the athletic type and women of the leptosomatic type have average brain weights, and men of the leptosomatic type and women of the athletic type have the lowest brain weights. These details might well be associated with the relative ratios between body weight/size and brain weight (Fig. 3.3). From about the age of 20, brain weight gradually decreases, on average by a total of

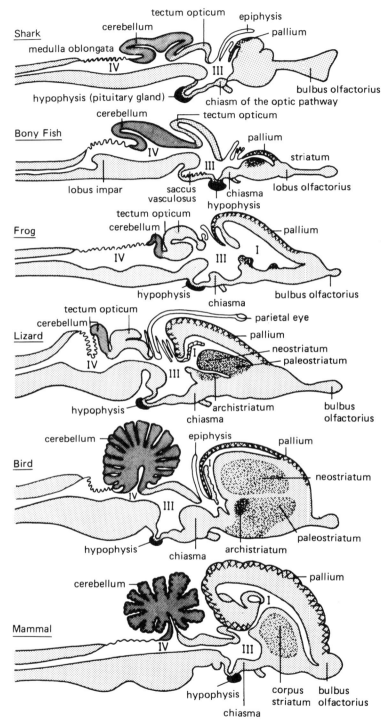

Fig. 3.4. Schematic comparison of longitudinal brain sections of phylogenetically different, highly developed vertebrates.

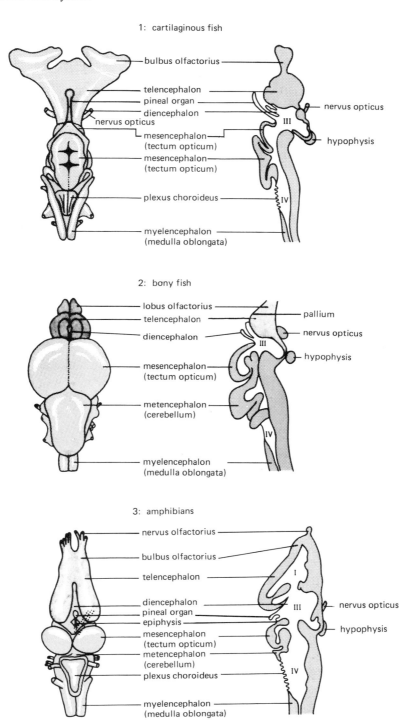

Fig. 3.5. Comparative overview (dorsal aspect and lateral section) of the brain formations in phylogenetically different, highly developed vertebrates: 1: cartilaginous fish; 2: bony fish; 3: amphibians; 4: reptiles; 5: birds; 6: mammals; 7: human.

4: reptiles

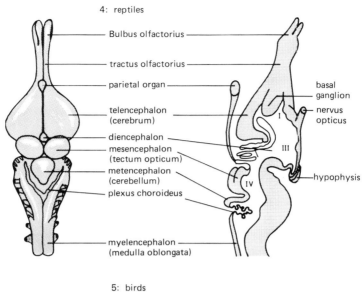

Bulbus olfactorius

tractus olfactorius

parietal organ

telencephalon
(cerebrum)

diencephalon

mesencephalon
(tectum opticum)

metencephalon
(cerebellum)

plexus choroideus

myelencephalon
(medulla oblongata)

basal
ganglion

nervus
opticus

hypophysis

5: birds

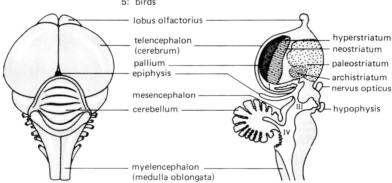

lobus olfactorius

telencephalon
(cerebrum)

pallium

epiphysis

mesencephalon

cerebellum

myelencephalon
(medulla oblongata)

hyperstriatum

neostriatum

paleostriatum

archistriatum

nervus opticus

hypophysis

6: mammals

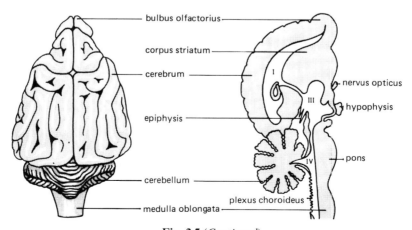

bulbus olfactorius

corpus striatum

cerebrum

epiphysis

cerebellum

medulla oblongata

nervus opticus

hypophysis

pons

plexus choroideus

Fig. 3.5 (*Continued*)

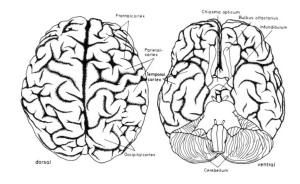

7. Human

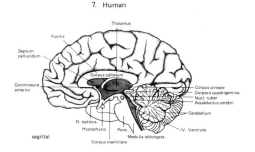

Fig. 3.5 (*Continued*)

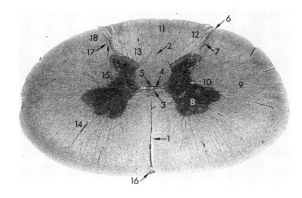

about 8% by age 90. This is not due to a decrease in the number of nerve cells, but to a general shrinkage in volume (dehydration).

At a weight gain of 88%, the hemispheres of the prosencephalon (forebrain) exhibit the greatest increase in weight of all the areas of the brain. For a size comparison of important brain areas, some average values for men in regard to weight (g) and volume (cm^3) are as follows: hemispheres, 1,200 g/l, 160 cm^3; cerebellum, 150 g/144 cm^3; medulla oblongata, 26 g/25 cm^3.

3.2.2 Comparative Overview of the Functional Morphology of the Major Sections of the Human CNS

3.2.2.1 Spinal Cord (Medulla Spinalis)

The spinal cord realizes the actual basic structural plan of the CNS, that plan remaining recognizable as such in many parts of the brain. In the human, it consists of a segmentally sectioned nerve tube averaging about 40 cm in length and weighing, on average, between about 34 and 38 g (Blinkov and Glezer, 1968). It is located in the spinal canal of the vertebral column. Cross sections reveal a shape in which the "*gray matter*" (*substantia grisea*), rich in ganglion cells and capillaries, is arranged in a butterfly pattern around the relatively small central canal (canalis centralis) which is filled with cerebrospinal fluid (liquor cerebrospinalis). The exterior of the spinal cord is wrapped in a thick mantle of "*white matter*" (*subtantia alba*) which consists predominantly of myelinated fibers (Fig. 3.6). The total volume of gray matter in the spinal cord amounts to about 5 cm^3, and the white matter amounts to about 23 cm^3. During development, the gray and white matter undergo different growth rates in the various regions. The spinal cord is divided nearly

Fig. 3.6. Cross section of the thoracic region of the spinal cord in the cow: 1) fissura mediana ventralis, 2) septum medianum dorsale, 3) commissura alba ventralis, 4) commissura grisea, 5) central canal, 6) dorsal lateral sulcus, 7) dorsal root fibers, 8) gray matter, 9) white matter, 10) formatio reticularis, 11) Burdach's column, 12) Goll's column, 13) fasciculus dorsalis, 14) fasciculus ventrolateralis, 15) columna lateralis, 16) arteria spinalis ventralis, 17) vessel in the region of the dorsal root, 18) zona terminalis.

in two by a deep fissure both on the dorsal and on the ventral sides (*septum medianum dorsale* and *fissura mediana ventralis*). Two horns grow dorsally and two grow ventrally from the gray matter during ontogeny, the *anterior horns* (*columnae anteriores*) and the *posterior horns* (*columnae posteriores*), such that a cross section reveals a shape strongly reminiscent of a butterfly (Fig. 3.6).

The spinal cord is subdivided by spinal nerves occurring in pairs. Mammals, humans included, have 31 pairs of spinal nerves which are distributed among five regions of the spinal cord (Fig. 3.7):

- The cervical part of the medulla (pars cervicalis), ca. 9.5 cm in length, with eight pairs of spinal nerves,
- The thoracic cord (pars thoracica), ca. 23.5 cm in length, with 12 nerves,

- The lumbar region of the cord (pars lumbalis), ca. 5 cm in length, with five nerves,
- The sacral region of the cord (pars sacralis), ca. 3 cm in length, with five nerves,
- The coccygeal region of the cord (pars coccygea), ca. 3 cm in length, with one nerve.

In the course of development from the newborn to the adult, the volume of the gray and white matter in the individual regions of the spinal cord increases 5- to 20-fold. As a rule, the number of spinal nerve pairs in vertebrates is dependent upon the number of vertebrae. The number established in the human, 31, is based on the fact that the growth in length of the spinal cord remains behind that of the spinal column. The so-called *cauda equina* (horse's tail) below the second lumbar vertebrae results from this. All of the more deeply set spinal nerves progress downwardly from here as a horse's tail within the

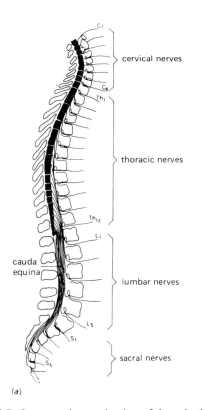
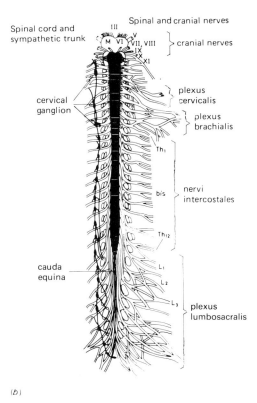

Fig. 3.7. Segmental organization of the spinal cord and spinal nerves in regard to the backbone. Lateral aspect with vertebrae (*a*), ventral aspect (*b*), subdivided into sympathetic trunk (*left*) and spinal nerve exit points (Reprinted with permission from Rohen, 1971).

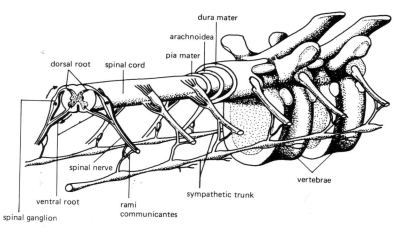

Fig. 3.8. Diagram of the backbone and spinal cord showing the latter's connections to the vegetative nervous system of the sympathetic trunk.

vertebral canal and group around the *filum terminale* (end fibers) at the coccyx. Thus, the spinal nerves issuing from the intervertebral foramina cover a considerable distance in the vertebral canal before they exit it.

Each spinal nerve emerges from an interior (ventral) and from a posterior (dorsal) root in the spinal cord (Fig. 3.8). The *dorsal roots* contain the sensory (afferent) pathways coming from the periphery of the body; the ventral roots, on the other hand, contain the effector (efferent) pathways. The cell bodies of the dorsal fibers are in particular spinal ganglia at the side of the spinal cord; the ventral pathways are motor ganglion cells in the anterior horns of the gray matter. Distal to the spinal ganglia, the dorsal and ventral nerve cords merge bilaterally for a short distance, only later to separate and diverge again. In contrast to the nerve roots, then, sensory and motor fibers run next to one another in some fiber paths.

All impulses running from the higher regions of the CNS to the trunk and extremities, or vice versa, are connected by descending or ascending pathways that run within the spinal cord, primarily in the white matter. Regarding the different effects mediated by the spinal cord pathways, not only is the position of the pathway in the spinal cord known, but its origin and terminus as well as the synaptic pathways in the higher regions of the CNS are known, also (see Sect. 3.2.2.2.2).

The most important *ascending pathways* within the spinal cord in the human (Fig. 3.9a) are

- The tractus spinothalamicus,
- The tractus spinobulbaris
- The tractus spinocerebellaris.

(The Latin name indicates the origin and terminus of the pathway).

The ascending pathways are formed without interruption by neurons of the spinal cord whose fibers extend to the thalamus, located in the diencephalon, and to various parts of the cerebellum.

- Phylogenetically, the *tractus spinothalamicus* is the oldest ascending nerve pathway of the spinal cord. It handles those nerve impulses in the human that carry the diffuse, primitive sensations of pain, temperature, pressure, and touch or that trigger instinctive "fight or flight" reflexes.
- The *tractus spinobulbaris* is phylogenetically younger, handles tactile sensitivity and proprioceptive sensation, and facilitates finer localization and differentiation of touch response of the skin. Moreover, it establishes the necessary connections between the mechanicoreceptors of the skin and muscles and the higher regions of the CNS, which are needed to coordinate complicated movements. Some of the impulses conducted to the posterior

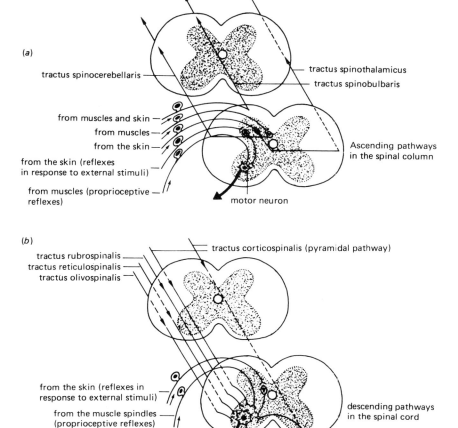

Fig. 3.9. Ascending (*a*), and descending pathways (*b*) in the spinal cord in the human.

white columns (funiculus posterior) of this system reaches the cerebellar cortex and in this way are involved in motor response.

- Phylogenetically, the cerebellar white column of the *tractus spinocerebellaris* (funiculus lateralis medullae spinalis) is a pathway old enough to be present in fish. It conducts impulses essential for coordinating motor responses from receptors of the muscles and skin to the cerebellum.

The most important *descending pathways* in the spinal cord (Fig. 3.9b) are

- The tractus corticospinalis (pyramidal tract),
- The tractus rubrospinalis,
- The tractus reticulospinalis,
- The tractus vestibulo- or deiterospinalis,
- The tractus olivospinalis.

Of these, the last four tracts are also referred to as the extrapyramidal system.

In addition to these pathways, there are also descending pathways of the vegetative nervous system whose development is not yet fully understood. The descending pathways conduct impulses from the brain to the motor neurons in the anterior horn that have a somatomotor function. The pyramidal pathways are so important in this regard that the other systems are referred to collectively as the extrapyramidal system.

- Phylogenetically, the pyramidal pathway makes its first appearance in mammals. It directly connects the cerebral cortex with the motor neurons in the anterior horn and makes possible the immediate voluntary influence of these cells. In this way, finely adjusted movements, such as those articulated by the human hand, can be directed by the higher relay centers.

The *extrapyramidal pathways* conduct impulses to the motor anterior horn cells that are not subject directly to the will. Apparently, voluntary movements are directed, in large part, over the pyramidal system without consciousness involvement:

- The *tractus rubrospinalis* originates in the mesencephalon. It conducts impulses of the higher brain centers extrapyramidally, particularly those of the basal ganglia of the cerebral hemispheres (see Sect. 3.2.2.4) and the cores of the cerebellum (see Sect. 3.2.2.2.2), to the motor anterior horn cells of the spinal cord.
- Among the fibers that run in the *tractus reticulospinalis* are fibers that affect yet other fibers beyond the control of the anterior horn neurons, those that affect muscle fibers and, thus, in turn, can influence processes in the proprioceptive reflex arcs. Other fibers conduct inhibitory impulses of the cerebral cortex.
- The *tractus vestibulospinalis* handles the crucial work of the vestibular apparatus via the Deiters nucleus to the skeletal muscles for reflex retention of the normal head position and body equilibrium.
- The *tractus olivospinalis* originates in the olivary nucleus of the medulla oblongata (see Sect. 3.2.2.2.1) and leads impulses that previously were relayed or routed from elsewhere in the brain to the olivary body, to the motor anterior horn cells.

Owing to the conductive *function of the spinal cord*, injury to it (for example, a crushing injury or severance) leads to both sensory and motor paralysis of the body areas lying below the point of injury. The only processes that do not cease to function are those that are controlled reflexively by specific centers of the spinal cord without

brain involvement, by the so-called autoapparatus of the spinal cord. These function without conscious control. The brain centers of the nerve fibers issuing from the specific segments through the anterior roots in the spinal cord are classified collectively as the *niveau centers*. The motor niveau centers would be the origination points of the motor nerve fibers that are located in the anterior horns of the gray matter. The origination points of the preganglionic sympathetic nerve fibers of the autonomic nervous system (see Sect. 3.3) are located in the lateral horns of the thoracic and lumbar segments. The more highly organized the CNS, the less capable of independent performance are the niveau centers. In young mammals, vasotonia can be restored in large part by autonomic activity of the niveau centers following severance of the spinal cord or destruction of the higher blood pressure center in the medulla oblongata. A specific autonomy can even develop for the inspiratory muscles; however, this is not the case in adult mammals and, thus, does not apply to adult humans. The origination points of some of the parasympathetic fibers of the autonomic nervous system are located in the sacral region of the spinal cord. These fibers innervate a series of abdominal organs. All of these centers, however, are also under the control of the higher regions of the brain, in particular, the diencephalon; they are also connected to the cerebrum, as evidenced by the fact that all of these body functions can be influenced voluntarily (autogenic training). However, conditions of paraplegia, in which the lower sacral region of the spinal cord has been severed from the rest of the CNS, reveal that most of the abdominal organs, for example the bladder and the intestines, will function reflexively, without involvement of the will. In dogs, for example, even pregnancy and birth can proceed solely under the control of the niveau centers. Such findings are corroborated in humans by the sensational instances of clinically brain-dead pregnant women who were kept alive by artificial means until their fetuses matured.

All other brain regions located in the head area of vertebrates derive from the original organization of the spinal cord. Before delving more deeply into the most important functional-morphological aspects of the individual

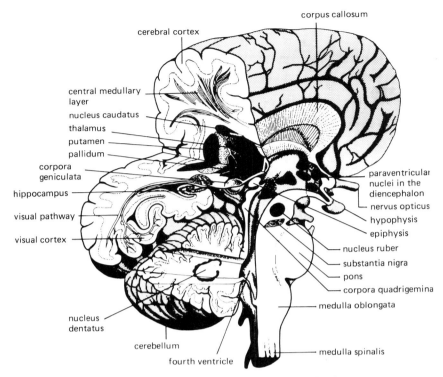

Fig. 3.10. Cutaway diagram of the human brain.

brain regions, attention should be directed to Fig. 3.10, a cutaway diagram that represents the topography of the most essential structures of the human brain. The great extent to which deviations in the basic organization of the spinal cord have occurred becomes evident from this diagram. The changes go hand in hand with extraordinary functional sophistication. Certainly, past neuropathological studies have explained many interrelationships of function in the human brain. However, the field of neurobiology has relied primarily on animal models for its research and continues to do so. In this regard, studies of the rat brain have proved extremely useful because most essential neuronal functional processes are readily observable using this subject. For this reason, the essential functional–morphological interrelationships of the rat's brain are presented in Fig. 3.11. These diagrams can be used as a reference model when discussing the characteristics of the human brain.

3.2.2.2 Rhombencephalon

The *rhombencephalon* connects anteriorly to the spinal cord, extends to the mesencephalon, and consists of two main parts, the *medulla oblongata* (*myelencephalon*) and the *metencephalon* with its *cerebellum*, *pons*, and *tegmentum*.

3.2.2.2.1 Medulla Oblongata Myelencephalon

The important functional regions of the 3-cm long medulla in the human are

1. The *formatio reticularis*, a net-like area of gray matter woven of many fiber bundles, which constitutes the continuous junction between the rhombencephalon and mesencephalon and performs crucial filtering functions for the brain's incoming and outgoing information.
2. The *pyramidal pathways* on the ventral side

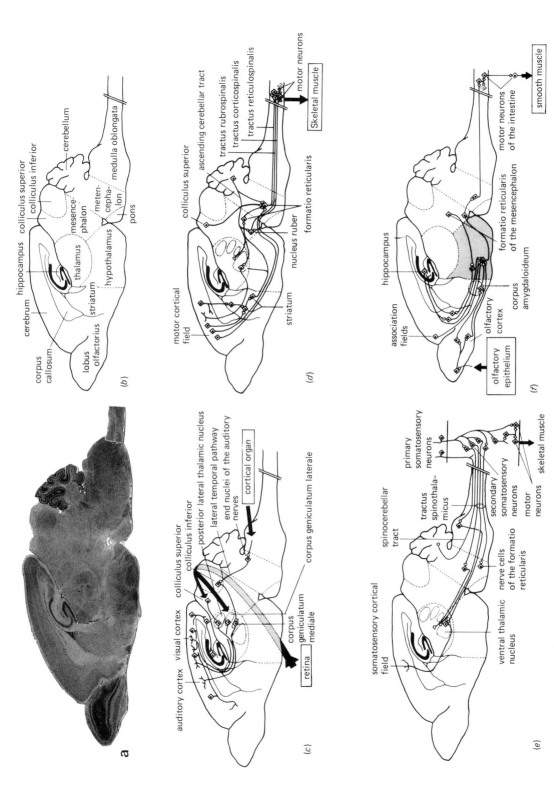

Fig. 3.11. Overview of the most important functional-morphological systems in the rat brain: (*a*) histological longitudinal section; (*b*) overview of the most important brain regions; (*c*) visual and auditory system; (*d*) motor system; (*e*) somatomotor system; (*f*) hypothalamus routing center.

59

TABLE 3.1. Overview of the cranial nerves.

Cranial nerve and number	Entry/exit region of brain	Quality	Innervates
I.	N. olfactorius (fila olfactoria)	Bulbus olfactorius	Sensory	Olfactory epithelium of the nose
Ia.	N. terminalis	Bulbus olfactorius	Sensory	Olfactory epithelium of the nose (in reptiles: Jacobson's organ)
II.	N. opticus	Diencephalon	Sensory	Retina
III.	N. oculomotorius	Mesencephalon	Motor	Eye muscle
IV.	N. trochlearis	Mesencephalon	Motor	Eye muscle
V.	N. trigeminus (3-way branch)	Rhombencephalon	Sensory and motor	Face
VI.	N. abducens	Rhombencephalon	Motor	Eye muscle
VII.	N. facialis	Rhombencephalon	Sensory and motor	Face (mimetic)
VIII.	N. statoacusticus	Rhombencephalon	Sensory	Inner ear (hearing, equilibrium)
IX.	N. glossopharyngeus	Rhombencephalon	Sensory	Pharyngeo- and motor-muscualture and palatine mucosa
X.	N. vagus	Rhombencephalon	Sensory and motor	Laryngo/pharyngeo-musculature, pharyngeal mucosa, intestines
XI.	N. accessorius	Rhombencephalon	Motor	Branch of the n. vagus
XII.	N. hypoglossus	Rhombencephalon	Motor	Musculature of tongue and hyoid

leading from the cerebellum to the spinal cord,
3. The laterally situated *olivary bodies* (*nuclei olivae*) whose nuclei belong to the extrapyramidal motor system (EPMS), and
4. Several origination points of cranial nerves V-XII (see Table 3.1).

The inner composition of the medulla oblongata is more complex than that of the spinal cord (Fig. 3.12). Of course, the gray matter also contains motor nuclei regions in its ventral *floor plates* and, primarily, sensory nuclei in the lateral *alar lamina*, but the gray matter no longer shows the butterfly pattern so typical of the spinal cord. It forms a rather diffuse reticulated body, the functionally crucial *formatio reticularis* in whose region cranial nerves V-XII originate and which constitutes the transition to the mesencephalon. The so-called *pyramids* emerge on the ventral side of the medulla. These contain the continuous pyramidal pathways which run caudally to the interior of the spinal cord. The so-called *olivary bodies*, whose nuclei are connected to other nucleus regions in the tegmentum of the mesencephalon, the cerebellum, and the spinal cord as the extrapyramidal motor system, emerge laterally from the medulla. Primarily, the roof of the medulla consists of epithelia of the *plexus choroideus* (*vascular plexus*, see Fig. 3.5.4).

The nucleus groups of the gray matter are widely seperated by the broadening of the central spinal cord canal in the rhombencephalon to the fourth ventricle resulting in their final positions being next to one another. Ascending and descending fiber bundles course through the region of the *myelencephalon* not only longitudinally, but also by criss crossing one another and, thus, moving to opposite sides.

They unite ventrally with telencephalic pathways and, together with their routing cells, form the pons, a massive bulb of fibers at the metencephalon.

As a conductive organ, the *medulla oblongata* contains

• somato sensory nuclei from which the tactile organs, gravity-sensory organs, and the ear (and in fish and amphibians the lateral line sense organs associated with sensing currents) are innervated:
• viscero sensory nucleus centers supplying nerves to the sensory organs (proprioreceptors) within the body:
• somato motor nuclei which innervate body and eye musculature:
• viscero motor nuclei for motor control of the glands and musculature of the foregut, or protogaster.

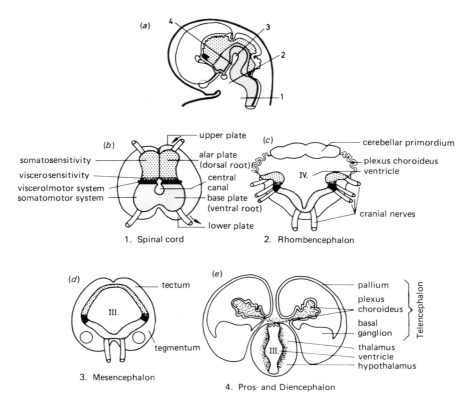

Fig. 3.12. Diagram of the embryonic human brain (*a*), representing the base and alar plate sections of the spinal cord (*b*), rhombencephalon (*c*), mesencephalon (*d*) and telencephalon (*e*). (Reprinted with permission from Rohen, 1971.)

A series of *brain centers* exists in the medulla oblongata and in sections of the rhombencephalon connected to the medulla oblongata. The origination areas of a series of *cranial nerves* lie in the niveau centers much as was described previously for the spinal cord, but the cranial nerves differ distinctly from the spinal nerves. The 12 cranial nerves can be divided into three groupings according to their function and the areas they innervate:

• sensory nerves (I, II, VIII) with somato sensory function,
• visceral nerves (V, VII, IX, XI) with viscero sensory and motor function in the throat and thorax regions,
• somato motor nerves in the head area.

The cranial nerves are distinguished according to their manner of origination. Sensory nerves and visceral nerves are outgrowths of the pros- or diencephalon. The others, including the spinal nerves, emerge from centers in the CNS that are formed very early in the course of development. The afferent cranial nerves terminate in corresponding entry areas of the CNS, whereas the efferent cranial nerves originate in the region of their source nuclei, which are simultaneously their points of issuance. The rhombencephalon is the main origination area of the somato motor cranial nerve group.

The 12 cranial nerves are listed in Table 3.1 together with their entry or exit area in the brain, their quality, and the region that they innervate.

There is a series of *motor* and *vegetative function centers* in the medulla oblongata superordinate to the niveau centers for the cranial nerves by which motor functions (e.g., sucking, chewing, swallowing, and vomiting) and vegetative functions (e.g., salivation and pupillary reaction) are controlled. Of particular significance are those centers of the medulla that control breathing and circulation (see Sect. 3.3).

Circulation centers regulate heartbeat frequency and coordinate the frequency with the demands placed upon the system at any given moment. This is achieved particularly via the centers of the nervus vagus. Blood pressure and peripheral blood distribution are regulated, in contrast, by the sympathetic nervous system.

The *breathing center* controls the motor response of the respiratory muscles and is especially sensitive to changes in CO_2 tension and ion concentration in the blood. It is located at the base of the medulla in the formatio reticularis. The caudally located sections of the breathing center regulate, among other things, the phylogenetically older phenomenon of gasping breathing common to amphibians and evidenced in the human only in exceptional instances when the higher centers cease to function. Other areas of the formatio reticularis serve as the inspiratory center, and still others function as the expiratory center. Located in the rostral areas, the pneumotactic center coordinates respiratory movement. There is also a coordination center for coughing and one for sneezing, as well as an especially important *spasm*, or *convulsion*, *center* which is activated by an increase in blood CO_2 or by an oxygen deficiency. Beginning in the spasm center, a general stimulus affects all centers of the CNS and leads to manifestations of spasm. In fatal instances, spasm and general paralysis lead to death by asphyxiation.

3.2.2.2.2 Metencephalon (Hindbrain)

The *pons* and the *tegmentum pontis* are located in the section of the rhombencephalon called the metencephalon (hindbrain). Above these is situated the *cerebellum* which is composed of two hemispheres connected at its lower sections by the three *pedunculae cerebelli* (peduncles of the cerebellum).

Pons. In the region of the metencephalon, the *pons*, ca. 2.5 cm long and 3 cm wide, serves a purely conductive function. Fibers from the prosencephalon are relayed here to the cerebellum. Motor efferent pathways of the pyramidal system and of the EPMS pass through the pons,

as do the afferent motor pathways and pathways of the auditory nerves. The higher regions of the breathing center lie in the formatio reticularis section of the pons.

Several cranial nerves have their nuclei of origin in the *tegmentum pontis*: V = nervus trigeminus, VI = nervus abducens, VII = nervus facialis, as well as the nuclei of termination of the cranial nerve nervus statoacusticus (VIII; see Table 3.1).

Cerebellum. The cerebellum, the dorsal differentiation of the rhombencephalon, plays an extremely important role in coordinating and regulating motor response and maintaining normal body position (Figs. 3.4 and 3.5). Anatomically, the cerebellum can be divided into two parts: the *vermis* (worm), which is the older part, phylogenetically, and the pair of hemispheres that, by far, occupy the most space. The cerebellum is connected to the rest of the brain by three *pedunculae cerebelli*, each ca. 1 cm long and 1.4 cm thick, and connected to the medulla oblongata and spinal cord by the posterior *brachia restiformia*. The cerebellum is connected to the pons and, from here, to the cerebrum by the *middle brachia pontis* and is connected to the mesencephalon by the anterior *brachia conjunctiva*.

The mature hemispheres of the cerebellum in the human are ca. 6 cm long, 10 cm wide, and 4.4 cm in height. In adults, their weight ranges between 136 and 169 g with an average volume of 162 mm^3. Of this, ca. 90% is cortical matter and ca. 10% is white fiber matter. At birth, the cerebellum in the human amounts to only 5 to 6% of the total brain weight, and the figure rises to ca. 11% in the adult. This level is attained by a particularly high growth rate in the cerebellum beginning just after birth and continuing through the end of the second year, at which time a ratio of 10.6% is attained, nearly the relative value for the adult. After about age 50, the absolute values for weight and volume decrease: however, within the framework of total brain reduction, the ratio between the weight of the cerebellum and the total brain weight remains relatively constant.

The most important *functions of the cerebellum* (maintaining balance, regulating muscle tone, coordinating movements) are performed in three seperate sections of the cerebellum that, phylogenetically, became differentiated successively. These sections are the *paleocerebellum*, the *archicerebellum*, and the *neocerebellum*

The *paleocerebellum* is the oldest section of the cerebellum. It is connected to the *nervus vestibularis*, which, in the human, becomes part of cranial nerve VIII and takes over coordination of data from the inner ear pertaining to equilibrium. The *archicerebellum* receives data regarding proprioceptive sensation. The *neocerebellum*, phylogenetically the youngest section, is the connection to the motor centers of the prosencephalon.

The cerebellum is a region of the brain that, similarly to the cerebrum (see Sect. 3.2.2.4.2), has undergone extremely specialized differentiation in the course of ontogeny. Its particular configuration is, however, less a characteristic of higher evolutionary development than it is connected to the way of life of individual groups. In vertebrates that move especially quickly, the cerebellum shows, in part, an even more pronounced ridging to increase the cortical surface than does the cerebrum. Apparently, differentiation of the cerebellum is correlated with the capacity of the movement apparatus. The greater the demands placed upon it, and the finer the resultant coordination of movement, the larger and structurally more differentiated this region of the brain becomes. Thus, in the case of amphibians and reptiles, which move predominantly by crawling, the cerebellum consists merely of a thin plate above the rhomboid fossa. In fish, on the other hand, and here particularly among "good swimmers," the cerebellum has developed into a significant structures. In mammals, and especially birds, the cerebellum is greatly enlarged by formations of folds. To the archaic *paleocerebellum*, there then evolved, phylogenetically, the development of the *neocerebellum* which, in birds and mammals, stands in conjunction with the motor centers of the cerebrum.

The number of nerve cells in the cerebellar cortex is extraordinarily large in spite of its modest size in comparison to the entire CNS. Even from a rough, general picture of a section of the cerebellum, it can be seen that the lightly colored matter in the convolutions is less pronounced than in the cortex (Fig. 3.13a). The cortex of the cerebellum is divided into three layers (Fig. 3.13b)

1. *The molecular layer (stratum moleculare)* in which the loosely stored star and basket cells are located. The *star cells* perform functions of association in this layer; the axons of the *basket cells* terminate in so-called fiber baskets at the perikarya of the Purkinje cells.
2. The *ganglion cell layer (stratum ganglionare)*, consisting of the very large perikarya of the *Purkinje cells* (see Fig. 3.13b) whose dendrites, in reticulated fashion, extend into the molecular layer and whose axons pass through the granular and the medullary fiber layers into the cores of the cerebellum.
3. The *granular layer (stratum granulosum)*, consisting mostly of small, densely stored *granular cells* whose dendrites are highly ramified and couple with neurites, the so-called mossy fibers.

On the whole, the cerebellum is a region of the brain that receives messages via afferent pathways from diverse sensory organs and brain regions to coordinate movement of the entire organism. It is capable of controlling all manner of movement via the efferent, extrapyramidal pathways (Fig. 3.14). Thus, impulses flow to the cerebellar cortex from the skeletal muscles (*tractus spinothalamicus*), from the vestibular apparatus of the inner ear (*tractus vestibulocerebellaris*), from the olivary body which receives impulses from the extrapyramidal motor system, and from the prosencephalon via the pons (*tractus corticopontocerebellaris*). The last is the largest afferent pathway to the neocerebellum and enables a virtual point-for-point transfer of the functions of the cerebral fields to the cerebellum.

All afferent impulses received by the cerebellum are conducted by the formidable dendrite nerwork of the large *Purkinje cells* (see Fig. 1.3d and 3.13b) to the cerebellar cortex whose axons form, in their totality, the efferent pathways. These first carry impulses to the nuclei of the cerebellum from where the impulses then reach the various nuclei of the extrapyramidal system

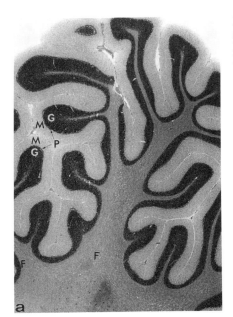

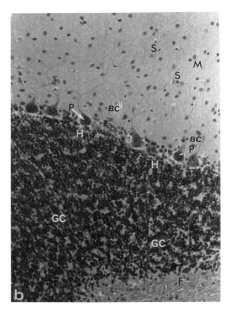

Fig. 3.13. Cross section of the cerebellum of a rhesus monkey: *(a)* overview, *(b)* sectional enlargement; M: molecular layer; G: granular layer; F: medullated nerve fibers; S: star cells; BC: basket cells H: horizontal cells; GC: granular cells; P: Purkinje cells (see Fig. 5.8).

(in particular, the *nucleus ruber* in the mesencephalic tegmentum, *nucleus deiteri*) and from here control the prosencephalic motor cells in the spinal cord and, thus, the skeletal muscles. It is evident from this relay process that the cerebellum never directly induces motor effects such as movements of the eyes or extremities. Rather, these effects are prompted indirectly via the motor centers or the cerebral cortex. The cerebellum is not an independent center, but an adjunctive one.

Cessation or interruption of cerebellar function, therefore, does not result in a discontinuance of voluntary motor response, but rather only in coordination disorders (ataxia) which time and training can reverse if the controlling functions are taken over by the higher centers of the pros- and mesencephalon.

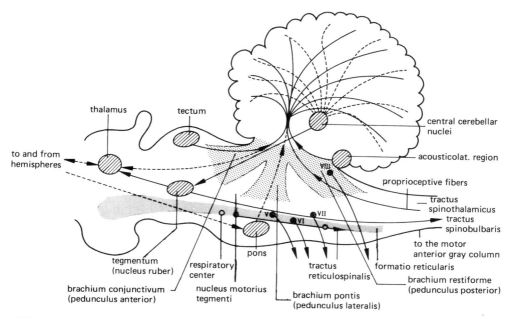

Fig. 3.14. Diagram of the most essential fiber connections in the mammalian cerebellum.

3.2.2.3 Mesencephalon (Midbrain)

The mesencephalon, only about 1 cm long in the human and weighing ca. 26 g (Blinkov and Glezer, 1968), adjoins the medulla oblongata in the continuation of the rhombencephalon primarily via the formatio reticularis. In this region, it surrounds the section of the *central canal* which connects cerebral ventricles III and IV in the form of the *aqueductus Sylvius* (see Fig. 3.4). The lumen of this connecting canal becomes enlarged with age. In the human, its cross section doubles in size between the ages of about 30 and 50 from about 1.6 mm^2 to 2.6 mm^2 in the anterior (rostral) region and from about 1.9 mm^2 to 4 mm^2 in the posterior (caudal) region.

Regarded topographically, the mesencephalon (see Figs. 3.4 and 3.5) consists of the mesencephalic roof (*tectum*), the ventrally situated *tegmentum* and the ventrally stored *pedunculi cerebri*. The quadrigeminal bodies (*corpora quadrigemina*), of which both anterior sections in virtually all higher vertebrates are the termini for the fibers of the optic nerve (*nervus opticus*), developed in mammals from the dorsally located mesencephalon of the lower vertebrates. The posterior

pair of the quadrigeminal bodies contains the end formations of the auditory pathway.

The *tegmentum* is primarily the terminal field of cranial nerve VIII (*nervous statoacusticus*) and is thus the auditory center. Furthermore, the important nucleus regions of the extrapyramidal system such as the *nucleus ruber,* the *nucleus reticularis*, and the nuclei of origin of cranial nerves III and IV (*nervus oculomotorius and nervus trochlearis*) are located here. Most of the pyramidal pathways dispatched from the cerebrum to the spinal cord pass through the *pedunculi cerebri*.

The *functional significance of the mesencephalon* in higher vertebrates (mammals) is that it bridges impulses of neighboring brain regions. It is of extreme importance because fiber masses, the pyramidal pathways in particular pass through it.

The function of the mesencephalon and, especially, the *tectum opticum*, has changed considerably during the course of phylogeny. In fish and amphibians, the tectum opticum remains primarily the visual center. Not only do the fibers of the cranial nerves III and IV (nervus oculomotorius and trochlearis), which

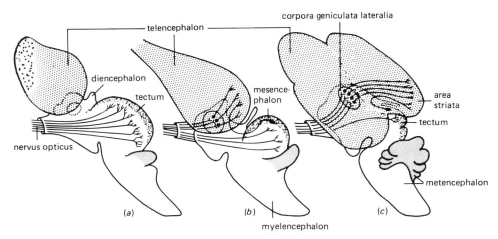

Fig. 3.15. Changes in path of the optic nerve (nervus opticus) fibers in fish (*a*), reptiles (*b*), and mammals (*c*).

control eye movement, enter the tectum opticum, but the fibers of cranial nerve II, the visual nerve (nervus opticus) enters, as well.

An essential association center, it is also an extremely important nervous processing center for many instincts and learning processes. On a higher level, especially in the case of mammals, however, its function is solely to relay, or route, impulses from neighboring brain regions.

The few fibers of the optic nerve that, in mammals and birds, still run directly from the retina to the anterior region of the corpora quadrigemina in the mesencephalon, the corpora bigemina, serve only to stimulate reflex responses such as the pupillary reflex, and no longer are associated with vision (Fig. 3.15).

3.2.2.4 Prosencephalon (Forebrain)

The prosencephalon consists of an unpaired midsection, the *diencephalon*, and the two cerebral hemispheres of the *telencephalon*, the surface area of which is so extraordinarily large in humans and higher mammals that it covers the other regions of the brain (diencephalon, tectum of the mesencephalon, and cerebellum) as a *mantle* and, thus, is also referred to as the *pallium*.

While, through special structural differentiation of the cerebellum, a center was developing in the region of the rhombencephalon to control the body's unconscious motor response, the *cerebrum*, a new, superordinate integration

system for numerous nervous functions, was being formed in the region of the prosencephalon. Viewed from the perspective of comparative anatomy, the development of the cerebrum represents one of the most impressive examples of higher development, for, in the course of vertebrate phylogeny, this brain region underwent the most extraordinary change in form and the greatest increase in volume of all the regions of the brain. This was attended by a concomitant change in its primary functions and a broadening and refinement of its role. In the human, the cerebrum is virtually overdeveloped (see Fig. 3.5) and it is to this region of the brain that we owe our preeminent standing in nature.

In the human, the original function of the prosencephalon as the rhinencephalon has been diminished and is now only of minimal significance. The fibers of the olfactory nerve (cranial nerve I, nervus olfactorius) run from the olfactory and taste cells in the area of the nose and mouth to the fila olfactoria region at the anterior lower side of the prosencephalon. In humans and birds, the eye is the primary sensory organ, whereas the sense of smell remains predominant in the majority of mammals. This is evidenced by the size of the olfactory lobe which is located in the prosencephalon.

During early stages of embryonic development, two large vesicles emerge laterally from the prosencephalic structure and push outward to form the eyes. Thus, in higher vertebrates, the eye structures become fixed in the *ophthalmen-cephalon*, the region of the prosencephalon which, in its final formation, is integrated into the unpaired region of the prosencephalon, the *diencephalon* (*interbrain*). Through the outward pushing of the eyes, there is formed a long-reaching connection to the brain, the optic nerve (*nervus opticus* = *fasciculus opticus*, cranial nerve II). Thus, the optic nerve should not be considered a nerve, but a forwardly positioned, advance region of the brain. Its axons (neurites) originate in the ganglia cells of the retinas of the eyes and ca. 1.2 million of these axons converge in individual bundles at the optic disc. Outside of the retina, the optic nerve—as with the brain—is wrapped in protective sheaths until it enters the brain. The optic nerve in the region of the eyeball is somewhat coiled, offering a degree of latitude in eye movement without danger of ripping the nerve. The optic nerves of both eyes run through the eye canals in the base of the skull in the area of the brain in front of the pituitary gland and into the optic nerve crossing field, or chiasmal field, the *chiasma opticum* of the diencephalon. In lower vertebrates, a total and complete crossing of fibers occurs here. In the human, about a 50% crossing of fibers occurs.

In the human, the fibers coming from the lateral retina areas do not undergo this crossing, whereas those fibers from the middle parts of the retina do cross over to the other side in the chiasma opticum. Fiber bundles of both groups lead to the *corpora geniculata* (lateral geniculate bodies) in the continuations of the right and left optic nerves. Still other fiber bundles branch off prior to the chiasma opticum to the hypothalamus, the region under the thalamus and, thus, connect the visual apparatus with the vegetative nervous system.

Developmentally, two regions begin to emerge during ontogeny directly after the optic vesicles take shape from the protrusions of the prosencephalic wall: the paired structure of the *telencephalon* and the unpaired middle region of the *diencephalon*. The original organization suggestive of the spinal cord, still clearly recognizable in the medulla oblongata, is no longer so well defined in the di- and telencephalon.

3.2.2.4.1 Diencephalon (Interbrain)

Essentially, thickened sides of the third ventricle of the brain constitute the diencephalon. Paired brain centers in the dorsal region of a primarily sensory nature form the *thalamus*. The ventral area of the diencephalon beneath the thalamus is the *hypothalamus* whose viscero motor functions are predominantly vegetative, effecting motor response in the intestines. The *hypophyseal stalk* (*infundibulum*) emerges from the floor of the hypothalamus and terminates in the *neurohypophysis*. It joins with the *adenohypophysis* which emerges from the palatine vault to form the *hypophysis* (*pituitary gland*), the unpaired superordinate hormonal gland (Figs. 3.4, 3.5, and 3.10). Further, the *chiasma opticum*, the crossing field of the optic nerves flowing into the diencephalon, is located in the floor of the hypothalamus.

The dorsal tegmentum of the diencephalon, the *epithalamus*, is configured variously among the different vertebrate groups. Its extremely thin-walled integument can be invaginated deeply in the first two ventricles of the cerebral hemispheres as the *tela choroidea*. Additionally, glandular appendage organs such as the *pineal bodies,* can project from the epithalamus. The *parietal eye* of the lower vertebrates, the *epiphysis* pineal gland) and *paraphysis,* an embryonic organ whose function remains unknown for the most part, belong to this grouping. Belonging to the epithalamus is the region of the *ganglion habenulae*, a paired brain center on the top side of the diencephalon whose purpose is to aid in the assessment and intake of nourishment. The connection between olfactory perception and mouth movements is controlled from here.

The *thalamus* in the human, covering about 3 cm and amounting to about 20 cm^3 and 1.5% of the total brain volume, is the central collection area for just about all information headed to the cerebral cortex. The fibers of the *tractus spinothalamicus* and *tractus bulbothalamicus*, which impart the senses of pain, touch, temperature, and proprioceptive sensation, terminate here. The *nucleus posterior* in the thalamus receives input from the eye and inner ear which then are projected from here to the association centers of the visual and auditory cortex of the cerebrum. All incoming impulses are filtered in the thalamus, modified and, to a degree, prepared for reflexes on a higher level. Through its connection to the extrapyramidal motor system (EPMS), the thalamus is involved directly in coordinating processes of movement, and through its connection to the hypothalamus, it is linked directly to the vegetative nervous system, as well, whereby it assists in controlling said system.

The *hypothalamus* represents the superordinate control system of the vegetative nervous system. Among the areas it regulates are blood pressures, blood sugar and water content, fat storage, heat regulation, and the sleep–wake rhythm of general metabolism. Typical vegetative motor activities such as trembling caused by cold, urination, and defecation are controlled from here. Of perhaps greater importance is the control of stimulus-response conditions and instinctive behavior. The preeminent role played by the hypothalamus in effecting vegetative-body and animal-mind events is founded in the fact that there are close connections between nervous and humeral control of body functions. Thus, the brain centers of the *nucleus supraopticus*, *nucleus preopticus*, and *nucleus paraventricularis neurosecretions*, which are located in the hypothalamus, produce *inhibiting* and *releasing hormones* (see Chap. 7.1.3.2) that reach the neurohypophysis by means of axonal transport through the infundibulum or reach the adeno-hypophysis via hypothalamic-hypophyseal portal circulation from where they control hormone production in the peripheral hormonal glands.

Understandably then, a brain region with these functions is especially important to the organism. Accordingly, no significant change in function has occurred here in the course of phylogeny. The structure, therefore, may be regarded as relatively conservative. It has, in all vertebrates, approximately the same function for which a certain mass of nervous substrate is required. The overall configuration of this structure is, thus, less a factor of the level of development than body size of an animal. Consequently, the hypothalamus in the human did not increase in size to the same extent that the other areas of the prosencephalon did. In relation to the total brain size of other, less highly developed mammals (for example, the mouse), it is smaller by a factor of 7.

The *pituitary gland (hypophysis)* (see Fig. 3.16) consists of an anterior lobe (*lobus anterior*), a middle lobe (*pars intermedia*), and a posterior lobe (*lobus posterior*). It is the relay, or routing, center for nearly all endocrine interactions. Hypofunction leads to dwarfism and a specific form of diabetes (diabetes insipidus), whereas hyperfunction leads to gigantism. Growth hormones and gonadotropic hormones are produced in the anterior lobe (*adenohypophysis*). The middle lobe consists mostly of epithelial cell ridges and special cysts that contain colloids and are coated with epithelium. The posterior lobe (*neurohypophysis*) is composed of neuroglial cells and nerve fibers. Products of the hypophysis enter the bloodstream via endocrine pathways or gain direct access to the diencephalon through the liquor cerebrospinalis. The secretions affect other endocrine glands and assure their harmonious interaction.

The following hormones are formed in the *adenohypophysis* (Fig. 3.16): (a) *corticotropin* (ACTH, adrenocorticotropic hormone) which controls the activity of the adrenal, or suprarenal, gland; (b) *somatotropin* (STH, growth hormone); (c) *thyrotropin* (thyroid-stimulating hormone, TSH) which regulates the thyroid gland; (d) *prolactin* (luteotropin, LTH; lactation hormone) which, in combination with progesterone and the follicular hormone stimulates milk producton; (e) various *gonadotropins* including the follicle-stimulating hormone (FSH); and (f) *luteinizing hormone* (LH) affecting both male and female reproductive glands. Additionally, there is a series of other endocrine substances such as the *interstitial cell stimulating hormone* (ICSH).

In the human, the *pars intermedia* amounts to only about 2% of the total hypophysis. It consists of residual follicles of the adenohypophysis cavity which contain colloids. *Intermedin*, the substance which controls metachrosis in fish, amphibians, and reptiles, is produced here.

As a neuroendocrinological object of examination, the *neurohypophysis* is composed of *pituicytes*, a specific neuroglia, and nerve fibers exceptionally low in medullary matter. The perikarya of these fibers are located in the

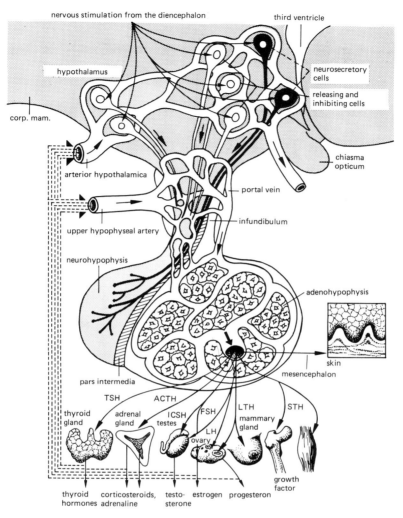

Fig. 3.16. Diagram of the composition and function of the hypothalamohypophyseal control system (pituitary gland) in vertebrates.

hypothalamus as the *nucleus supraopticus, paraventricularis,* and *preopticus*. In them are formed the coupled neurosecretions *oxytocin* and *vasopressin* which progress through the *tractus hypothalamo hypophysialis* into the neurohypophysis by means of axonal transport (see Chap. 9.1) where they are stored and released as in a neurohemal organ. Although physiologically very different, both hormones are closely related chemically. They are peptides of 9 amino acids which contain a disulfide bridge to which the histochemical dye *Gomori* with chrome alumhematoxylin reacts to show neurosecretions. The

granular axonal swellings in the neurosecretory tract induced by the releasing factors are known as *Herring bodies*.

Oxytocin regulates contraction of the smooth uterine muscles and, thus, stimulates labor. Furthermore, it promotes milk secretion by contracting the myoepithelial cells around the mammary alveoli.

Adiuretin has an antidiuretic effect in that it promotes the reabsorption of water in the distal region of a nephron in the kidney. Its effect is induced by the activation of *adenyl cyclase* in the distal tubuli of the kidney. It also has a vasomotor effect in that it regulates vascular tone. Apparently, regulating the release of hormones is dependent upon osmotic blood pressure in that the actual osmoreceptors are activated in the hypothalamus.

In addition to the pure neurosecretions formed by the nuclei of the hypothalamus, so-called *releasing and inhibiting factors* are also formed here in certain ganglion cell groups. These factors are secreted into the adenohypophysis via *hypothalamic-hypophyseal portal circulation.* After they are formed, they are secreted mostly into the blood vessels situated there. They then move through the infundibulum to the distal region of the adenohypophysis. The result is that these agents, which also can be called hypophysiotropic hormones, can effect releasing or inhibiting responses in the various cells of the adenohypophysis.

The following nomenclature has been suggested for the known releasing and inhibiting hormones. Their specific effect can be derived from their designation:

> *Corticotropin releasing hormone* (CRH)
> *Luteinizing hormone releasing hormone* (LH-RH)
> *Follicle-stimulating hormone releasing hormone* (FSH-RH)
> *Thyrotropin releasing hormone* (TRH)
> *Somatotropin releasing hormone* (SRH)
> *Prolactin release inhibiting hormone* (PRIH)
> *Melanocyte-stimulating hormone* (MSH)
> *Melanocyte release inhibiting hormone* (MRIH)

The chemical structures of these hormones have not been identified in all cases, but, generally, they are short chain polypeptides. TRH, for example, is a tripeptide, whereas CRH, FSH-RH, and LH-RH are each a decapeptide. The secretion mechanism of these hypophyseotropic hormones might be influenced by feedback mechanisms or neuronal influences in which case neurotransmitters such as noradrenaline, dopamine, or serotonin would play an important role.

The interaction between the hypothalamus and hypophysis as well as the effects of the pituitary hormones are presented in Fig. 3.16. The perikarya of corresponding centers of the hypothalamus (i.e., the nucleus supraopticus)

are activated to increase production of neurohormones by stimulation of the CNS. After they are synthesized, and, depending on the substance group (releasing factors), they reach either the neurohypophysis (oxytocin, adiuretin) or the hypothalamic-pituitary portal system of the adenohypophysis by means of axonal transport (see Chap. 9.1). Although the releasing factors, facilitated by the neurohemal structures, reach their peripheral sites of action (uterus, kidney) via the bloodstream, the releasing and inhibiting hormones arrive at the various cell types of the adenohypophysis via the short-circuited hypophyseal portal circulatory system. Here, they induce the release of various pituitary hormones that trigger reactions generally known at their sites of action.

3.2.2.4.2 Telencephalon

Certainly, it is the vertebrate *telencephalon* that has undergone the most profound changes in the course of phylogenesis (Figs. 3.4, 3.5, and 3.10). In fish, it is essentially only an olfactory center. In higher vertebrates, especially mammals, it evolved into the highest associative and integrative control system of the entire brain. This functional change might well have been the result of a general change in the telencephalic basal ganglia, indeed, primarily in the *pallium* (Fig. 3.17). In primitive vertebrates, the gray matter is located deep beneath the brain's surface. It is derived from, and still is differentiated only slightly from, the spinal cord. Called the *paleopallium*, its role is solely olfactory in nature. Beginning at the level of the amphibian, the paleopallium becomes differentiated from other sensory organs by additional pathways. It becomes a superordinate integration area, the so-called *archipallium*. It is situated laterally and basally. In vertebrates, the archipallium develops further into the *hippocampus* and the basilar, striated bodies, the *corpus striatum*. The latter is an important association center whose pathways run from and to the tegmentum of the mesencephalon. With increasingly higher development from the level of the reptiles, the paleo- and archipallium shift to beneath the surface of the cerebral hemispheres and form between them the *neopallium* which further develops into the stratiformed *cortex*.

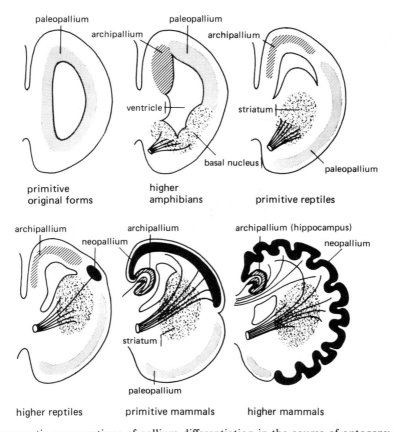

Fig. 3.17. Diagrammatic cross sections of pallium differentiation in the course of ontogeny in vertebrates.

The pallium of birds can be seen to develop progressively as it assumes ever greater associative functions. The striated body, the striatum, shifts further inward. In mammals, this process of *cerebralization* reaches its zenith as the neopallium (*neocortex*) and evidences an exceptionally great increase in surface area through extensive ridging. This is attended by inner cytostructural differentiation.

In its ultimate form in humans, the telencephalon, consisting of two hemispheres, divides into the telencephalic basal ganglia and the *pallium*, the main mass of which is formed from the *cortex* located at the surface.

On the basis of its ascending and descending fiber systems, the massive *basal ganglia* section of the cerebrum, or *colliculus ganglionaris*, can be subdivided into the region of the *caudate nucleus* (nucleus caudatus), the *putamen*, the *claustrum*, and the *amygdaloid nucleus* (corpus amygdaloideum) whereby the nucleus caudatus and the putamen together form the so-called *corpus striatum* which constitutes a unified, extrapyramidal motor center.

The structures of the *rhinencephalon*, formations that are localized forward in the floor of the hemispheres in higher forms, are particularly ancient parts of the prosencephalon that were of paramount importance early in the evolution of vertebrates. They consist solely of cortical matter that remains simply structured. Each of the two nasally situated, distended *olfactory bulbs* (bulbi olfactorii) leads with a *peduncle* (pedunculus olfactorius) to the actual *olfactory cortex* (cortex olfactorius) located directly under each half of the cerebrum.

The *hippocampus* is yet another special region of the prosencephalon that emerges quite early in the development of vertebrates as a differentiation of the archicortex. It is located first at the

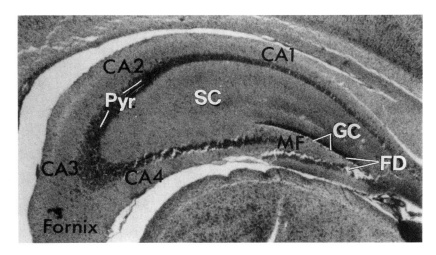

Fig. 3.18. Cross section of the hippocampus of a Tupaija (tree shrew). FD: fascia dendata; GC: granular cells; MS: mossy fibers; SC: Schaffer collaterals; Pyr: pyramidal cells.

open boundaries of the cerebral cortex and becomes involuted into a long, cylindrical body as a result of an arciform growth pattern. An histological cross section of the hippocampus (Fig. 3.18) reveals a structure with an extremely characteristic organization of components, i.e., together with the adjacent *fascia dentata*, the individual regions form an S-shaped structure. The lower part of the fascia dentata is positioned around the C-shaped structure (horn of Ammon) of the actual hippocampus. Both parts comprise one functional unit. Typical of archicortical areas, the C-shaped hippocampus has only a single cell bed and can be subdivided into four regions (*cornu Ammonis* regions) with individual cell differentiations and separate functions (see Chap. 5 and Figs. 3.18, 5.9).

The amygdaloid nucleus complex (corpus amygdaloideum), fornix and cingulum (see Fig. 3.5.7) are closely connected to the hippocampus and form one functional unit, the *limbic system*. These structures have branched off from the rhinencephalon in the course of phylogenetic development and no longer are connected functionally to it. The limbic system has vegetative functions and is associated with emotional processes that are triggered by various sensory perceptions.

The other great region of the cerebral cortex, the *neocortex*, which comprises nearly 70% of all

the nerve cells of the entire CNS, is phylogenetically and ontogenetically the youngest region of the brain. Its primordium and further development lead from the reptiles to the line of mammals and constitutes the prerequisite for the special development in humans. To understand the composition and function of the prosencephalon and, especially, the cortex, which is so utterly crucial in the human and is superordinate to all sections of the nervous system, special attention is given here to the development of the *cerebral cortex*.

Composition and Development of the Cortex. Essentially, the cortex emerges in three stages during ontogeny. It originates from a cell body layer (*matrix zone*) near a ventricle. The matrix zone has a superficial fiber zone (*marginal zone*). First, the matrix zone widens through intensive cell division. Ultimately, cells leave it and migrate to the primary fiber zone, giving rise to a zone rich in cell bodies called the primary cortex or zonal layer in the primary marginal zone. The more cells that migrate from the matrix zone, the thicker the cortex becomes. Differentiation, which leads to a layer formation, does not occur until later. The first fine differentiation occurs in humans in the third or fourth embryonal month; the final differentiation does not occur until after birth under functional conditions. The swarming process of the nerve cells from the matrix layer concludes in the second semester of pregnancy. Fibers grow from the nerve cell bodies of the primary cortex that, by and large, form the medullary layer. As a result of the elapsed swarming process, *gray matter* lies outside of the cortex and *white matter* lies within it, a state of affairs quite different from the usual in the regions of the brain. The myelin sheaths around the fibers mature in various periods of time, first in the older, conservative structures and later in the fiber sheaths belonging phylo- and ontogenetically to the younger systems. Fine differentiation should progress in the fiber region until the age of 35.

The axons growing from the cortex can remain either as association fibers within the same hemisphere, or they run to the opposing hemisphere as *commissural fibers*, or they exit the cortical region as *projection fibers* and establish contact with other regions of the brain.

Both cerebral vesicles of the prosencephalon growing out to the cerebrum are connected in their middle area to the diencephalon which, even later, remains quite small. Thus, growth is limited to forward, rearward, and lateral downward directions. The hemispheres, therefore, proliferate during embryonic development in those directions until they are impeded by the skull cap. Then, the direction of growth turns back in the forward direction. The inwardly involuted regions of the *hippocampus* develop through a downward, arcing movement of the surfaces of both hemispheres.

The various outgrowth zones and directions result in a four-lobed segmentation in each *cerebral hemisphere*: the *frontal lobe* (*lobus frontalis*), *temporal lobe* (*lobus temporalis*), and *occipital lobe* (*lobus occipitalis*), which are connected by way of the island region, a special structure, to the *parietal lobe* (*lobus parietalis*).

The neuronal matter for this extensive growth originates in the matrix zone from which the cells migrate, inducing formation of the lobes. In the next stage of cell growth, rather intensive cell accumulation occurs at specific locations within the lobes that give rise to local prominences that eventually become the *gyri* (convolutions). Growth of the walls is retarded, on the other hand, in some places within these prominences. This later results in the formation of *sulci*

(furrows). Thus, the furrow relief of the cortex is a result of an increase in mass in certain regions attended by simultaneous growth retardation in adjacent areas.

This sequence of developmental stages in the first months of embryonic development shows that the human brain is first smooth-walled (*lissencephalic*) and then, as early as the eighth month, is mostly furrowed (*gyrencephalic*), i.e., predominantly subdivided into longitudinal furrows. The furrowing begins at the boundaries of the lobes and at the phylogenetically oldest cortical regions (primary furrows) and tend to run diagonally to the predominant growth direction of the telencephalon. The primary furrows are

- *Sulcus centralis* between the frontal and parietal lobes where the primary motor and sensory cortical fields later develop,
- *Sulcus parieto-occipitalis* between the parietal and occipital lobes,
- *Fissura calcarina* on the inside of the occipital lobe in the future primary optical projection center of the cortex,
- *Sulcus cinguli* on the inside of the telencephalic vesicle as separation of the future rhinencephalic region from the actual new brain, and
- *Fissura lateralis cerebri* between the frontal and temporal lobes in the region of the future speech and auditory centers which are developing here.

In the course of the brain's further maturation, secondary furrows eventually emerge within the individual lobes which also tend to run longitudinally and are relatively constant in form. Numerous tertiary furrows branch off from these secondary furrows which vary in form and number and are responsible for the complexity and individuality of the cortical relief.

Associated with the formation of furrows is an extensive increase in surface area, for the cortex, regardless of its organizational level or size, remains the same in thickness (ca. 1.5 to 4.5 mm). Thus, *gyrencephalization* provides for a greater relay area in the hemispheres. According to Blinkov and Glezer (1968), the surface area of various cortical regions in the cerebrum of an adult amounts to 105 cm^2 for the occipital area,

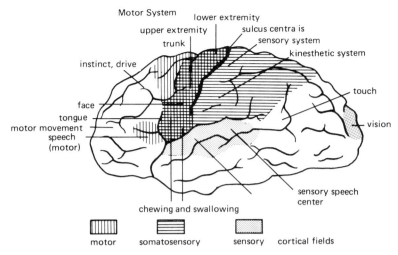

Fig. 3.19. Motor, somatosensory, and sensory projection fields of the cerebral cortex in the human.

79 cm² for the lower parietal area, 72 cm² for the upper parietal area, 197 cm² for the temporal area, and 208 cm² for the frontal area yielding a total surface area of about 660 cm². In general, the individual furrows are not topographic boundaries of specific cortical regions. Nevertheless, functional-morphological analyses show that motor representation centers lie chiefly in the frontal regions, whereas sensory centers for the coordination of sensory perception lie in the caudal regions (Fig. 3.19). In comparison to other central nervous structures, however, we know relatively little about the functions of the cortex.

The *structure of the neocortex*, as a whole, is replete with regional differentiations. Thus, it is best to divide the cortex into more or less uniform areas, or fields, that can be described histologically and can be distinguished readily from one another. With this in mind, Brodmann (1925) distinguished about 40 cortical fields. In spite of many points of agreement, Ecomo divided the cortex into almost 80 fields. An examination of the function and cytoarchitectonics of these fields showed that about $\frac{1}{3}$ of the areas distinguished in the human as projection fields are associated with defined actions and reactions in the periphery. As yet, no correlation with external stimulus reactions has been found for $\frac{2}{3}$ of the fields. For this reason, these fields also are referred to as inter- or association fields,

the presumption being that they serve this end. The physical relationship of the various body sections in the region of the motor or sensory part of the cerebral cortex is shown in Fig. 3.20. It can be seen that the relative size of the brain areas does not correspond to the structures they innervate, but rather to the nerve cells they innervate or the number of receptor and motor units.

The extraordinary complexity and difficulty involved in finding clear organizational criteria, as reflected in the discrepant assessment regarding the number of brain regions and their demarcation, also impede any attempt to construct a general pattern of association among the neuronal elements on a cellular level from the variety of manifestations (see Chap. 5). Hence, examination of cortical structures is always limited to a basic type, and findings cannot be applied to the wide variety that remains insufficiently researched.

Cells and fibers are arranged in the *cerebral cortex*, which covers the entire surface of the cerebrum, in such a manner that a clear layering becomes evident. In the course of ontogenetic and phylogenetic development, it undergoes increasing differentiation to the extent that a six-layered cortex results.

A detailed microscopic analysis of the cortex reveals considerable differences in cell forms, quantity and organization. Those regions of the

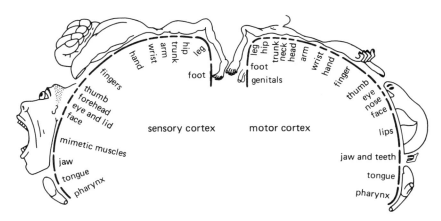

Fig. 3.20. Representational fields of body sections in the sensory and motor cortex of the telencephalon. (Reprinted with permission of Macmillan Publishing Company from The Cerebral Cortex of Man by Wilder Penfield and Theodore Rasmussen. Copyright 1950, Macmillan Publishing Company; copyright renewed 1978 Theodore Rasmussen.)

cerebral cortex that are uniform are designated as areas (areae). These can be distinguished clearly from one another by virtue of their specific structure, or architecture, by various methods of representation:

- *Cytostructure* (differentiation according to cell type),
- *Myelostructure* (varying organization of myelinated nerve fibers),
- *Chemostructure* (specific behavior of nerve cells vis-à-vis histochemical treatment),
- *Gliastructure* (glial formation), and
- *Angiostructure* (specific arrangement of the blood vessels).

The *neocortex* is divided from the outside to the inside into the these six layers (Fig. 3.21):

I. *molecular layer* (*lamina zonalis*), consisting of scattered, small, horizontally oriented cells and tangential *associative fibers*;
II. *outer granular layer* (*lamina granularis externa*), composed of densely packed granular cells whose axons terminate in the same layer;
III. *outer pyramidal layer* (*lamina pyramidalis externa*) of pyramidally shaped cells whose neurites, departing basally, form the pyramidal *projection pathways* and are wrapped in a myelin sheath even within this layer;
IV. *inner granular layer* (*lamina granularis interna*), formed similarly to layer II;

however, pronounced in the region of the visual cortex;

V. *inner pyramidal layer* (*lamina pyramidalis interna*), composed in part of large *pyramid cells* (see Fig. 1.3c), but also of neurons that are aligned horizontally and especially pronounced in the visual cortex;
VI. *spindle cell layer* (*lamina multiformis*), composed of multiformed cells, the larger ones lying predominantly outside, the smaller ones lying predominantly within. Their neurites course into the inner medullary layer as well as into the outer cortical layers lying in the opposing direction.

Regarding the course of the fibers within the cerebral cortex, it has been established that ascending fibers from the thalamus terminate, greatly ramified, as axodendritic synapses in layer IV, whereas intracortical associative fibers terminate in levels II and IV. Primarily descending axons of the pyramid cells, which are routed at the same time via collaterals within the cortex, exit layers III, IV, and V. The individual *cortical fields* of the different regions of the cortex can be subdivided cytologically by specific cell type (see Fig. 3.21), i.e., the motor areas mainly contain large pyramid cells whereas the sensory areas consist of the more simply configured granular cells.

With regard to routing, the extraordinarily complex cytostructural composition of the

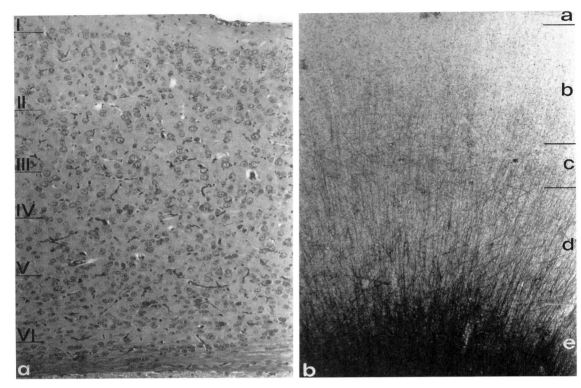

Fig. 3.21. Cross section of the cerebral cortex of a mouse: (*a*) hematoxylin and eosin (H & E) combination dye reveals cell bodies; (*b*) silver impregnation of the fibers from Weigert-Pal. I: molecular layer; II: outer granular layer; III: outer pyramidal layer; IV: inner granular layer; V: inner pyramidal layer; VI: multiform layer; *a*: pia mater; *b*: outer main layer; *c*: Baillarger's bands; *d*: inner main layer; *e*: medulla.

cerebral cortex (see Figs. 3.19 and 3.20) enables all centers to be connected with one another, regardless of their location. *Projection pathways* connect the cerebral cortex with more deeply located structures of the CNS, *association pathways* facilitate neuronal cross-linking, and *commissural pathways* control the flow of information between the two hemispheres. One other phenomenon is that of crossing, or *decussation*, particularly of pyramidal pathways and most of the other ascending and descending pathways, as well, to the side of the body opposite the cerebral hemispheres, i.e., as compared to the position of the representation fields. Thus, the right half of the body is innervated by the left hemisphere, and vice versa. Functional coordination and equalization of both hemispheres occurs in the crossing, or decussation, region via connecting commissures,

particularly in the so-called trabecular (*corpus callosum;* see Fig. 3.5.7).

Based upon the unique composition and *function of the cerebral cortex*, some capabilities that have their early beginnings in animals are nearly perfected in the human. These capabilities elevate the human being above all other members of the animal world. Consider *language* and its resultant *verbal communication*. The associated capability of using the spoken word as a vehicle of communication independent of the person who uttered it must be viewed as a basis of culture. It is preserved and reproduced, giving rise to the possibility that its meaning can be disseminated and received through the millennia. This occurs in oral and in written form ("*extracerebral chains of association*"), whereby, in the millions of years of human development, the invention of writing, 8,000 to 10,000 years ago at the earliest, is relatively recent. Yet, its impact lead to an explosive increase in information storage.

It has been recognized for more than 100 years that various regions of the brain work synergistically to create the ability to *speak and comprehend language.* Meanwhile, the process that enables this, and the regions of the brain that are active in the process, have been revealed to researchers to a great extent. Reception of a spoken word occurs via the *primary auditory center.* The impulse is conducted to the *sensory language* or *speech center* where it is processed into meaning. The motor *speech center* becomes activated to facilitate a *verbal repetition* of this word. In doing so, it conducts the impulse to the motor cortical field where the muscle movements necessary for speech are activated and their execution is coordinated.

The path leading to *articulation of a written word* (or a *word read*) is even longer: the *primary visual center* is activated, the optical impression is processed by the *reading center* in the *gyrus angularis* (the middle part of the inferior parietal lobule) where the visual form is translated into acoustic form which is then further processed by the sensory speech center and is finally articulated.

If what is read is to be answered in written form, then the path from the speech center via the reading and visual centers is made more complicated and is extended to involve the sequence control mechanism and a triggering of movements necessary for the act of writing.

Crossing, or decussation, of most *nerve fibers* departing the cortical fields results in the phenomenon that the left side of the face and body is controlled by the opposite, or right, hemisphere, and vice versa. This relationship of opposing sides and hemispheres is especially true of the primary motor and somatosensory cortical fields such as for movement of the hands and feet. The eye and ear, in addition to having a strong relationship of opposing sides, also are involved by commissural fibers extending to the hemispheres of the same side. Essentially, the division of sensory and motor functions of the hemispheres is symmetrical. Certain specialized functions, however, lead to functional asymmetry, i.e., they predominantly engage only one

hemisphere. The coordination of the hemispheres then ensues via the various secondary connections in the brain. For example, *speech capacity,* or the ability to recognize a melody and to recall verbal information, is localized asymmetrically.

Visual recall of people and things is associated with the function of the regions of the cerebrum in the area under the temporal and occipital lobes.

Emotional reactions are also important within this context. Apparently, they are represented especially strongly in an asymmetrical manner. *Emotions* are associated with the *limbic system* deep within the brain. Furthermore, the right hemisphere contributes greatly to the emotions, whereas the left is only minimally involved. The right hemisphere controls not only appropriate emotional reactions to given stimuli, but is also responsible for registering and comprehending the feelings of others, for example, those feelings that accompany a statement. Meaning, however, without emotional background, is comprehended by the left hemisphere.

Recent investigations have shown that functional *asymmetry* also manifests itself anatomically. Thus, despite the general tendency of functional structures to be symmetrical, anatomical differences have been found in several regions between the two hemispheres. For example, *Heschl's convolutions* in the sensory speech center, are larger in their posterior region in the left hemisphere than in the right and, in fact, this is the case even before birth. One can conclude that language performance of the left hemisphere is related to factors of anatomy and is not the consequence of language development in childhood. It was recently discovered that a change in the organization of the tissue corresponds to the enlargement of the Heschl convolutions. There is a special cell layer in this region the dimensions of which are larger in the left hemisphere than in the right by a factor of 7.

Right-handedness seems to be based upon asymmetries expressed in the form of an increase in the size of the right temporal lobe and left parietal and occipital lobes when compared to their opposing sides. These increases are expressed as subtle recesses on the inner surface of the cranium and are evidenced even in fossilized *hominids* including the *Neanderthal.*

Additionally, various asymmetries, especially in the region of the Sylvius sulci (*fissura sylvii*), are associated with right- and left-handedness, but particularly among the right-handed. Whether or not this tendency in right-handers is hereditary is still an open question. Perhaps one day, comparative computerized tomographic examinations of family members will offer further information in this regard.

Most functional areas have been located by way of lesions that have led to a loss of specific body function or mental capability that corresponds to a particular representation center in the neocortex. Thus, for example, injury to the region of the lower convolutions of the frontal lobe (lobus frontalis), the so-called *Broca speech center*, leads to *motor aphasia*, i.e., the inability to speak even though the muscles involved in the process of swallowing, etc., can be summond to action voluntarily and the capacity to comprehend language remains intact. If, however, injury occurs to the *Wernicke speech center* located on the sensory side, posteriorly in the upper temporal lobe (lobus temporalis) and, thus, adjacent to the primary acoustic center, then the resultant impairment is one of *sensory aphasia*, i.e., the afflicted individual can hear words but is incapable of processing their content into meaning.

The loss of regions involved with vision can have similar effects. Failure of certain cortical fields can lead to *psychanopsia* (psychic or *"soul blindness"*) even though the eyes remain perfectly functional. The sequence of variables that leads to impairment or loss of normal functions is extremely complex; however, it is often the only key to understanding, or gaining insight into, the manner in which the cerebrum functions.

3.3 Vegetative Nervous System (Sympathetic and Parasympathetic)

In conjunction with the endocrine system, the *vegetative, visceral,* or *autonomic nervous system* (NS) in vertebrates controls, in conjunction with the endocrine system, the body's organ functions by regulating the function of the smooth musculature of the intestines, the musculature of the heart, and the functional activity of the glands. It regulates essential processes such as maintaining the inner milieu through heat regulation, water equilibrium, and ion balance; it controls metabolism and energy regulation, and coordinates the circulatory system, respiration, digestion, and reproduction.

The vegetative NS is divided into two parts that function antagonistically, the *sympathetic nervous system* and the *parasympathetic nervous system*, which originate in different regions of the central nervous system (Fig. 3.22). The peripheral, fundamental elements of both vegetative systems consist of two neurons that come into synaptic contact with each other in a peripheral ganglion.

The *sympathetic nervous system* forms a ganglial chain connected via the *asympathetic trunk* (*truncus sympathicus*) laterally on both sides of the spinal column in the region of the thoracic and upper lumbar spinal cord. The ganglia are connected to the nerve cell bodies that are located in the lateral horns of the spinal cord by means of *preganglionic fibers* that exit the anterior roots of the lateral horns of the spinal cord, the *rami communicantes albi* (white communicating branches). These fibers are relayed to postganglionic fibers, the *rami communicantes grisei* (gray communicating branches) in the ganglia (Fig. 3.23). These are considerably longer than the former and course to the peripheral sensory organs along with the spinal nerves. In addition to the spinal nerves, the sympathetic fibers of the tenth cranial nerve accompany them en route.

In contrast to the sympathetic nervous system, the *parasympathetic nervous system* originates partially from the brain stem and partially from the sacral region of the spinal cord. Although the routing of the preganglionic fibers of the sympathetic nervous system already has occurred in the immediate vicinity of the spinal cord, in the sympathetic chain, the axons of the parasympathetic nervous system are considerally longer and are routed to the organs being innervated or to the parasympathetic ganglia which, similarly, lie distant from their point of origin. The parasympathetic nervous system does not have its own guidance system. Rather, its fibers run in the pathways of the cranial

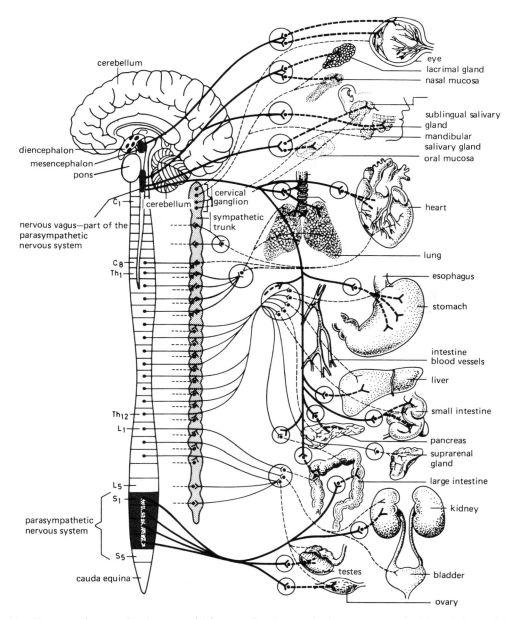

Fig. 3.22. Diagram of vegetative (autonomic) innervation (sympathetic, parasympathetic) and the position of the vegetative nuclei of origin in the human body.

nerves, for example, the *nervus oculomotorius*, the *nervus facialis*, the *nervus glossopharyngeus*, and, in particular, the *nervus vagus* (Fig. 3.22).

The two parts of the vegetative nervous system function antagonistically. Generally speaking, the parasympathetic system serves to control the function of the inner organs whereas the sympathetic system brings its particular influence to bear under conditions of stress. For example, one effect of the parasympathetic nervous system is to slow heart beat frequency whereas the sympathetic system increases pulse to meet momentary demands.

Numerous, diverse functional characteristics

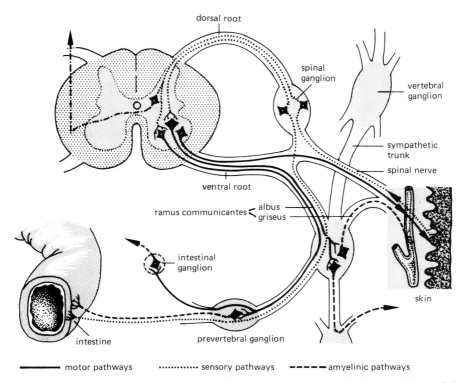

Fig. 3.23. Possible reflex routes between sympathetic and somatic neurons in the periphery of the body.

can be distinguished. Of utmost importance is the release of glycogens from the liver stimulated by the sympathetic system resulting in an increase of the body's capacity to perform work. The parasympathetic nervous system, on the other hand, suppresses glycogen mobilization, promotes rest and reversion to the body's former state, intensifies and slows digestion, and controls the intensity of heart and circulatory function.

As is the case in all nerve cells, stimulus conduction occurs in the vegetative system by means of *neurotransmitters* (see Chap. 7). Although impulse conduction at the synapses of the preganglionic fibers takes place with the help of *acetylcholine* in both vegetative systems, this transmitter is used only in the parasympathetic system, not in the sympathetic system, at the postganglionic synapses to the sensory organ. *Adrenaline* and *noradrenaline* serve signal conduction in the sympathetic system (as do certain other catecholamines in the CNS). Thus, the functional difference between the para-

sympathetic and sympathetic systems rests in their differing transmitters. This is extremely important as it applies to pharmacological considerations in as much as their *mimetica* (substances that imitate the effect of particular system transmitters) and their *lytica* (substances that inhibit system transmitters) are used clinically for the treatment of hyper- or hypofunction in many different organs:

- *Sympaticomimetica* include noradrenaline, adrenaline, ephedrine, isoproterenol, and the amphetamines Benzedrine and pervitin;
- *Sympaticolytica* include ergotamine, yohimbine, phentolamine;
- *Parasympaticomimetica* include acetylcholine, muscarine, pilocarpine, carbachol, cholinesterase inhibitors (physostigmine and neostigmine);
- *Parasympathicolytica* include atropine and scopolamine.

The *manner in which the vegetative NS functions* appears to be extremely complex. It has been discovered that sympathetic postganglionic synaptic contacts exist not only with the sensory organs, but also with the peripheral sympathetic ganglia. Thus, the postganglionic parasympathetic neuron is not only influenced by the acetylcholine of the preganglionic synapse, but also by the sympathetic endings. This leads to an integrative function of the parasympathetic neuron. Rather than simple impulse conduction, the system is capable of modulation in that it permits some of the necessary regulatory and integrative functions to be carried out directly in the peripheral area without the assistance of the CNS.

Stimulation of the sympathetic nervous system affects the entire system and one stimulus can trigger a whole series of reflexes (Fig. 3.23).

- *Viscerovisceral reflexes* (from the viscera to the viscera) are triggered peripherally; they affect the triggering organ itself and regulate, for example, motor response or secretion.
- *Viscerocutaneous reflexes* (from the viscera to the skin) are triggered when the stimulus from an inner organ is conducted in part to the skin via the rami communicantes grisei and certain affected areas of the skin react to this impulse, for example, with increased blood flow.
- *Visceromotor reflexes* (from the viscera to the motor response system) conduct the stimulus from an inner organ to a motor neuron in the spinal cord. This results in so-called abdominal defense, or resistance, in instances of inflammation processes. Reflex paths of this sort play an important role in medical diagnosis (appendicitis). If, on the other hand, the stimulus originates in the tactile organs (pain, pressure, temperature), then the progression of reflexes can elapse in the reverse direction as well (outside to inside) and trigger cultivisceral reflexes.
- *Cutivisceral reflexes* (from the skin to the viscera). The viscera, in turn, can trigger viscerovisceral reflexes. Thus, it is possible in this way affect motor response or secretion in the inner organs by changing skin temperature (in response to weather, etc.)

Part of the stimuli of internal and external organs that normally are conducted vegetative-reflex-ively without conscious control can be conducted to superordinate brain regions of the cerebral neocortex via the *tractus spinothalamicus*. However, due to the diffuse reactions of the sympathetic nervous system, they do not enable further information to be imparted about their specific point of origin. Thus, these afferent stimulus patterns are projected to the affiliated skin area in which the response, for example, pain, is felt as a result of the cortical fields being stimulated.

The inner organs, then, have their narrowly delineated *cutaneous projection fields*, or *Head's zones*, in the skin (Fig. 3.24) in which symptoms of illness of the inner organs can be recognized. As such, these zones are especially important in medical diagnosis. The integration of all vegetative components of our organism is carried out by nucleus centers in the hypothalamus of the diencephalon (see Sect. 3.2.2.4.1). These nucleus centers regulate the automic nervous system not only via the nerves, but also, and primarily, humorally by way of neurosecretion mechanisms.

3.4 Derivatives of the Placodes

Just as the neural crest developed during ontogeny in the region of the spinal cord and gave rise to a number of derivatives (see Chap. 2.1), so do *placodes* develop in various places of the embryonic ectoderm in the head area in vertebrates during this same period. These placodes innervate a number of sensory organs. The *labyrinth vesicles* and their sensory neurons emerge from the dorsal-lateral placodes as does the *lateral line system*. The *epibranchial placodes* above the dorsal boundary of the branchial fissures form the neurons of the *taste buds*. The rostral *ophthalmicus placode* supplies the neurons of skin sensitivity for the *ganglion ophthalmicum*. The olfactory placodes, located most anteriorly, send their basal axons (*fila olfactoria*) deep within to connect with the *tractus olfactorius* which runs to the prosencephalon.

The individual components of the nervous system become differentiated from these separate embryonic structures (neural plate, neural crest,

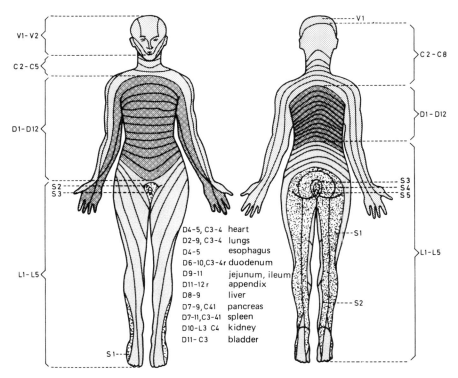

D4–5, C3–4   heart
D2–9, C3–4   lungs
D4–5   esophagus
D6–10,C3–4r   duodenum
D9–11   jejunum, ileum
D11–12 r   appendix
D8–9   liver
D7–9, C4l   pancreas
D7–11,C3–4l   spleen
D10–L3 C4   kidney
D11– C3   bladder

Fig. 3.24. Segmental and peripheral skin innervation areas as cutaneous projection fields of inner organs (Head's zones).

and placodes) and connect with one another and with the secondary sensory cells and effectors during ontogeny to form one great, functional entity.

3.5 Nonneuronal Structures in the Nervous System

3.5.1 Neuronal Sheaths

Nerve tissue is protected by special sheaths of mesodermal tissue, i.e., developmentally, this tissue originates in the mesoderm. In the region of the CNS, this tissue is the meninges (see Sect. 3.5.3), wheras in the peripheral nerve tissue it is the neural sheath (Fig. 3.25). Within the neural sheath is found an additional type of sheath called the *endoneural sheath* (or reticular fiber sheath, or fatty sheath). An endoneural sheath surrounds each peripheral nerve fiber that is in contact with a myelin sheath. This last variety of sheath surrounds the fiber like a tube (see Chap. 1.2.4). Each rather large peripheral nerve is separated, along with other fibers, into its own individual bundle (*fasciculus*) by the connective tissue of these endoneural sheaths. The outer connective tissue matter of the *endoneurium* solidifies into a very hard layer, the perineurium. The number of axons within the *perineurium* varies. They are separated into individual bundles by the loose, vascular connective tissue of the endoneurium. These nerve bundles, which are surrounded by a perineurium, are, in turn, gathered together into a nerve cord by the *epineurium,* a rather rigid lining. The nerve cord is visible macroscopically. These connective tissue sheaths constitute a formidable diffusion/metabolic barrier in the peripheral nervous system, as do the meninges in the central nervous system, and provide

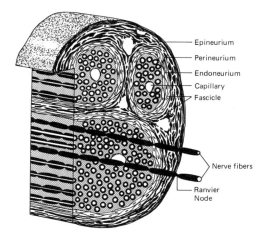

Fig. 3.25. Diagram of the composition of the neural sheath of peripheral nerves.

mechanical protection, as well. The blood vessels needed to sustain the nerves are located within the sheaths.

3.5.2 Ependyma and Circumventricular Organs

The wall lining of the *central canal* (ventricle and spinal cord canal) is composed of *ependymal cells* which join to form an epithelial cell formation. A great number of them, the *spongioblasts*, are ciliated (Fig. 3.26). It is through the pulsation of their cilia that the flow of the liquor cerebro-spinalis (cerebrospinal fluid, CSF) is directed into the *cerebral ventricles* in the direction of the lumbar spinal canal. In animals of lower developmental level ontogenetically and phy-

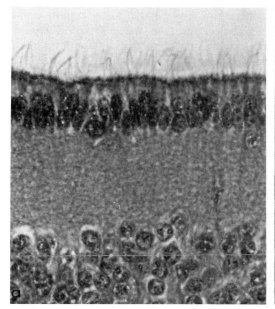
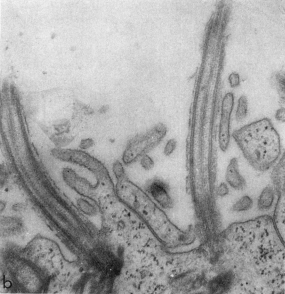

Fig. 3.26. Ciliated ependyma from the region of the second ventricle of a lamprey (*a*); electromicrograph of cilia of an ependymal cell of a goldfish (*b*).

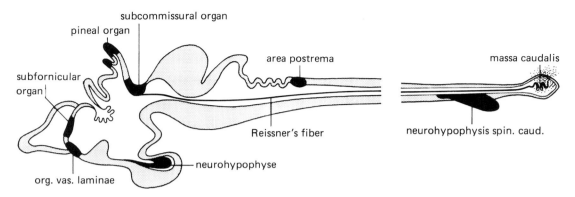

Fig. 3.27. Diagram of the position of circumventricular organs and the path of the Reissner fiber in the CNS of vertebrates.

logeneticallly, the ependymal cells, with their radial processes (as seen from the central canal), permeate the wall structure of the neural tube. Only later in the course of ontogenetic development is the connection to the outer wall interrupted. If the ependymal cells also lose their connection to the inside of the neural tube, then they leave the epithelial group and become pure glial cells. The ependymal cells retain their capacity to divide. Thus, one cell descendant might stay in the epithelial group whereas another might depart this group, migrate to the periphery and become differentiated either into a neuroblast or a glioblast (see Fig. 2.5).

In addition to normal differentiation of the ependyma as a boundary of the cranial ventricles and spinal canal, the ependymal glia can undergo special differentiation, especially in the region of the third and fourth ventricles, resulting in organ-like, densified special structures rich in blood vessels, the so-called *circumventricular organs* (Fig. 3.27). Since these organs are highly secretory, one presumes that they perform a special function in regulating osmotic equilibrium between blood and CS fluid. The circumventricular glial organs contain chemorecptors that react with great sensitivity to changes in the pH of the blood and CS fluid. For this reason, they appear to control vegetative reactions that regulate the blood's acid-base equilibruim. Additionally, they may play a substantial role in CS fluid circulation and in regulating water

inflow to the cerebrospinal fluid (under pathological circumstances, formation of edemas, or hydrocephaly).

The most important circumventricular organs are the *subfornicular organ* in the region of the diencephalon and the *subcommissural organ* which, aside from the human and dolphin, occurs in all vertebrates and lies in the region of the commissura posterior. A long, homogeneous fiber, *Reissner's fiber* (Fig. 3.28), whose function is still not fully understood, extends from its ependymal region to the spinal canal.

There are also periventricular organs in the diencephalon that are especially prevalent in the area of the thalamus and the infundibulum as well as in the *area postrema*, located at the entrance of the spinal canal at the base of the rhomboid fossa.

3.5.3 Meninges

The actual nervous tissue of the brain and neural tube is bounded by the ectodermal glial cortical layer called the *membrana limitans gliae*. To the outward side of this cortical layer are the *meninges* (Fig. 3.29) which consist of mesodermal connective tissue. The former is the soft meninges, the *meninx primitiva* in lower vertebrates and the *leptomeninx* in higher vertebrates. In cyclostomes and teleosts, the *meninx primitiva* is divided into a basal *endomeninx* and a *intermeningeal tissue* which contains fat and fills out the wide cavity

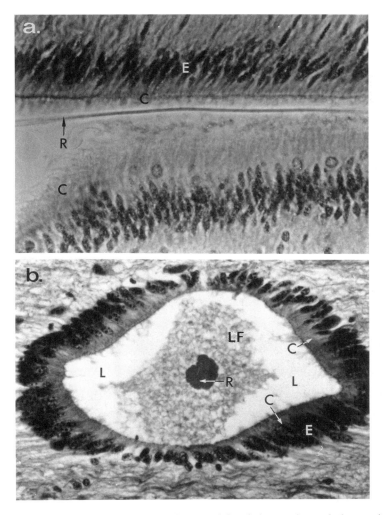

Fig. 3.28. Reissner's fiber: longitudinal section in the ventricle of the myelencephalon and spinal cord of a lamprey (*a*); cross section of the central canal of a cow (*b*). R: Reissner's fiber; LF: leaked CS fluid; C: cilia border; E: ependymal nuclei with numerous nucleoli; L: lumen of the central canal.

between the relatively small brain and the cranial bone.

In quadrupedal vertebrates (Fig. 3.29b), a "hard" meninx, the *pachymeninx* or *dura mater*, is present in addition to the leptomeninx, the latter consisting of soft connective tissue. The pachymeninx is split by the perichondrium of the cranial bone and vertebral canal. The cavity between the dura mater and the perichondrium resulting from the split, the *cavum epidurale,* is filled with veins or sinus cavities or even with solid tissue. This connective tissue can shift, or

be displaced, only in the region of the flexible vertebral canal.

In mammals (Fig. 3.29c), the meninges have undergone one additional differentiation. The leptomeninx is subdivided such that three meningeal layers exist beneath the membrana limitans gliae: first, the *pia mater,* the soft nerve membrane rich in blood vessels which is so closely connected to the membrana limitans gliae and permeats all sulci and recesses lying on the surface of the brain and spinal cord; then, the heavily reticulated, nearly vessel-free arachnoid

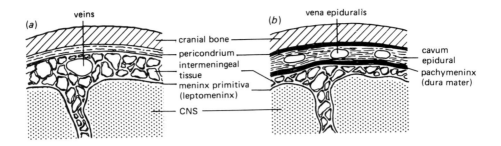

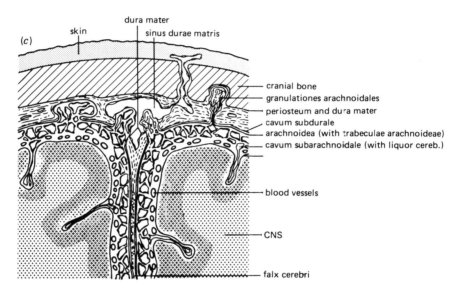

Fig. 3.29. Composition of the meninges in fish (*a*), amphibians (*b*), and mammals (*c*).

membrane, the *arachnoidea,* which encompasses a cavity pervaded by fibrous septa and trabeculae, the *cavum subarachnoidale.* Above it lies the hard *dura mater* which is seperated from the arachnoidea by the subdural cavity. The latter is filled with serous fluid whose pressure keeps the dura mater taut. A recess in the dura mater, the *falx cerebri,* forms a septum that separates the two hemispheres from one another. The cavum subarachnoidale is expanded in many places into so-called *cisterns.* The *cisterna cerebellomedullaris* and, with it, the entire subarachnoidal cavity above the fourth ventricle, is connected with the inner cavities of the CNS via openings in the dorsal covering of the rhomboid fossa (*foramen of Magendie,* foramen Luschka = apertura ventriculi terminalis medul-

lae spinalis, apertura ventriculi quarti cerebri) and, thus, contains cerebrospinal fluid (Fig. 3.30).

One function of the meninges is to protect sensitive nerve tissue. Also of functional significance, the vascular plexi of the leptomeninx in the region of the *plexus choroidei* (see Sect. 3.5.4) play an important part in the production of cerebrospinal fluid. The meninges also help in equalizing above-normal intracerebral pressures through the connection of the subarachnoidal cavity with the fourth ventricle. The leptomeninx and subarachnoidal cavity envelop not only the outer area of the CNS, but they also accompany the vessels and are drawn at their entrance points into the brain and spinal cord in funnel-like fashion. In this way, perivascular, sleeve-like cavities known as *Virchow-Robin cavities* emerge

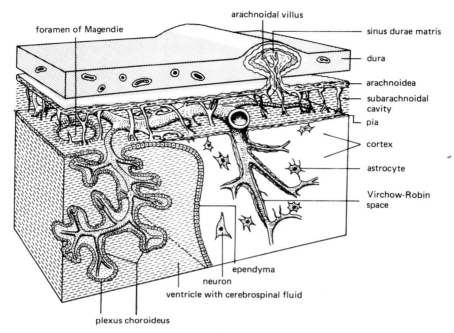

Fig. 3.30. Diagram of the relative positions of the meninges, plexus choroideus, blood vessels, and cerebrospinal fluid in the CNS of mammals.

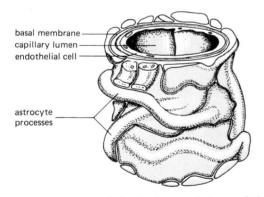

Fig. 3.31. Diagram of the basic morphology of the blood-brain barrier (bbb).

filtration of matter and thereby maintain a constant ion environment for the neurons of the CNS.

The leptomeninx fulfills one other essential function, that of a defense organ: in lymphoreticular tissue, for example, lymphocytes, histiocytes, or macrophages can be mobilized upon specific stimulation.

3.5.4 Blood Vessel Networks (Plexus Choroideus)

between neural tissue and the endothelium of the blood vessels (Fig. 3.30). These mesodermal sheaths effectively seperate the blood vessels from the neural tissue along the entire length of the vessels and contribute to the formation of the blood–brain barrier (BBB, Fig. 3.31), a metabolic barrier system the function of which is to absorb differentiated secretions for the nerve tissue from the circulatory system during the

The outer wall of the neural tube in the dorsal region of the brain consists, in part, of the original, single layer of epithelial cells, the upper plate, referred to as the *lamina tectoria* or lamina epithelialis. The vessel-laden pia mater is attached here from the outside in lamellar fashion as the *telae choroideae* and fuses with it into a vascular plexus, the *plexus choroideus* (Fig. 3.32). Such plexi are found in various parts of the cerebral roof (plexus choroideus ventri culi lateralis, pl. ch. ventr. tertius, pl. ch. ventr. quarti) and project, in mammals, into the cerebral

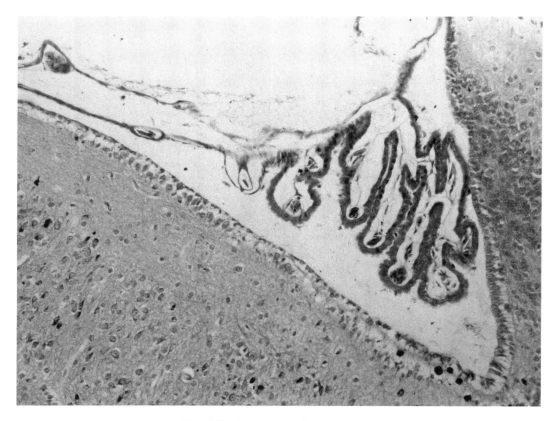

Fig. 3.32. Plexus choroideus of a frog.

ventricles as tree-like, ramified formations with abundant surface area (Fig. 3.33). The highly flexuous capillaries of the plexus are enlarged in a manner reminiscent of the renal glomerulus such that substances easily reach the telae through ultrafiltration or diffusion from the decelerated bloodstream and, from here, cross the blood-brain barrier (BBB), i.e., they are able to cross into the cranial ventricles under the controlling secretory function of the lamina epithelialis, assisted by active transport processes.

The function of the plexus, apparently with assistance of the ependyma, is primarily related to secreting the clear cerebrospinal fluid (CSF; rate of production 0.5 ml per minute), which completely fills the interior cerebral cavities (ventricles), the spinal canal, and the subarachnoidal cavity and, thus, rinses the brain from all sides (Fig.3.33).

3.5.5 Cerebrospinal Fluid

The cerebrospinal fluid (*CSF, liquor cerebrospinalis*, or *1. cephalospinalis*) is important on a number of levels but, even today, is not thoroughly understood. In humans, there exist considerable differences in the *composition of the CSF* and blood plasma that show the CS fluid not to be only an ultrafiltrate of the blood plasma (Table 3.2). The lesser concentrations of potassium, glucose and, above all, protein (1/200 of the plasma) and cholesterol (1/500 of the plasma) in the CSF are especially characteristic. The electrolyte milieu of the CS fluid and the brain is held in the strictest ion equilibrium by the activity of the blood-CSF barrier and the BBB despite changes in blood electrolyte content. The composition of the extracellular liquid remains rather constant throughout the organism;

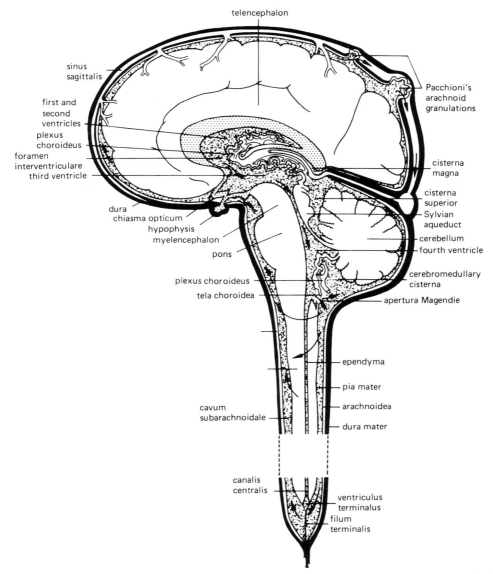

Fig. 3.33. Topographical diagram of the meninges, plexus choroideus, and cerebrospinal fluid in the CNS of a mammal. Fluid flow is indicated by arrows.

however, the neurons, which are especially sensitive to ion change, receive an additional means by which to exchange substances and derive the benefit of an additional protective factor.

The differences in concentration between CS fluid and plasma, however, are not always the same for early ontogenetic stages or in phylogenetic comparison (Table 3.3). It has been determined from these findings and corresponding morphological facts that it is likely that the CS fluid originally performed a nutritive role similar to that of the lymph for the other organs of the body. This function becomes clear, for example, in the Amphioxus (Branchiostoma), which is on a lower phylogenetic level and in which the central canal remains connected to the intestinal tube via the so-called *canalis neuren-*

TABLE 3.2. Concentrations of various substances in the cerebrospinal fluid and blood plasma

Substance		CS fluid	Plasma	CS fluid/plasma
Na^+	(mval/kg H_2O)	147.0	150.0	0.98
K^+	(mval/kg H_2O)	2.9	4.6	0.62
Mg^{2+}	(mval/kg H_2O)	2.2	1.6	1.39
Ca^{2+}	(mval/kg H_2O)	2.3	4.7	0.49
Cl^-	(mval/kg H_2O)	113.0	99.9	1.14
HCO_3^-	(mval/kg H_2O)	25.1	24.8	1.01
P_{CO_2}	(mm Hg)	50.2	39.5	1.28
pH		7.326	7.409	—
Osmolarity	(mosmol/kg H_2O)	289.0	289.0	1.00
Protein	(mg/100 ml)	20.0	6000.0	0.003
Glucose	(mg/100 ml)	64.0	100.0	0.64
Inorganic P	(mg/100 ml	3.4	4.7	0.73
Urea	(mg/100 ml)	12.0	15.0	0.80
Creatinine	(mg/100 ml)	1.5	1.2	1.25
Uric acid	(mg/100 ml)	1.5	5.0	0.30
Lactic acid	(mg/100 ml)	18.0	21.0	0.86
Cholesterol	(mg/100 ml)	0.2	175.0	0.001

(Data in part from Davson, 1967).

TABLE 3.3. Average protein concentrations of serum and CSF of various vertebrates.

	Average protein concentration (mg/ml)	
	Serum	Cerebrospinal fluid
Fish	31	47
Frog	33	1.3
Chick	31	1.9
Human	70	0.5

(Reprinted with permission from Zucht and Rahmann, 1974.)

tericus during early ontogeny. In this case, the CS fluid represents a kind of central nervous circulatory system with whose assistance the entire metabolic process of the CNS must be carried out, including humoral regulative remote actions, because there exists no direct access to blood circulation. In higher vertebrates, the nutritive component is taken over increasingly by the blood vessels that permeate the CNS such that the CS fluid assumes other functions. For example, the CNS has an extracellular space of only about 4 to 6% in contrast to that of the other organs with about 17 to 20%. Accordingly,

the CS fluid might be ascribed the essential function of being a "lymph drain"; the CS fluid system can absorb proteins, catabolites, and even cellular debris that are no longer usable and remove them from the CNS in the region of the lumbar since there are no actual lymphatic vessels in the CNS or meninges. The CS fluid flows from the plexus choroidei, through the ventricle and central canal of the spinal cord into its caudal region (Fig. 3.33). It also flows through the apertures of the fourth ventricle to the pia mater and to the subarachnoidal cavity and connects with the venous sinus of the dura via the arachnoidal villi (*Pacchioni arachnoid granulations*). Its connection to the lymphatic system of the venous circulatory system via the connective tissue sheaths of the brain and spinal nerves that penetrate the meninges is of particular importance, as is its connection to the lumbar region of the cord.

The *Reissner fiber* (RF; Fig. 3.28, Sect. 3.5.2) should be mentioned once again in this regard. It represents the secretory product of a special ependymal differentiation beneath the commissura posterior at the boundary between the diencephalon and mesencephalon, the subcommissural organ (SCO), which is present even in the most primitve chordates (tunicates) and cyclostomes, and is uniformly configured in all the vertebrates excluding insectivores, whales (cetacean), and humans. One hypothesis has been put forward that suggests that the SCO and the RF serve to detoxify the CS fluid by virtue of the fact that SCO secretions to the CS fluid form complexes that are twisted together into strands by the cilia of the SCO and are removed from the brain in the form of the RF. According to this view, the SCO-RF system would be, in effect, the brain's own excretory system. In contrast, more recent findings indicate that the liquor cerebrospinalis is not simply a kind of "*rubbish dump for CNS catabolites,*" but that it may govern humoral control functions. The presence of various hormones in the CS fluid might explain this: the octapeptides *vasopressin* and *oxytocin* have been detected, hormones known to be formed in the hypothalamus of the diencephalon that reach the neurohypophysis via neurosecretory pathways through the infundibulum of the hypophysis. In this regard, primary attention was directed to specific cell groups of the circumventricular organs, and it was discovered that they secreted many products into the CS fluid, resulting in the concept of *cerebrospinal neurocriny* (Sterba). In any case, the physiological-chemical component of the CNS seems to be of paramount importance even though it is not understood in many of its finer details. Nonetheless, autoradiographic and radiochemical studies have revealed a secretion of proteins from brain structures that do not belong to the circumventricular system, as, for example, from the visual layers of the optic tectum of teleosts into the CSF. This suggests that CS fluid proteins must not be exclusively serogenic in origin (on the part of the plexus choroidei), but—even if only to a minimal degree—can originate in nerve tissue itself.

The CSF also performs an important mechanical, or hydrostatic, function. CS fluid and brain have nearly the same specific weight (the brain weighs 1,400 g in the air and only about 50 g in CS fluid). If the brain were not practically weightless in the CS fluid mantle that surrounds it from all sides, if it were held in place only by blood vessels, nerves, and the trabeculae arachnoidae, then even the most mimimal shock to the brain would have dire consequences. The liquid envelopes in both media, nerve tissue and

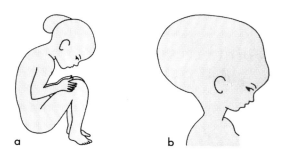

Fig. 3.34. Liquor fistula *(a)* and hydrocephalus *(b)*.

CSF, absorb blows to the head area, preventing localized injuries except in extreme instances.

Also, the pulse wave of the CNS arteries, which could potentially damage nerve tissue function, is neutralized in the arachnoidal and Virchow–Robin cavities in that arterial blood pressure of the relatively thin-walled arteries is transferred to the veins without nervous, vasomotor control by way of CS fluid pressure against that of the pulsating blood.

Production of the CS fluid begins very early in embryonic development, together with brain development, as a structure-forming element. Interruptions of the flow, for example, by blockages of the *aqueductus Sylvius* between the third and fourth ventricle, or by accumulation of CSF at the brain's surface can lead to *hydrocephalus internus* or *hydrocephalus externus* (Fig. 3.34).

The CSF is of extreme clinical-diagnostic importance. In cases of various disorders of the CNS, especially inflammatory changes, suspicion of tumors, syphilitic illnesses, etc., the composition of the CS fluid can important diagnostic evidence. To this end, CS fluid samples are obtained directly from the ventricles, from the lumbar region by tapping the dural sack, suboccipitally from the cisterna cerebellomedullaris, or by opening the skull (*trepanation*). The samples are then tested for weight, cell content, protein, sugar, electrolyte content, etc. Replacing the cerebrospinal fluid with air allows the ventricles to appear in vivid contrast in x-rays (*Pneumocephalography*, when tumors are suspected). Violent headaches result, however, since the brain stretches from its state of suspension. Consequently, this method has been replaced widely by modern computer tomography.

4
Evolution and Architecture of the Nervous System in Invertebrates

4.1 Evolution of Nerve Cells: General Remarks

The formation of nerve cells, i.e., elements whose function within an organism is to conduct impulses, was immensely important for higher development (*anagenesis*) in animals during their *phylogeny*. It was, indeed, the development of nerve cells that enabled animals to develop their own system of communication which, in turn, gave the entire organism the capacity to act and react. Higher development in the organisms of multicellular animals (*metazoa*) was coupled with ever-advancing development and refinement of the nerve cells into nervous systems. The more differentiated the composition and functional modality of the nervous system of a species was, the more complex and adaptive did its performance become. That meant that it had all the more positive advantages of selection.

The distinctive attribute of nerve cells is their ability to conduct impulses. Furthermore, they are also involved in the trophic secretion of synthesized products, or neurosecretions. A key factor in discussing which of these two functional components "came first," i.e., the trophic or the conductive, was the discovery of the phenomenon of neuronal transport. When nerve cells first began to develop during early ontogeny in the simplest metazoa, the formation of long cell processes might well have had two equally important functions: (a) to conduct impulses efficiently and (b) to supply nerve fiber terminals with substances synthesized in the cell body via

effective transport within the fibers. That means that the greatest advantage of *selection* might have been precisely a linking of functions as just described.

Just such a fine differentiation in cells occurred in the platyhelminths (flatworms). In the type of cell that was principally conductive in nature, the neurosecretory functional component was reduced to the secretion of transmitter substances. Accordingly, conductivity plays a very small part in cells that were essentially neurosecretory in nature. Furthermore, development of the nervous system and the neuro-endocrine (neurosecretory) organs paralleled one another even on a relatively low evolutionary level (polychaetes, marine annelids). The nervous system in invertebrates serves the same purpose as it does in vertebrates: it coordinates the processes that take place within the body as well as the organism's responses to its environment. Nonetheless, the invertebrate nervous system is fundamentally different from the vertebrate system and it can manifest itself variously within the great diversity of invertebrates. Only the main types are represented here.

4.2 Organization of the Nervous System in Invertebrates

Although a genuine nervous system does not emerge until the phylogenetic level of the *metazoa,* or multicellular organisms, many monocellular organisms (*protozoa*) have cell structures that function in a complex manner,

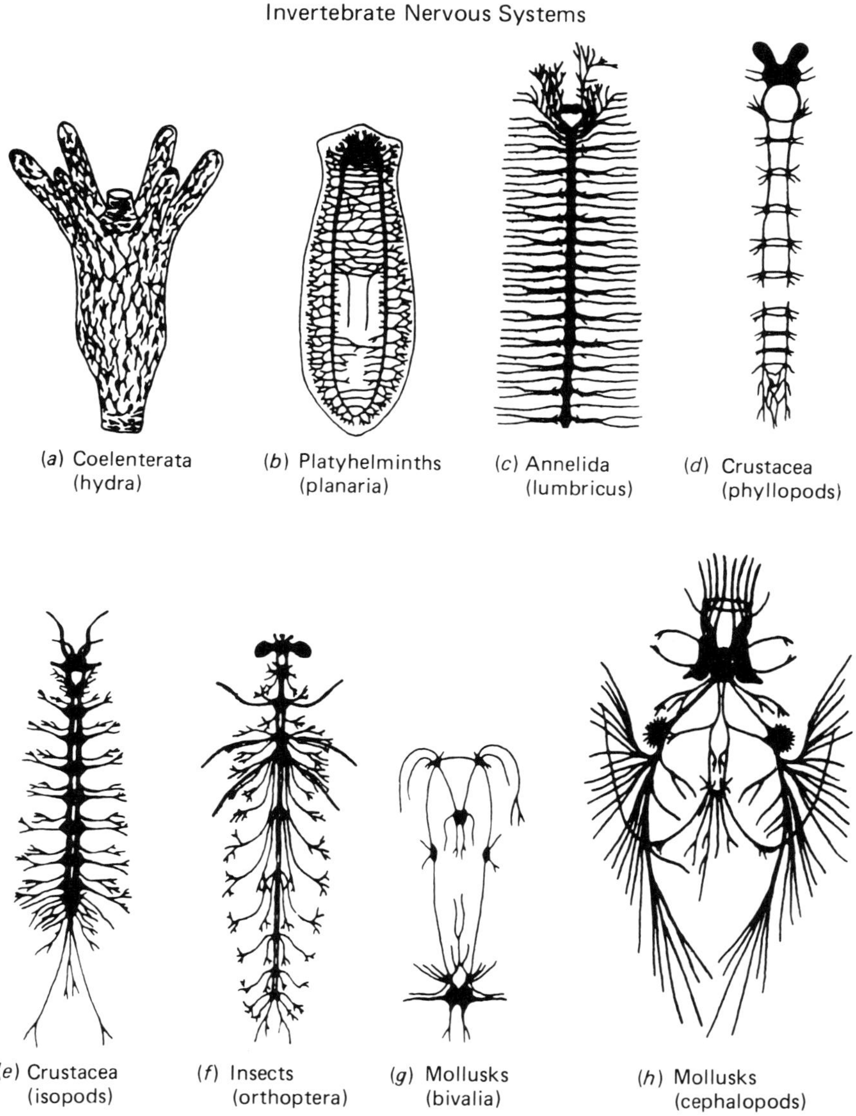

Invertebrate Nervous Systems

(a) Coelenterata   (b) Platyhelminths   (c) Annelida   (d) Crustacea
    (hydra)   (planaria)   (lumbricus)   (phyllopods)

(e) Crustacea   (f) Insects   (g) Mollusks   (h) Mollusks
    (isopods)   (orthoptera)   (bivalia)   (cephalopods)

Fig. 4.1. Overview of the forms of nervous systems represented by invertebrates.

i.e., that enable the organisms to undertake directed movement and react to outside stimuli through changes in membrane potential. However, structures of this sort do not enable a mono-cellular organism to engage in associative acts such as conditioned reflexes that would lead to the formation of memory (see Chap. 11).

The comparative morphology and physiology of the nervous systems of extant *metazoa* (Fig. 4.1) reveal that this organ system, despite its diversity of form, might well have one and the same origin. Of course, the evolution of nervous systems cannot be derived solely from fossil analysis since the less solid nerves rarely leave traces during fossilization. However, comparative developmental anatomy does show that, in the course of evolution, the nervous systems in animals were modified to a far lesser extent than were the other organs. Both the composition of the nerve tissue (from nerve and glial cells) as well as the physiology of the nervous systems, especially in regard to their electrical and chemical properties, are extraordinarily similar in vertebrates and invertebrates. The nerve cells in all instances are ectodermal in origin, i.e., they all derive from the same outer blastodermic layer. They manifest themselves in three forms: as *sensory nerve cells* of the epidermis (for example, olfactory epithelium), as subepithelial or more deeply lying *nerve plexi* (intramural nerve cells of hollow organs) and as *centralized nervous systems* (CNS).

The *coelenetrates* (Fig. 4.1a) exhibit the most simply organized of the nervous systems. It is only minimally centralized, partially as a ringed system. The individual nerve cells in hydrozoa and a few actinia form a diffuse nervous system that is distributed over the entire body. A stimulus that is perceived at any location of the body produces a stimulation that spreads equally over the entire nerve network in all directions. After numerous repetitions of the same stimulus, the state of stimulation continues: a kind of *facilitation* sets in. Nevertheless, clear evidence of actual learning in coelenterates has not been shown to date.

Although the nerve cells in lower *platyhelminths* (*flatworms*) are organized in a kind of *diffuse nerve network* similar in kind to that of the coelenterata, the higher flatworms, for example, the *Turbellaria sp.* that live in running waters and oceans, show a conglomeration of nerve cells in the head amounting to a *brain*. Nerve fibers run from sensory organs (primitive eyes, tentacles, olfactory fossa) to this brain. Others exit it as nerve funiculi (medullated fiber nervous system; Fig. 4.1b) to service the crawling and swimming organs. Flatworms are useful because of their simple orientation behavior. (Their importance as laboratory animals in learning and memory experiments is addressed in Chap.10.)

Nerve cells began to consolidate into so-called *ganglia* as segmented organisms developed. Ganglia are concentrations of many nerve cell bodies attended by a fabric of nerve fibers, the *neuropile*. It is through the neuropile that individual ganglia establish contact with each other via collateral nerve fiber strands (Fig. 4.1c–h). Injections of fluorescent dyes help reveal

that species-specific, tree-like ramifications extend from the axons of these types of ganglia, i.e., that identical synaptic connections exist between homologous nerve cells of different individuals of one and the same species.

Each body segment in segmented animals (*articulates*), to which both *annelids* and *arthropods* belong, has its own pair of ganglia that are connected to *commissures* (nerve threads) (Fig. 4.1c–e). Each ganglion pair facilitates reflex functions within the body segment as well as interactions with adjacent segments that are connected to the ganglion pair via *connectives* (Fig. 4.2). In this way, the nervous system of the articulates is organized in the form of a brain that lies dorsally to the esophagus and

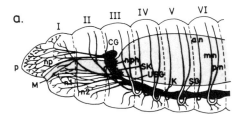

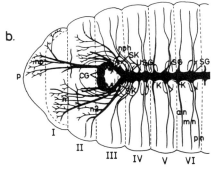

Fig. 4.2. Diagram of the nervous system in annelids (lumbricus, or earthworm): (*a*) lateral aspect; (*b*) dorsal aspect. I–VI: segments 1 to 6; p: prostomium; M: oral aperture; CG: cerebral ganglion, brain; EC: esophogeal connectives; HEG: hypoesophogeal ganglion; SG: segment ganglion; C: connectives; an: dorsal branch of an anterior segment nerve; mn: ventral branch of a middle segment nerve; pn: dorsal branch of a posterior segment nerve; n1: nerve branches of the esophogeal connectives to segment I; n2: nerve branches of the posterior, ventral esophogeal connective areas to segment II; np: nerves to the prostomium, head; nph: nerve plexus of the esophagus.

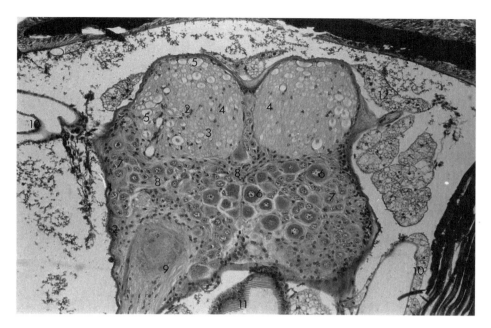

Fig. 4.3. Cross section of the subpharyngeal ganglion (mandible ganglion) in the American cockroach (*Periplaneta americana*): 1) neurilemma, 2) glial cell nuclei, 3) neuropil, 4) pharyngeal connectives, 5) giant fibers, 6) giant cells (motoneurons), 7) ganglial cells, 8) commissure, 9) mandible nerve, 10) spiracles, 11) salivary gland secretory duct, 12) fat bodies.

a *"rope-ladder" nervous system* that lies ventrally to the intestinal tube. This nervous system is extremely well suited for neurophysiological experimentation in that the ganglia, which consist of only a few cell bodies, are readily accessible and, thus, lend themselves well to measurement. Functional associations, therefore, are obtained relatively easily. The ganglion chains of grasshoppers and crickets (Fig. 4.3) serve especially well in this regard. Similarly, the field of neurobiology has gained valuable understanding from the *giant fibers* of earth-worms and leeches regarding conductivity in individual nerve fibers. These data play heavily in the analysis of reflex arcs. Beyond the rope-ladder nervous system, there emerges on the level of the arthropods *superganglia* or *brains*. These are, in fact, a substantial accumulation of numerous individual ganglia in the anterior body segment that are appreciably more complex in organization than a segmental ganglion pair. An upper pharyngeal ganglion of this sort, which is situated dorsally to the intestinal tube, has become a brain (*cerebrum*)

in *insects,* for example, comprised of a *protocerebrum* that receives information from the eyes via optic lobes, a *deutocerebrum* that accepts impulses from the antenna, and a *tritocerebrum* that innervates the anterior intestinal tract and the head region (Fig. 4.4). A histological cross section of the brain of a bee (Fig. 4.5) reveals the complex organization of various cell body regions, neuropil networks, and nerve fiber threads. Of paritcular significance are the *antennal lobes,* which balance olfactory stimuli, and the *optic lobes.* The *corpora pedunculata,* neuropil networks shaped like mushroom caps, are ascribed complex associative routing functions.

The electron microscope has enabled re-searchers to discern astounding differences in dimension regarding the cytologic organization of the insect brain as compared to that of vertebrates. For example, the diameter of nerve fibers in the brain of an ant varies from between 0.1 and $0.8\,\mu m$ whereas the variance in the cortex and cerebellum of a cat ranges between 0.2 and $3\,\mu m$. The size of the synaptic vesicle in the

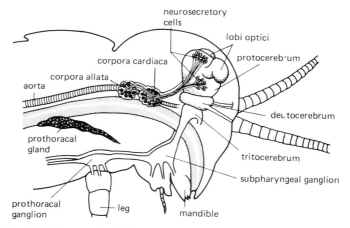

Fig. 4.4. Diagram of the insect brain and endocrine regulating system.

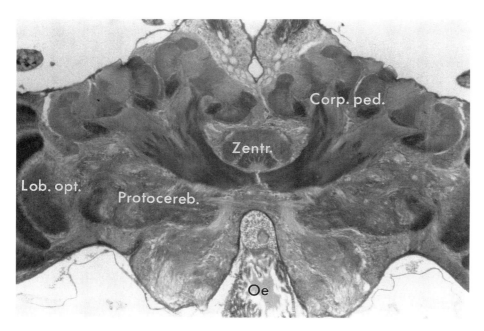

Fig. 4.5. Histological cross section of a bee's brain; protocerebrum with corpora pedunculata, optic lobes, and central body. Oe: esophagus.

ant amounts to 20 nm. The same is about 50 nm in the cat. Although myelin sheaths are completely absent in the insect brain, they are present in vertebrates and are from 0.1 to 0.2 μm in thickness. It follows from these data that the nerve fibers in the insect brain are emplaced considerably more densely than in vertebrates, a factor that may contribute substantially to the former's outstanding neuronal performance capacity.

Generally speaking, a *neuroendocrine regulating system* for the control of bodily functions is an important final element in the nervous sustem in arthropods (Fig. 4.4). In this regard, the *corpora cardiaca,* which are located behind the brain, function as control centers. They are supplied with *neurohormones* that are formed in neurosecretory cells in the protocerbrum. The corpora cardiaca, formed as neurohemal organs, are in contact with the *corpora allata,* an organ group that is functionally both neuroendocrine and endocrine in nature. Both organs have important control functions, in combination with the *prothorax gland* (which produces the hormone *ecdysone*), in processes of metamorphosis and, thus, in the process of ecdysis in particular. (The neuroendocrine and endocrine regulating system of arthropods is often viewed analogously to the hypothalamus-hypophysis system in vertebrates.)

The nervous system of the *mollusks* comprises in its totality all structural types of the other invertebrates from the most primitive level of development in flatworms to the central nervous system in cephalopods (cuttlefish), which is comparable to the insect brain. On the basis of body sections, pedal (foot), visceral (intestinal), pleural (lung), abdominal, and cerebral (head) ganglia are developed in mollusks (Fig. 4.6). Individual neurons with a diameter of more than 1 mm occur in *opisthobranchs* (gastropods) (for example, of *Aplysia sp.* and *Tritonia sp.*). They are especially well suited for neurophysiological experiments in conductivity, injection experi-

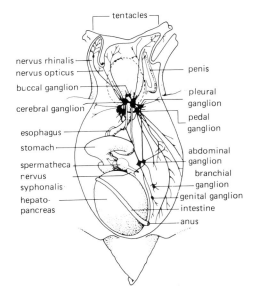

Fig. 4.6. Diagram of the nervous system of an opisthobranch snail.

ments using fluorescent dyes to depict the topography of individual nerve cells in the total organism, and ion or neuropharmacological solutions for the analysis of neuronal membrane functions.

Among the mollusks, the brain of the *cephalopods* represents the most highly developed nervous system of the invertebrates. There are calculated to be approximately 10^8 neurons in the *octopus,* a cuttlefish. (By way of comparison, the human prosencephalon contains about 10^{11} neurons!) In the octopus, they are arranged in differentiated brain structures of which the optic lobes predominate (Fig.4.7). In accordance with the great complexity of organization of the nervous system, the cephlopod exhibits an extensive inventory of behavior the documentation of which cites performance of complicated acts of learning and memory (see Chaps. 10 and 11).

The nervous system of other invertebrate phyla such as the *echinoderms* or *hemichordates* is far less differentiated than that of the articulates and mollusks. In adaptation to the organization of the total body, it is reduced in echinoderms to a ringed nervous system from which individual radial strands are dispatched. It is a product of the secondary radial symmetry of its body shape and, above all, a result of its less vagile, somewhat sessile manner of existence. It does not attain the level of an actual central nervous system. The spectrum of behavior exhibited by these forms is commensurately less differentiated.

The following parallels can be drawn between the nervous systems of vertebrates and invertebrates when they are examined with regard to comparative phylogenetic aspects:

The foremost characteristic of the vertebrate brain, i.e., the development of a hollow spinal tube or brain composed of dorsal ectodermal material and filled with cerebrospinal fluid, can be found in some extant invertebrates belonging to *deuterostomes* such as the *acraniates* (Branchiostoma) and *tunicates*. The *enteropneusts* also have a nerve plexus located on the outside of the integument with medullary threads that are enlarged in certain areas. The so-called collar medulla deriving from this is regarded as an homologous structure to the spinal cord in

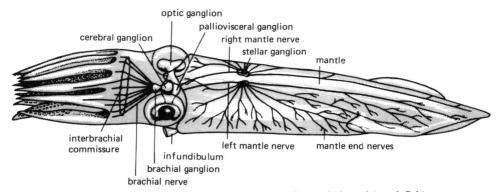

Fig. 4.7. Diagram of the nervous system of a cephalopod (cuttlefish).

chordates (tunicates, acraniates, and *vertebrates).* Even primitive protostomes such as the *bryozoa* have a similar cerebral organ, which is the reason that they are viewed by some systematists as old forms of proto- and deuterstomes.

Other similarities in makeup between the vertebrate and invertebrate brain are those which are regarded not as homologies, but rather as manifestations of convergence: formation of the rhinencephalon as the most anterior region of the brain in vertebrates as well as in nemertines, articulates, and pulmonates (lunged snails); differentiation of the spinal cord in vertebrates and of the abdominal nervous cord in invertebrates in motor and sensory areas; partially body-oriented segmentation of the nervous system which is primary in articulates, however limited in vertebrates to stem plates located in the somites and the spinal nerves branching from the spinal cord.

5

Principles of Circuitry in Neurobiological Information Processing

Nerve cells never occur in isolation within an organism. In addition to their close association with glial cells (see Chap. 1.2), they are always connected to other neurons and receptor or effector cells in the region of their nerve fiber endings via the formation of synaptic contacts. The integration of nervous systems into highly complex functional systems, i.e., into information and control centers, increases proportionately to the animal's higher developmental phylogenetic standing.

Presently, a precise analysis of the relay architecture that underlies these centers is feasible only on the lowest level of circuitry as insurmountable difficulties still stand in the way of deeper understanding. The primary obstacles are

- The immense complexity of the systems under scrutiny, for example, the several hundred billion neurons that make up the human brain,
- The vast number of nerve fiber endings on each individual nerve cell (as many as 10,000 terminals per neuron!),
- The diverse modes of synaptic association between two neurons (excitatory–inhibitory, electrical–chemical), and
- In the case of chemical transmission of impulses, the great diversity of transmitter substances.

An analysis of neuronal circuitry based on today's level of understanding will prove fruitful only if the broad spectrum of research methods available to neurobiology is brought to bear, for example, the methodologies of electrophysiology, neuroanatomy (using isotope and dye tracers), electromicroscopy, and even neurocybernetics, the last contributing computer-simulated models of circuitry. Several relatively simple, generally corroborated principles of organization and performance have been explained on the basis of electrophysiological and neuroanatomical findings (Fig. 5.1): in vertebrates, a circuit consists of at least two neurons and an effector cell (a muscle or gland cell, for example). In this instance, one neuron reacts sensorily (afferently, conveying impulses toward the nerve center) and the second neuron reacts motorily (efferently, conveying impulses toward an effector) (Fig. 5.1a). The efferent neuron, therefore, receives impulses from the effector cell and conducts them, after synaptic relay, to the motor neuron which, in turn, conducts impulses back to the effector cell. This is referred to as a *monosynaptic reflex arc.*

Thus, synaptic connections are initiated between those neurons that conduct impulses by way of their axon terminals from one region to *relay neurons* in another region. In this latter region, *interneurons* are targeted, as well. These conduct their impulses back to the relay neurons (Fig. 5.1b). The equalized impulses are then conducted by the relay neurons, whereby reverse connections to the interneurons are again possible. Thus, various synaptic connections can be made within one such *synapse triad* consisting of an input neuron, an interneuron, and a relay neuron. If several of these synapse triads are linked in series or in parallel, as is the case in

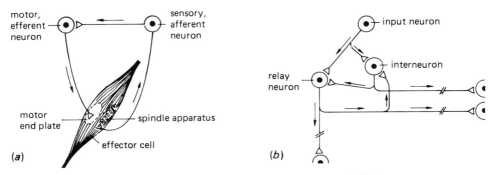

Fig. 5.1. Simple neuronal circuit (*a*); synapse triad (*b*).

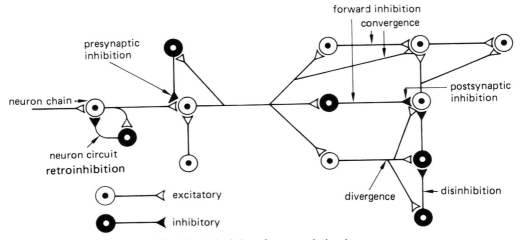

Fig. 5.2. Principles of neuronal circuitry.

the retina (see Fig. 5.7), then the result can be a system of circuitry that is extremely complex.

5.1 Neuronal Circuitry

In view of the extraordinary significance that the various principles of circuitry hold for the processing of neuronal information, additional neuronal circuiting possibilities will be discussed in the following section (see Fig. 5.2 for a general overview). The modes of connection by which *neuronal circuits* can be linked with one another are numerous and depend upon whether the axon terminals establish synaptic contact with the soma, the dendrites, or even other axons of a receiving cell (axosomatic, axodendritic, axoax-

onal synapses; see Chap. 1.1.3.4). Additionally, the type of transmitter substance secreted by the presynapses and the response mode of the postsynaptic membrane fall into two distinct categories, either excitatory (i.e., stimulating synapses) or inhibitory (i.e., inhibiting synapses; EPSP or IPSP, see Chap. 6.3.3).

Divergent circuitry (Fig. 5.3a) refers to the relay of *one* neuron (or one receptor cell) to *several* receiving neurons. *Dale's rule* applies here. It states that all axon terminals of a given neuron secrete the same transmitter and, thus, they all function in the same mode, i.e., either in an excitatory or an inhibitory fashion. Divergent circuitry is instrumental in rendering equal afferences accessible to various regions of the CNS. Peripheral receptor cells *exemplify* diver-

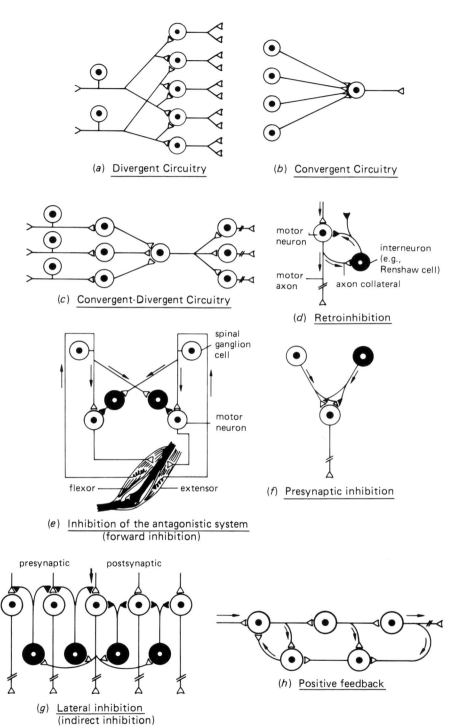

(a) Divergent Circuitry   (b) Convergent Circuitry

(c) Convergent-Divergent Circuitry

(d) Retroinhibition

(e) Inhibition of the antagonistic system
(forward inhibition)

(f) Presynaptic inhibition

(g) Lateral inhibition
(indirect inhibition)

(h) Positive feedback

Fig. 5.3. Overwiew of simple neuronal circuitry systems.

gent circuitry. Their afferent, sensory fibers enter the spinal cord via the dorsal roots, become ramified there, and innervate numerous spinal cord neurons.

In the case of convergent circuitry (Fig. 5.3b), the terminal ramifications of a number of neurons converge at one nerve cell. This results in a spatial and temporal integration of impulses from many cells or receptors at one receiving nerve cell. In apes, for example, the number of axon terminals believed to converge at one motoneuron in the spinal cord has been put at up to 19,000, and the number of Betz cells in the motor cortex of the prosencephalon is estimated to be upward of 60,000.

Convergence–divergence circuitry refers to that phenomenon in which both of the previously mentioned principles occur in linked fashion with one another. Each individual neuron receives convergent information via numerous postsynaptic input impulses from various other neurons or receptor cells. It integrates these data and disseminates them further to a great number of receiving cells via a multitude of axonal, presynaptic terminal ramifications. The retina (see Fig. 5.7) exemplifies this type of circuitry. It maintains divergent connections with numerous receiving neurons in the photoreceptor cell whereas, in contrast, any of its neurons receives afferences from a

multitude of photoreceptors. Additionally, in the case of mossy fibers and Purkinje cells, a circuit is activated in the cerebellum that is patterned on the principle of divergence–convergence (see Fig. 5.8). Temporal and spacial pattern discrimination result from this circuit.

There are two basic forms of simple *inhibitory neuronal assemblies* in the CNS of vertebrates. One is presynaptic in nature, the other postsynaptic.

In the instance of postsynaptic inhibition, inhibitory synapses address the cell body of the neuron that is to be inhibited. Their effect is to inhibit the depolariation of the postsynaptic membrane, that condition of depolarization having been brought about by the excitatory effects of other synapses. Various modes of circuitry are possible:

1. *Reverse,* or *recurrent inhibition* (Fig. 5.3d): in the sense of negative feedback, axon collaterals of an excitatory neuron activate an inhibitory interneuron that reacts upon the original neuron in an inhibitory manner by means of inhibitory synapses. The *Renshaw cell,* which functions as an inhibitory motoneuron, has come to exemplify this mode of avtivity in motoneurons in the spinal cord (Fig. 5.4) by reducing the rate of discharge in these motoneurons. With high frequencies of

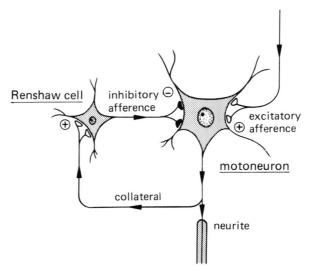

Fig. 5.4. Diagram of postsynaptic inhibition of a motoneuron by a Renshaw interneuron (*left*); excitatory afferences (*right*) enable stimulation.

up to 1,000 impulses/second and durations of up to one second, correspondingly long inhibitory postsynaptic potentials (IPSP) are triggered by the release of the transmitter glycine at the motoneuron. A similar reverse feedback was described relative to the *hippocampus* whereby basket cells, functioning as interneurons, were stimulated by axon collaterals of the pyramidal cells whose soma activity they, in turn, inhibit.

2. *Afferent collateral forward inhibition* (inhibition of the antagonistic system; Fig. 5.3e) refers to the phenomenon whereby an inhibitory interneuron is stimulated directly (i.e., in a "forward" direction) by an afferent neuron. This is a principle that underlies antagonistic inhibition. An example of such an assembly is the inhibition of extensor muscle neurons simultaneously to the activation of flexor muscle neurons.

In the case of *presynaptic inhibition* (Fig. 5.3f), inhibitory synapses of an inhibitor neuron address the presynaptic endings of excitatory cells where, under such influence, secretion of the transmitter is inhibited. Thus, during such axoaxonal inhibition, it is not the postsynaptic membrane that becomes hyperpolarized, but rather the depolarization of presynaptic nerve endings is inhibited, i.e., fewer excitatory postsynaptic potentials (EPSP) can be triggered at the same frequency of the action potential. Presynaptic inhibition essentially reduces weaker impulses and, thus, contributes both to the filtering of signals and to the intensification of signal contrast.

Lateral inhibition (environmental inhibition; Fig. 5.3g) designates the condition in which the inhibiting interneurons react pre- or postsynaptically not only upon the excited cell but also upon adjacent cells of like function that have not been stimulated at all, or only weakly stimulated. This results in an especially strong inhibition of the area in the vicinity of the active cells. Examples of lateral inhibition can be found in the retina where it serves to enhance or generate contrast.

Positive feedback (Fig. 5.3h) refers to the commonly held claim that there also must exist positive feedback assemblies in the CNS. It would lead to a circuiting of impulses from the feedback of an impulse to cells that are already stimulated (*impulse oscillation circuits*). The existence of this mode of circuitry is still disputed. It would be conceivable that such neuronal chains, once stimulated, could sustain their induced activity over a rather long period of time. Several authors ascribe the phenomenon of *short-term memory* to a circuiting of impulses that originates in this manner in positive feedback circuits (see Chapts. 10.1 and 11).

5.2 Reflex Circuitry

Analysis of the neurobiological fundamentals of reflex processes has contributed greatly to the understanding of neuronal circuitry. Indeed, it has led to the understanding that the regualtion of body functions results, in part, from a hierarchical graduation in the linkage of neuronal pathways. When one speaks of reflexes one refers to modes of behavior in which a direct, involuntary, and stereotypical response of efferent nerve fibers is brought about in a predetermined manner by the stimulation of afferent fibers. Anatomically, reflex sequences are rooted in reflex arcs that receive impulses via afferent fibers and, subsequent to synaptic relay, pass them on to efferent systems.

In the most basic sense, *axon reflexes* (Fig. 5.5a) are capable of controlling vegetative functions (vascular dilation, for example) in that a single receptor (a pain fiber, for example) can lead received impulses directly to the effector via collaterals of an individual axon, i.e., unassisted by an interneuronal synapse. Consider that, subsequent to surface irritation of the skin, axon reflexes of pain receptors that lie within the skin are instrumental in processes of localized vascular dilation (rubefaction, or reddening of the skin) subsequent to surface irritation of the skin. A further example is offered by the so-called *Lewis reaction*: under conditions of extreme coldness, vasoconstriction of the skin, which is induced thermally, is interrupted periodically in order that tissue damage be avoided.

A *monosynaptic reflex arc* (Figs. 5.1 and 5.5b) is one in which the beginning (receptor) and end (effector) of the reflex pathway lie within the same

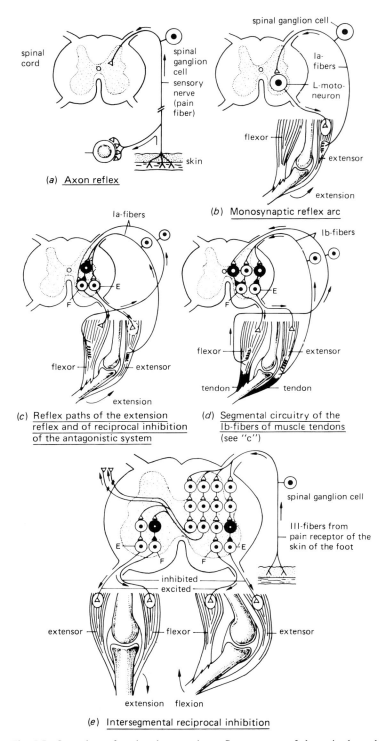

Fig. 5.5. Overview of variously complex reflex systems of the spinal cord.

organ. There exists only one synapse between the afferent and efferent neuron. Using the patellar tendon reflex as an example, if the quadriceps (extensor) muscle is extended suddenly by a blow to its tendon, then the muscle spindles are affected by the blow, as well. Their extension effects a stimulation of the so-called Ia-fibers of spinal ganglion cells which run via the posterior root to the anterior horn of the spinal cord and, there, directly stimulate the L-motoneurons of the same muscle. The last then leads to a contraction of the extensor muscle within 3.5 msec and causes the leg to "jerk" up (extend). To achieve an effective straightenting of the leg, the L-motoneuron of the affiliated flexor must be inhibited. This occurs within the framework of a *reciprocal inhibition* of the antagonistic system (Fig. 5.5c) via an inhibiting interneuron that is stimulated by an axon collateral of the Ia-fiber of the extensor spinal ganglion.

The reflex response can be completed in various ways, depending on the muscle tension of the leg (Fig. 5.5d):

• Normally, the muscle spindle of the extensor is contracted which causes a reversal of the impulse in the Ia-fibers;
• In cases where the tendon receptor shows greater than normal tension, its Ib-fibers inhibit the associated α-motoneuron via an interneuron (autogenous inhibition);
• Conversely, the Ib-fibers stimulate the α-motoneuron of the opposing flexor muscle; and, finally,
• The collaterals of the α-motoneurons can inhibit each other reciprocally in reverse fashion via a Renshaw cell which functions as an interneuron (see Fig. 5.3d); thus, an uncontrolled buildup of neuronal activity is avoided.

Although, in the proprioceptive reflex system, the beginning and end of the reflex arc lie in the same organ, the receptors involved in a *polysynaptic reflex arc system* are located apart from the sensory organ (*reflex in response to external stimuli*). The defense reflex (flexion reflex) might well serve to exemplify a poly-synaptic reflex arc system. In response to a pain stimulus at the sole of one foot, it triggers both a flexion in the joints of that stimulated leg and,

simultaneously, effects. a straightening of the joints of the contralateral leg. In this instance, the afferent impulses are conducted from the pain receptors in the skin to the spinal cord by means of the spinal ganglion cell and from there are conducted further (Fig. 5.5e):

1. Via excitatory interneurons to the flexor motoneurons (F) which stimulate the ipsi-lateral flexor muscle and thereby raise the leg,
2. Via inhibitory interneurons to the extensor motoneurons (E) which cause the ipsilateral extensor to relax,
3. Via excitatory interneurons to the extensor motoneurons of the opposite side which stimulate the collateral extensor and extend this leg (crossed stretch reflex),
4. Via inhibitory interneurons to the flexor motoneurons of the opposite side which cause the collateral flexor muscle to relax and, lastly,
5. To other ascending and descending seg-ments of the spinal cord, since not all extensor and flexor muscles served by just one segment.

In contrast to the proprioceptive reflexes, reflex time in response to external stimuli is not constant. Indeed, it decreases exponentially with increasing stimulus intensity. Such a reduction, however, is not achieved by accelerated impulse conduction. Rather, it is a factor of reduced synapse time that results from the cumulative effect of a superinflux of impulses. Parallel to this, response time, as a rule, also decreases with increasing stimulus intensity and the brain receives the stimulus information, as well. The possible modes of reflex circuitry in vertebrates are summarized in Fig. 5.6, this time based on the regulation of the skeletal-musculature motor system. *Regulation of motor response* is possible on three or four different levels of neuronal circuitry: in the spinal cord (see Fig. 5.6), in the mesencephalic and thalamic regions and, finally, in the prosencephalon. Circuiting within this last area, however, is subject to conscious control and, thus, in the strictest sense, no longer fulfills the definition of a true reflex process.

The stimulation impulses arriving from the muscle spindles of the periphery return either to the motor end plates in the muscle fibers as

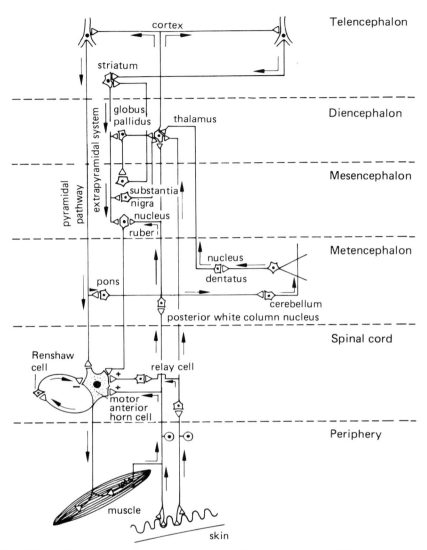

Fig. 5.6. Modalities of interconnection among neuronal reflex arcs as evidenced in sensosomatic innervation on various CNS levels.

proprioceptive reflexes with only one synaptic
relay via the motor anterior horn cells of the
spinal cord, or they can ascend to the cerebral
cortex via the sensory pathways. In this latter
process, they are relayed first in the nuclei of
the posterior cord (white column) of the medulla
oblongata (see Chap. 3.2.2.2.1) and, then, relayed
once again in the thalamus (see Chap. 3.2.2.4.1).

The impulses from skin receptors are conducted either in the spinal cord, as in response
to external stimuli, or they reach the diencephalon from where they are conducted further
to the cortex of the prosencephalon or the corpus
striatum.

In descending pathways, impulses from the
pyramidal cells in the cortex (see Fig. 3.21) run
through the *pyramidal pathways* directly to the
motor anterior horn cells in the spinal cord.
Parallel to this, afferences can course over the
pathways of the extrapyramidal system which
originates in the corpus striatum of the cerebrum.

Collateral ramifications produce additional
alternatives for connecting the afferent and
efferent systems on the various levels.

5.3 Examples of Central Nervous Circuitry Systems

One dividend of the increasingly sophisticated laboratory techniques of recent years in the fields of neuronanatomy and, especially, neurophysiology has been the ability to track individual neuronal circuit systems in various regions of the brain and sensory organs, despite the enormous complexities of such systems. Especially impressive within this context is the circuitry represented in the retina, the cerebellum, the hippocampus, and the neocortex. Because the functional modalities of these kinds of neuronal circuitry systems might contribute significantly to the understanding and eventual construction of artificial electronic networks, they are surveyed briefly in the following pages. To better comprehend the simplified circuitry diagrams that are presented here, the topographical details are discussed first with the help of schematic drawings. The histological features and secondary functional–morphological interrelations of the individual structures already have been discussed in Chap. 3.

5.3.1 Retina

The retina (Fig. 5.7) in vertebrates has a six-layered configuration. There are three layers that are abundant in cell bodies (sensory cell layer, layer of interstitial cells arranged in a bipolar and horizontal manner, and ganglion cell layer representing the origin of the visual nerve fibers). These are separated by interposed fiber layers (Fig. 5.7a). The cell bodies of the sensory cells that aggregate in the second layer form two cell processes that differ morphologically and functionally. Indeed, both the *retinal cones,* which facilitate color perception, and the *rods,* which serve to distiguish brightness, are situated in the first layer, the layer closest to the light. Impulses of the photoreceptor cells are relayed in the third and fourth layers to interneurons, i.e., *bipolar cells* (BC) and *horizontal cells* (HC). The impulses of several bipolar cells are integrated by a *fourth layer, a ganglion cell layer,* which conducts the information to the diencephalon (in mammals) via a long axonal process via the optic nerve. Since the horizontal cells form cross-connections between these cells in the region of the bipolar layer, each ganglion cell receives data from a larger number of sensory cells. As a rule, however, each sensory cell conducts to several ganglion cells. Only in the *fovea centralis,* the location of sharpest visual acuity on the retina, are there exactly as many ganglion cells as sensory cells. In all other areas of the retina, the number of ganglion cells is substantially lower than that of the sensory cells, which is estimated to be 120 million rods vis-à-vis 6 million cones. In their totality, the neurites of the ganglion cells, which total about

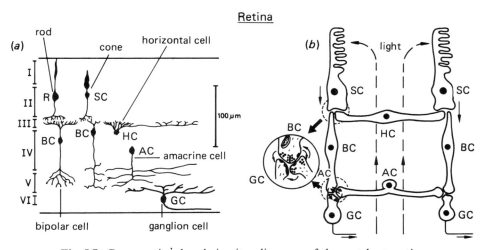

Fig. 5.7. Cross-sectional and circuitry diagrams of the vertebrate retina.

1 million, comprise the *nervus opticus* which conducts its impulses to the brain (to the mesencephalon in lower vertebrates, to the diencephalon in mammals). The high degree of *convergence* in neuronal circuitry is documented to be 126:1. In contrast, *divergent* access to an immense number of receiving/processing neurons predominates in the CNS. This high rate of reticulation of the individual elements of the retina serves to equilibrate and moderate impulses. Functionally, this results in the eye's powers of temporal, spatial, and spectral resolution. The formation of synapse triads (Fig. 5.7b) very well might be an essential component in the development of such fine-tuning among the impulse activities of the various cell systems in the formation of *synapse triads* (Fig. 5.7b).

Apparently, these are arranged, on the one hand, between sensory cells (SC), bipolar cells (BC), and horizontal cells (HC), and between *amacrine* cells (AC) and ganglion cells (GC), on the other. Extraordinarily complex neuronal circuitry results from circuits of several of these synapse triads linked in sequence or in parallel.

5.3.2 Cerebellum

The cerebellum (Fig. 5.8) is the component of the CNS that is responsible for the functional complex of "sensory–motor coordination" (see Chap. 3.2). The integrative tasks associated with it correlate with the fact that the cerebellum is closely linked to numerous other regions of the brain (see Fig. 3.14). The demands that the system, in its entirety, places upon the cerebellum are managed in a fixed, geometrically arranged network of neuronal functional elements (see Fig. 3.13). To best address the activity by which the cerebellum manages the system-wide demands placed upon it within its highly structured network, it is advisable to review the microscopic anatomy of the cerebellar cortex. It is, in fact, the cortex of the cerebellum that is the actual functional component in this regard (Fig. 5.8a).

Viewed macromorphologically, the *cerebellar cortex*—at least in mammals—is highly convoluted (see Fig. 3.13). It reveals a distinct two-layered organization. There are 10^{10} to 10^{11} granular cells densely packed in the *granular cell layer*. The axons of these cells ascend into the *molecular layer* that lies above. Here, they bifurcate and form the parallel fiber system. The *Purkinje cells* lie in the interface region between the granular and molecular layers and are a particularly characteristic element of the cerebellar cortex. The dendrite system of these cells spreads out in layer form and, it should be noted, is oriented at a right angle to the parallel fiber system.

In addition to the topographic arrangement, Fig. 5.8b offers a greatly simplified diagram repre-

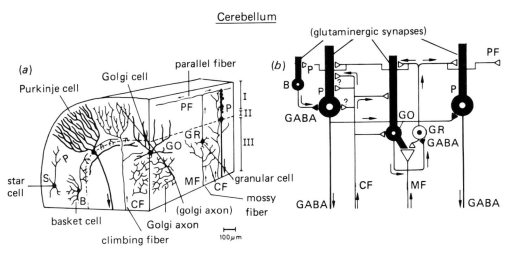

Fig. 5.8. Cross-sectional and circuitry diagrams of the cerebellum.

senting several modalities of linkage that are presently known to exist between the various types of neurons in the cerebellum.

Essentially, the cerebral cortex receives two differing types of impulse input, e.g., one via the *climbing fibers* (CF) whose cell bodies are located in the lower olivary bodies, and the other via the *mossy fibers* (MF) from deeper cerebellar regions. Both input systems have an excitatory effect upon the activity of the receiving Purkinje (P) cells.

Through their activity, the Purkinje cells, in turn, represent the essential output channel of the cerebellum. Since the transmitter released at their terminals consists of γ-aminobutyric acid (GABA), the Purkinje cells have an inhibitory effect upon the receiving systems.

Thus, the incoming impulses from the climbing fibers activate the Purkinje cells which dispatch their inhibitory impulses to the output channel and, on the basis of collateral cross-connections, also can have a self-inhibitory effect.

The incoming impulses from the *mossy fibers* (MF) primarily stimulate the *granular cells* (G) which are intercircuited among the Purkinje cells and whose broadly ramified fiber outgrowths, the *parallel fibers* (PF), stimulate both the Purkinje cells, via glutaminergic synapses, and the *basket cells* (B) which are located between them. The basket cells, in turn, exert an inhibitory controlling effect upon the Purkinje cells. The activity of the granular cells, however, is controlled, in turn, by GABA-ergic inhibitory signals from the *Golgi cells* (GO). Ultimately, the climbing fibers exert excitatory control over the Golgi cells.

The phenomenon by which an excitor input is relayed via an excitatory relay system (G) to the output channel (P) may be relevant to the extraordinary increase in the *degree of divergence* in the system overall: one mossy fiber *in natura* forms about 40 input structures that manifest themselves in so-called *rosettes*. Each of these rosettes establishes synaptic contact with about 20 granular cells. Each granular cell is connected to between 100 and 300 Purkinje cells via its parallel fiber system. These numbers lead to the divergent circuitry ratio of 1:100,000–300,000 (Ito, 1984).

In addition to distinguishing itself through this great degree of divergence in the input channel of the mossy fibers, the cerebellum also is characterized by equally formidable *convergence* in the output channel of the Purkinje cells: each Purkinje cell has as many as 100,000 dendritic spine processes. One parallel fiber terminal ends at every spine process. Thus, ca. 100,000 granular cells converge upon each Purkinje cell.

This extraordinarily great divergence/convergence ratio, nonetheless, does vary as it is a product of less than permanent circuitry. It is subject to change in both of its aspects. The Golgi cell associated with a given input "rosette" can control the flow of mossy fiber impulses to granular cells in a local circuit, i.e., one that is located in the granular layer. Furthermore, it can influence the grouping of the activities of the parallel fibers in a more extensive circuit such as one that is located in the molecular layer. Thus, it can affect the relay of granular cells to Purkinje cells.

By and large, the limitations of a relatively "fixed" plan of circuitry are that (a) the diverse modalities of interaction that result from the great number of constituent elements present an obstacle to their detailed analysis, and (b) the interconnections of the various component elements, which have been portrayed as essentially static in nature, might exist in reality as plastic-dynamic functional units.

5.3.3 Hippocampus

The *hippocampus* (Fig. 5.9) is a vital component of the *limbic system* (see Chap. 3.2.2.4) and one of the most essential integrating centers of the vegetative nervous system. Other cortical structures of the telencephalon belong to the vegetative nervous system, as do the amygdaloid and septum nuclei, the hypothalamus, and parts of the thalamus. As discussed in Chap. 3.2.2.4, the hippocampus represents a forebrain area that, regarded phylogenetically, became differentiated to the *archicortex* early through lateral displacement. These migratory movements induced a C-shaped arrangement of cells in the cross section of the prosencephalon in mammals in the lower portion of which an additional, denticulated cell formation, the

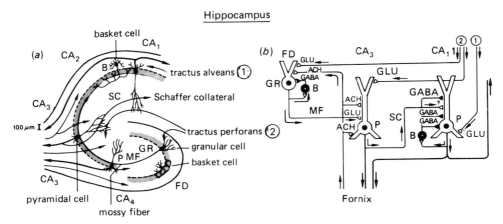

Fig. 5.9. Cross-sectional and circuitry diagrams of the hippocampus.

fascia dentata, emerged (Fig. 5.9a). The C-shaped hippocampus structure can be divided into four *cornu ammonis* regions (CA_1 to CA_4). The dominant cells in the hippocampus, as in the cortex, are the *pyramidal cells* (P) which are found here in dense cell formations. They represent the most essential output channel. The pyramidal cells are entwined in a web of intrinsic *basket cells* (B) whose axons form a basket-like, inhibitory network around the perikarya of the pyramidal cells. *Granular cells* (G) form a single, extremely compact cell layer in the *fascia dentata* (FD) section of the hippocampus. Their neurons—in like manner to the pyramidal cells of the CA-section—are surrounded by basket cells whose effect is inhibitory.

Functionally, the single-layered hippocampus is directly connected to the multilayered cortex (Fig. 5.9a). Essentially, it receives two input channels from the cortex, one via the *tractus alveans* which directly controls the CA_1 region, and the other via the *tractus perforans* which, as the most important input system, both directly stimulates the CA_2 to CA_3 regions and, additionally, also excites the fascia dentata whose granular cell processes, the *mossy fibers* (MF), indirectly control the pyramidal cells of the CA_2 to CA_4 regions.

The output of the pyramidal cells can be divided regionally, as well. The CA_4 pyramidal cells project back directly into the fascia dentata. The CA_3 cells divert their impulses collectively via the *fornix* fiber fasciculus as the main output

channel after they branch from collaterals, the so-called *Schaffer collaterals* (SC), which establish a connection to neurons of the CA_1 region. The pyramidal cells of the CA_1 and CA_2 regions project in the direction of the fornix whereby the CA_1 neurons send back still other projection fibers to the entorhinal cortex. All output fibers that are united in the fornix and originate in the cell layer of the hippocampus (CA_1 to CA_4) are directed toward two areas: the septum nuclei of the limbic system and the respective collateral formations of the hippocampus.

Figure 5.9b offers a schematic rendering of the topographic details that have been discussed here. Excitatory signals traverse both input canals (i.e., the tractus alveans and the tractus perforans) by means of glutaminergic synapses from the cortex to the pyramidal cells (P). On the one hand, these fire into the appropriate output channels (back to the cortex or via the fornix to the septum). Then, too, they equalize the activity between them by excitatory information received via their Schaffer collaterals (SC), or the inhibitory GABAergic basket cells (B). Furthermore, the activity of the CA pyramidal cells is controlled by the granular cells (G) of the fascia dentata (FD) which are controlled by the cortex, as well, yet are stimulated via cholinergic fibers of the fornix and which, additionally, regulate themselves via inhibitory collateral basket cell (B) systems.

All of the neuronal elements that are involved in the paths of signals may be viewed as lying within a single plane, much as depicted in Fig. 5.9a. It seems perfectly reasonable, therefore, to conceptualize the entire formation of the hippocampus as consisting of a multitude of elementary layers that are oriented in parallel and whose circuitry is in parallel, as well. As such, each of these individual units could represent a complete processing module whose task could be the sequential processing of intricately queued patterns of information.

5.3.4 Neocortex

Chapter 3.2.2.4.2 establishes clearly how formidable a task it would be to depict a general circuitry diagram for the neocortex (Fig. 5.10) that would serve to clarify fundamental processing operations. Indeed, this is due to the immensely complex composition and arrangement of the various neuronal elements in the neocortex.

A section of the basic type of *isocortex* is depicted in Fig. 5.10a. It shows a very well–ordered pattern of layering and forms large portions of frontal, parietal, and temporal areas. Also, it is clearly distinguished from the more differentiated cortical sections of the somatosensory and somatomotor fields. The homotypical isocortex is composed of six layers (see Chap. 3.2.2.4.2):

1. The *molecular layer* (*lamina molecularis*) which is poor in cells, and composed of tangentially oriented fibers with an associative fuction,
2. The *outer granular layer* (*l. granularis externa*) which is rich in cells,
3. The *outer pyramidal cell layer* (*l. pyrmialis externa*) whose dendrites ramify in the molecular layer and whose axons establish contact with subcortical structures,
4. The *inner granular layer* (*l. granularis interna*) with small, irregularly shaped cells,
5. The *inner pyramidal cell layer* (*l. pyramidalis interna*) with its pyramidal cells which become ramified dendritically in the molecular layer and whose axons extend centrifigually to the medullary layer,
6. The *spindle cell layer* (*l. multiformis*) which also contains some pyramidal cells but which is comprised predominantly of spindle-shaped cells with radial axons.

The topographically layered configuration of the isocortex facilitates associative connections of the most diverse inputs and outputs from the sensory projection areas as well as via collaterals from afferent pathways. The structures that are involved in associative processes exist in immensely complex reciprocal relationships with one another, i.e., their activity is induced reciprocally. Figure 5.10b shows that the circuitry connecting the various areas of the

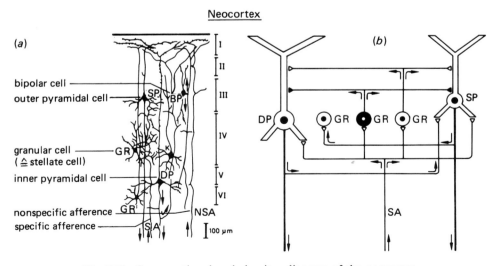

Fig. 5.10. Cross-sectional and circuitry diagram of the neocortex.

associative apparatus accommodates this type of activity, e.g., between the cortical and thalamic centers (*the thalamocortical impulse circuit*), between the cortex and the formatio reticularis (*the corticoreticular circuit*) and, above all, between various areas of the cortex (*the corticocortical circuits*). Specific impulses from afferent pathways or from adjacent regions of the cortex are conducted to the various types of neurons of the inner and outer *pyramidal cells* (DP, SP) as well as to the *granular cells* (G) situated between them. The granular cells, some of which have inhibitory, others of which have excitatory transmitters, respectively inhibit or stimulate the activity of the pyramidal cells. The latter regulate their own activity and that of the granular cells via collateral ramifications. Numerous cortical neurons are characterized by the fact that their axons merge into bundles that establish a radial path that leads them to the medullary layer. These bundles originate high in the third layer. Most fibers of the radial fascicles are axons of the pyramidal cells of the third and fifth layers. They are also axons of the pyramidal and spindle cells of layer 6. It is these radial fascicles that lend radial organization to the deeper layers of the cortex and divide it into *columns* of cell-rich cortical radii which are the basis for the concept of histological columns put forth here (see Chap. 3.2.2.4.2 for the concept of functional cortical columns). The fibers that form radial fibers can be subcategorized even further. The axons of the cells in the fifth and sixth layers represent "genuine" projection fibers, i.e., the fibers from layer 6 project to the thalamus and those from layer 5 project to other subcortical centers. In contrast, the fibers of the neurons of the third layer return to the cortex and, thus, are involved in the transmission of information internally within the system. As such, they form the system of ipsolateral coticocortical fibers. This consists of the U-fiber system which has only a short effective range, connects the adjacent gyri, appears to be grouped into rather large bundles in the medullary layer, and, furthermore, organizes remotely situated areas into functional complexes. Additionally, there are fibers involved in the radial fascicles that are attributed to the system of contralateral corticocortical fibers and

connect homotopic fields of the two hemispheres. In addition to the fiber systems comprising the radial fascicles, there exist horizontal fiber networks in layers 3, 4, and 5. Essentially, collaterals and recurring collaterals of the pyramidal cells of layers 5 and 6 comprise these systems whose influence is upon their immediate areas. They run briefly in horizontal fashion within these fiber networks and then ascend vertically through the cortex into layers 2 and 3.

The impulses circuiting (reverberating) in the neuronal formations discussed above suggest that information processing in the cortex can be regarded in a manner similar to the process by which computers resolve *alogrithms,* or repeating calculations. Nevertheless, very few specifics are known about the actual mechanisms involved in the associative neuronal processes in the brain.

5.4 Outlook

In light of the circuitry priciples discused above within the context of neuronal information processing, one can conclude that an increasingly complex formation of spatially arranged neuronal networks emerges in the course of continued individual differentiation based upon the formation of the simplest neuronal assemblies during the earliest stages of embryonic development. The emergence of functional, neuronal units of integration may derive from the fact that individual, centrally located regions are controlled and connected by peripheral receptor populations in the sense of more or less straight-line circuitry. These may become highly stimulated in the course of a functional event. However, their activity is communicated to inhibitory neuronal systems which suppress the weaker impulses of surrounding neurons. The discharge patterns of neuron populations that result from this exhibit specific stimulatory peaks and inhibitory valleys. Insofar as it pertains to the formation of *memory* and the reactivation of previously stored memory content, one notes that an impulse pattern, which corresponds to that recollection in the manifestation of an engram, is established in the process of remembering. Two conclusions can be drawn from this: (a) in the process of neuronal information storage, each individually gathered bit of information is stored over wide areas in the nervous system, and (b) many bits of information are layered upon one another in storage within each part of a system that is involved in the processing of information (see Chap. 11.4).

Knoweledge of the aspects of these recently discovered neuronal assemblies within individual substructures of the brain, albeit only cursorily addressed here, have aroused the interest of information scientists and computer specialists alike. They are attempting to examine if and to what extent the functional modality of the human brain can be explained in physical-technical terms relative to carrying out its most essential tasks, beginning with the gathering, processing, and outputing of information, and extending to goal-oriented behavior, free expression of will, and creativity. The first positive beginnings in this direction have been realized in plausibility studies (Gerke, 1987; Palm, 1988). Indeed, these works should inspire future research in this area.

6

Electrophysiological Aspects of Information Processing

A characteristic common to all living things is that their cellular organization is comprised of myriad inner and outer membrane structures. These membranes serve to seperate from one another cell cavities, or compartments, which contain substances in aqueous form differing from one compartment to the next. An essential quality of living membranes is that of *semipermeability*, i.e., the property whereby a membrane is pervious only to specific substances, especially ions. This results in an asymmetrical distribution of the differentiated substances at the inner and outer layer of the membrane.

Correspondingly, measurable *electrical potentials* and/or *currents* results wherever an imbalance of ions occurs in the living organism or wherever ions are transported by membranes. The magnitude and extent of potentials and electric currents are cell specific and are controlled actively.

The outer membranes of nerve cells have developed in a unique manner. Essentially, they have the specialized capability to conduct rapidly alternating electrical signals in quick succession within the cell and to conduct these signals to other cells with specificity. These transient signals—*generator potentials*, *action potentials*, and *synaptic potentials*—all originate in the same manner, i.e., from rapid changes in the electrical properties of the cell membrane through which the *resting potential* is increased.

Therefore, we must concern ourselves first with the resting potential of a membrane in order to understand how changes in potential can be conducted as signals. It is precisely this nerve cell capability that underlies each and every reaction within an organism and, thus, is the basis for overall behavior. Indeed, electrical impulses conduct information from the outside to the brain and then, subsequent to central processing, carry information from the brain to the peripheral region, where the impulses trigger reactions.

An examination of the various bioelectrical phenomena enables conclusions to be drawn vis-à-vis the condition and functional modality of the entire organism and its individual organs. Electrophysiological methods of examination are readily applied in diagnostic medicine, for example, in *electroencephalography* (*EEG*), electrocardiology (*EKG*), and *electroretinography* (*ERG*), Furthermore, artificial electrical stimulation can be applied to stimulate a nerve in order to elicit specific responses at the effectors. This is extremely useful in scientific research. In the area of medicine, this approach is referred to as *electrotherapy*.

This chapter is intended to introduce the reader to the fundamentals and basic phenomena of nerve stimulation and stimulus conduction. When considering the function of nerve cells, electrical and chemical components are inextricably linked. The issues under discussion will become clear only when these two aspects of the neuronal process are clarified.

6.1 Resting Potential of Membranes

6.1.1 General Remarks

The *semipermeable membrane* is an essential constituent part of all living cells. It is a selective seperating medium between two fluid cavities that serves to induce *differences in potential,* since the two cavities in question contain diffused ions as well as a share of non diffused ions. The prevailing electrical potential of a fluid cavity can be determined by means of measuring electrodes. These electrodes are applied to both sides of a membrane. The electrode on the inner side of the membrane must be particularly fine so that it does not cause undue damage to the cell. The signals from the cells are amplified via the electrodes and rendered visible by a cathode-ray oscilloscope. To calculate the difference in potential between the two sides, the extracellular value is set at zero and the potential of the inside is compared to it. All nerve cells carry electrical charges on their outer membrane. Thin clouds of positively and negatively charged ions are spread over the intra- and extracellular surface; indeed, they extend only ca. 1 μm from either side of the membrane. The cytoplasm within the cell and the extracellular fluid, in contrast, are electically neutral, although they are conductive.

If the nerve cell is at rest, then a characteristic quantity of positive ions collects on the outside of the membrane and a specific quantity of ions collects on the inside. The quantities vary according to the type of nerve cell. At rest, the membrane maintains the separation of charges by means of mechanisms that will be discussed in detail. This results in the *resting potential* of the membrane (V_R). Resting potentials of nerve cells are specific; they show voltages of between -40 and $-75\,mV$, and in certain instances as great as $-100\,mV$.

All signals that originate in, elapse in, or emanate from a nerve cell are based upon changes of this resting potential. The notion of *membrane potential* (V_m) is more general. It refers to the nerve cell's particular potential at a given moment and in a given circumstance and is defined as

$$V_m = V_i - V_o$$

(membrane potential = inside potential − outside potential).

Membrane potential is directly proportional to the charges that are separated by the membrane.

The primary factor contributing to differences in membrane potential of the nerve cell is the unequal distribution of the four major ions, Na^+, K^+, Cl^-, and organic anions, (A^-), within the outer and inner cell cavities. Whereas Na^+ and Cl^- are concentrated outside of the cell, K^+ and A^- are found primarily within the cell. The organic anions predominantly consist of negatively charged amino acids and proteins.

Measurements of *ion distribution* have been gathered largely from the giant axons of cuttlefish, these axons being particularly large and accessible. The distribution of the most important ions at this nerve cell membrane is represented in Table 6.1.

For comparison, the ion distribution in human nerve cells is listed as well.

6.1.2 K+-Ion Equilibrium Potential as Evidenced in Glial Cells

K^+ and Na^+ move about freely within and outside of the cell and follow the concentration gradients in their spatial distribution. Their

TABLE 6.1. Distribution of ions at nerve membranes of the giant axons in cuttlefish (A) and in human nerve cells (B).

A. Giant axon of the cuttlefish

Ion	Internal concentration (mM)	External concentration (mM)
K^+	400	20
Na^+	50	440
Cl^-	52	560
A^-	385	–

B. Human nerve cells

Ion	Concentration per unit area of interior plasma membrane per μm^3	Concentration per unit area of exterior plasma membrane per μm^3
K^+	100,000	2,000
Na^+	10,000	100,000
Cl^-	2,000	1,000,000
Anions	107,000	

movement is impeded at the cell membrane which is virtually ion-impermeable. Ions are able to pass through the membrane only via specialized membrane pores that are composed of protein molecules and referred to as *channels*. A wide variety of *channel types* exists in the membrane, each of which is specialized to accommodate the passage of a specific type of ion. Selection is determined according to the size, charge, and hydration shell of the ions. The permeability of the membrane to the individual ions is dependent upon the number of differing channel types it evidences. Although there are channels in nerve cells for K^+, Na^+, and Cl^-, *glial cells* in contrast, have channels only for K^+, i.e., the membrane of these cells is permeable only to K^+.

Accordingly, the resting potential of the membranes of glial cells is determined only by K^+ ions. Thus, the mechanism of K^+ transport through the membrane in glial cells can be observed and understood without interference brought about by the movement of other ions. Just as it is the case in nerve cells, the interior of the glial cells has a high concentrations of K^+ and organic anions (A^-). However, high concentrations of Na^+ and Cl^- occur in the extracellular space (Fig. 6.1). K^+ tends to follow the concentration gradient and to diffuse out of the cell through the channels. In the process, the negative anions are left behind. This results in a surplus of positive charges on the outside of the cell and a negative surplus within the cell. Through electrostatic attraction of the surplus cations on the outside and negative anions within, the charged particles become distributed as a light mist over the surface of the membrane.

Two opposing forces are now in effect. The chemical concentration gradient drives the K^+ from the cell and raises the ion difference. Electrostatic attraction resulting from the growing electrical potential difference tends to draw the K^+ back into the cell. Thus, the K^+ flows out in accordance with the concentration gradient as long as the electrostatic attraction is so great that remigration and migration are equal and, accordingly, the outflow is stopped.

When K^+ is in equilibrium, i.e., when inflow and outflow are in balance, the membrane potential of glial cells is $-75\,mV$. Once it is established, this *equilibrium potential* can be maintained indefinitely without further expenditure of energy. *Membrane potential,* whereby an ion is in equilibrium, can be calculated by using the thermodynamic equation developed by W. Nernst in 1888:

$$E_k = \frac{R \cdot T}{Z \cdot F} \cdot \ln \frac{(K^+)_o}{(K^+)_i}$$

Whereby E_k is the value of the potassium

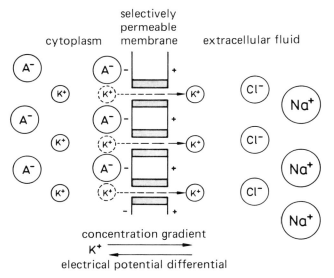

Fig. 6.1. Resting potential of a membrane is formed by virtue of the outflow of K^+ from the cell relative to the K^+ concentration gradients since the membrane is selectively permeable only to K^+.

equilibrium membrane potential, R is the gas constant, T is the temperature in centigrade, Z is the valence of (K^+), F is the Faraday constant and $(K^+)_o$ and $(K^+)_i$ are the K^+ concentrations outside and inside the cell, respectively.

If one were to use this equation to calculate the *potassium equilibrium potential*, for example, in the *giant axon* of the cuttlefish, then: $Z = 1/R \cdot T/Z \cdot F$ at 25°C amounting to 26 mV, the transformation constant for ln to the base $\log_{10}$ is 2.3. Accordingly,

$$E_k = 26\,\text{mV} \cdot 2.3 \lg_{10} \frac{20}{400} = -75\,\text{mV}.$$

The *Nernst equation* can be applied to any ion. For Na^+, the resulting value is $+55$ mV; that of Cl^- is -60 mV.

6.1.3 Ion Equilibrium Potential for K^+ and Na^+

The resting potential of the nerve cell is not determined as easily as it is for the glial cell. Essentially, it, too, is determined by the K^+ equilibrium, but the membrane is also permeable to a certain degree to K^+ and other ions. This means that channels are present for other ions, as well. The flow of the sodium diffusion, for example, disrupts the K^+ resting potential. Sodium is brought into the cell by means of a dual dynamic, i.e., the concentration gradient and the negative inner charge. Applying the Nernst equation, the membrane equilibrium of Na^+ is $+55$ mV. This means that, with a membrane resting potential of -75 mV, the sodium equilibrium is 130 mV removed from equilibrium. Thus, very strong forces of electrical attraction draw Na^+ into the cell.

The membrane is depolarized by the flow of Na^+ into the cell and the membrane potential is shifted in the direction of the Na^+ *equilibrium potential* (E_{Na}). Accordingly, more K^+ is required now on the outside to achieve a balance. The difference in potential between the inside and outside can be maintained in this manner for quite some time. Nonetheless, after a while substantial changes occur in the ratio of K^+ and Na^+ inside and outside that lead to a diminution of the membrane potential. For this reason, sodium is actively removed from the cell under expenditure of energy and, in a corresponding reaction to this, K^+ is reacquired. The reintroduction of K^+ and Na^+ follows their electrochemical gradient and uses adenosine triphosphate (*ATP*) in the process. The resting potential of a nerve cell membrane is balanced out by passive flows of ions of Na^+ and K^+ and by active, energy-expending counterflows. It maintains a delicate balance in this way. The *sodium ion pumps* of nerve cells alone use 15% of the oxidative energy of the neuronal metabolic process. A properly functioning ion pump is necessary for the survival of the nerve cells and, thus, for the survival of the total organism.

The Na^+ and K^+ pumps are interrelated. For each unit of energy expended, three sodium ions are removed from the cell and two potassium ions are reintroduced. Some authors are of the opinion that the $Na^+ : K^+$ ratio in this regard is 1:1. Only the first part of the process, the sodium ion pump, is energy consumptive. The potassium return of the second stage apparently represents a kind of equilibration on the part of the carrier molecules and progresses without further infusion of energy.

6.1.4 The Significance of Cl^- for the Resting Potential

The nerve cell membrane is pervious to Cl^- as well as to K^+ and Na^+. However, Cl^- is present in inverse proportion to K^+. Substantially more Cl^- is found extracellularly than within the cell since the negative charges there derive predominantly from large, impermeable protein molecules or SO_4^{2-} which are dependent upon the high concentration of K^+. The Cl^- permeability factor of most nerve cell membranes is quite high. Cl^- diffuses freely into and out of the cell. In most nerve cells, it is not actively pumped in either direction. Accordingly, it establishes its own equilibrium at the membrane. Since the resting potential of the nerve cell is determined by actively controlled K^+ and Na^+ ion concentrations, $Cl-$ has no effect on the resting potential as it is always distributed passively.

6.1.5 Quantifying Membrane Potential: The Goldman Equation

If the resting potential of a membrane (V_m) is influenced by more than one variety of ion, then each variety of ion affects the V_m. Its quantity is determined by the concentrations inside and outside the cell as well as by the permeability of the membrane to the ions present. The *Goldman equation* determines these quantitative relationships; however, it can be applied only if V_m is constant, i.e., when conditions of constancy prevail vis-à-vis the resting potential:

$$V_m = \frac{R \cdot T}{F} \ln \frac{P_K(K^+)_o + P_{Na}(Na^+)_o + P_{Cl}(Cl^-)_i}{P_K(K^+)_i + P_{Na}(Na^+)_i + P_{Cl}(Cl^-)_o}.$$

According to the equation, the effect of one type of ion upon the membrane potential is heightened proportionately to its concentration and to the degree of membrane permeability. In extreme instances, the effect of ions upon only one predominant type of ion (K^+, for example) is reduced. The equation, then, corresponds to the *Nernst equation* if the remaining varieties of ions do not influence the membrane potential.

6.1.6 Membrane Properties and Voltage-Dependent Ion Channels

The neuronal membrane, ca. 8 to 10 nm in thickness, consists of a double lipid layer in which mosaic-like proteins are embedded (see Fig. 7.5). Membrane lipids are hydrophobic at their centers, yet hydrophilic in the region of their head groups. In contrast, the ions of intercellular and extracellular space are hydrophilic and encased in hydration sheaths that considerably extend the proportional scale of the ions as well as their charge effect within the space, this by virtue of the dipolar character of water. This enlargement of the individual ions by the water sheaths (which, as a rule, do not readily become dissociated) in conjunction with the hydrophobic nature of the membrane impede casual passage into or out of the cell.

Thus, the membrane possesses specialized passage points (channels) that are composed of specific protein molecules and, therefore, are not hydrophobic. Glycoproteins, for example, have

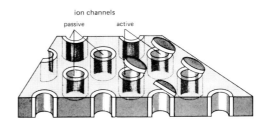

Fig. 6.2. Diagram of the two types of ion channels in the neuroplasm membrane: permanently opened channels enable passive entry of ions; actively restrictive channels regulate ion transport.

been identified as constituents of Na^+ channels. The configuration and the diameter of the channels are the primary factors in determining which types of ions, inclusive of their hydration sheaths, may pass through. The width of the channels, therefore, is crucial as is the number of differing channels per unit of surface area. These factors determine the quantitative ratio of ion varieties that may pass through the membrane.

The membrane has two kinds of *ion channels* (Fig. 6.2): those that are constantly open and allow passive migration of ions, and those that are opened or closed actively and regulate a controlled migration of ions through the membrane. The latter category is comprised of voltage-dependent channels and chemical-dependent channels. *Voltage-dependent ion channels* open and close relative to the magnitude and direction of the membrane potential. They occur predominantly in nerve fiber membranes and can have a density of up to 1,000 per 1 nm^2. *Chemical-dependent ion channels* occur mostly in postsynaptic membranes and react when a transmitter molecule combines with the receptor of a channel protein. Changes in the configuration of channel proteins can occur that lead to the opening or closing of the channel. Both types of ion channels, i.e., voltage-dependent and chemical-dependent ion channels, are composed of proteins.

In the case of *voltage-dependent channels*, changes in configuration occur in the channel proteins relative to the strength of the field at the membrane that causes the channels to open or close. Indeed, the *all-or-nothing principle* prevails here. It has been established, based on the examples of Na^+ and K^+, that each opened channel allows a precise quantity of ions to migrate before it closes.

Complex circumstances underlie the voltage-dependent ion channels in the membrane of the cell body of a nerve cell. It has five different types of channels for three varieties of ions (Na^+, K^+, and Ca^{2+}). The channels open variously and remain open for varying intervals of time. Thus, the migration of each ion is regulated in this manner. By and large, a sequence of individual signals is the response to a specific stimulus.

Essentially, the nerve cell expresses the intensity of a stimulus by utilizing the principle of *frequency modulation*. Signal amplitude remains unchanged in the process.

The in- and outflow of ions through the channels can be rendered visible and measured with the help of radioactive isotopes. However, these movements elapse so rapidly—and so many ions are involved—that, for all intents and purposes, it is not the number of ions that one measures, but rather the fluctuations of membrane potential that arise from the electrical currents generated in the process. Ion movements through the membrane are conducted by means of electrodes and measured as electric current. Three factors coordinate to generate these currents, or signals, through the nerve cell: the ion-specific channels, the concentration gradient of the ions involved, and the capacity of the membrane to store electrical charges (membrane capacity; Fig. 6.3).

Experimental studies of the kinetics of ion channels during stimulation of the nerve cell are carried out by utilizing a *voltage clamp*. This method has been instrumental in explaining in great detail the manner in which membrane potentials are built up and how changes in the various membrane currents elapse in relation to the potential.

6.2 Action Potential

6.2.1 Action Potential Defined

The nerve cell reacts actively to a subtle, short-lived change in membrane permeability or conductivity in relation to one or more types of ions. It does this by manifesting local fluctuations in potential at the stimulated location of the cell membrane. A buildup of so-called *local potential* occurs at these spots. If the stimulus is weak, then the local potential dissipates very quickly and the region is returned to its resting potential. If weak stimuli persist over a certain period of time and result in a minimum threshold rate and minimum threshold duration, or if the initial stimulus is sufficiently great that it crosses a minimum threshold intensity ($-50\,mV$), then it can bring about an increase in the resting potential and briefly convert the membrane charge from negative to positive. This change, brought on by an action in the cellular event, is referred to a *action potential.*

The action potential for any given cell type occurs in a typical and constant manner and is a functional consequence of its physico-chemical properties. In principle, action potentials can be triggered in all living cells whose "all-or-nothing reaction" proceeds in the same manner every time. However, it is especially the case in nerve (and muscle) cells that action potentials are precipitated so dramatically that they can be shown quite clearly.

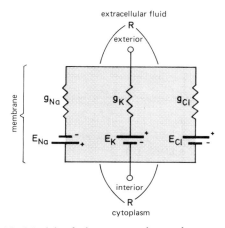

Fig. 6.3. Model of the neuronal membrane as an electrical circuit predicated upon ion-specific membrane channels (Na^+, K^+, Cl^-) linking the cytoplasm and the extracellular fluid. R: resistance; g: conductivity of the channel; E: equilibrium potential.

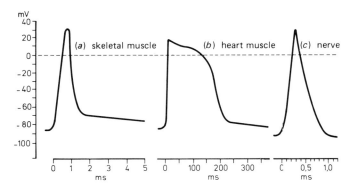

Fig. 6.4. Three examples of action potentials conducted intracellularly: skeletal muscle cell (rat, *a*), heart muscle cell (cat, *b*), and peripheral nerve cell (cat, *c*). Note the great differences in the periodicity of the potentials.

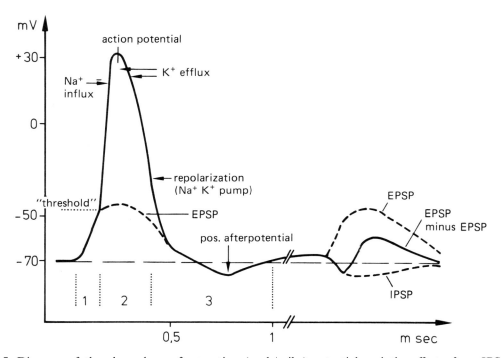

Fig. 6.5. Diagram of the chronology of an action (peak/spike) potential and the effect of an IPSP on an action potential that is built upon an EPSP (see sections 6.3.3.2 and 6.3.3.3).

Three instances of action potentials, each representing a different cell type, are given in Fig. 6.4. The differences lie not in the amplitude, but in the duration of the total event.

Figure 6.5 shows the individual phases of an action potential. It is triggered when the membrane is depolarized to about − 50 mV. The membrane is unstable at this threshold value. Thus, the action potential sets into motion complete depolarization of the membrane (*depolarization phase*). Within 0.2 to 0.5 msec, a rapid increase crosses the zero potential to peak at a *spike* of up to + 30 mV upon the return of the electrical field of the membrane. Finally, the electrical charge of the membrane of most cell types in the *repolarization phase* returns just as quickly, i.e., in a fraction of a millisecond. The return process can even lead to an *overmodulation* as the negative resting value is exceeded. The ultimate return to the resting potential then ensues in slower *after potentials*.

The after potentials can be *hyperpolarizing*, i.e., they can extend beyond the resting value into the negative field, or they can be *depolarizing*, i.e., diminished in a positive direction. The capacity to hyperpolarize is an essential property of the nerve cell membrane, one which will be discussed later at length.

6.2.2 Membrane Currents and Ion Shifts During Action Potential

What causes the *cell membrane* to *reverse its charge* during an action potential? It results from a displacement of ions in the region of the membrane. The resting potential is based on the high K^+ conductivity of the membrane and the potassium equilibrium potential was calculated to be about -75 mV according to the Nernst equation (see Sect. 6.1). However, if the membrane potential is diminished through *depolarization* by -40 to -50 mV to the general rating of the threshold value, then the conductive properties of the membrane correspondingly change for Na^+. Suddenly, the flow of sodium ions into the cell is increased. This decompensates the membrane even further which results in a further increase in the rate of inflow. Finally, the content of Na^+ in the cell attains a value 100 times that of the resting potential. The sodium content exceeds that of potassium during stimulation, a condition that, when of appropriate duration, results in the reversal of the membrane potential into the positive to a value of $+30$ mV (Fig. 6.5). Attainment of this value can be explained as follows. First, the condition of Na^+ conductivity is not sustained long enough for the membrane to achieve a complete transfer of charge that would result in sodium equilibrium ($+55$ mV). Secondly, K^+ conductivity is increased as Na^+ conductivity is increased. This results in a substantial countercurrent. Finally, the outflow of positive K^+ charges from the cell during the

repolarization phase offsets and surpasses that of the Na^+ inflow and the membrane potential again becomes more negative. Following this, the Na^+ ions actively are purged from the cell once again by action of the ion pump while, correspondingly, K^+ ions are reintroduced into the cell. Within 1 msec of the onset of the depolarization phase, the opened, active Na^+ ion channels close, become deactivated and do not reopen during the recovery period (*refractory time*), which is of specific duration. They open again only when the resting potential of the membrane has been reestablished.

6.2.3 Conducted Action Potential

In contrast to a *local potential*, which is restricted to a limited region of the neuronal membrane because it does not attain the threshold value, a peak, or spike potential (action potential) is conducted over relatively great distances. This extended range of conduction results from the fact that the electric currents that flow at the appropriate location at the membrane affect adjacent areas by stimulating them to depolarize the membrane and trigger an action potential. In a chain reaction, a wave of action potentials spreads over the entire length of the nerve fiber in a process referred to as *stimulus conduction* (Fig. 6.6). In principle, as has been shown under cell culture conditions (in vitro), the stimulus can propagate itself in all directions. Thus, the essence of stimulus conduction is that it is an extension, or continuation of membrane depolarization which is induced by small circuits from the initial location of the primary depolarization.

A brief *recovery*, or *refractory period* follows every action potential during which the resting potential of the affected membrane region is reestablished. The cell cannot be stimulated immediately subsequent to the spike (absolute refractory time). A phase then follows in which a reaction can be forced by an increased expenditure of energy, i.e., by unusually intense stimuli. The total duration of the recovery time amounts to only a few milliseconds during which period the remaining K^+ ion channels, which are involved in the spike, are opened and the last Na^+ channels are closed.

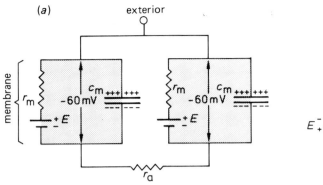

Fig. 6.6. Diagram of impulse conduction along a nerve fiber. Two adjacent membrane regions of an axon, which are connected by axoplasm, are represented by *a* and *b*, respectively. In *a*, both membrane regions are at rest; in *b*, an action potential spreads from left to right. The *broken lines* indicate the direction in which the current is flowing. r_m: membrane resistance; r_a: axon resistance; c_m: charge capacity of the axon.

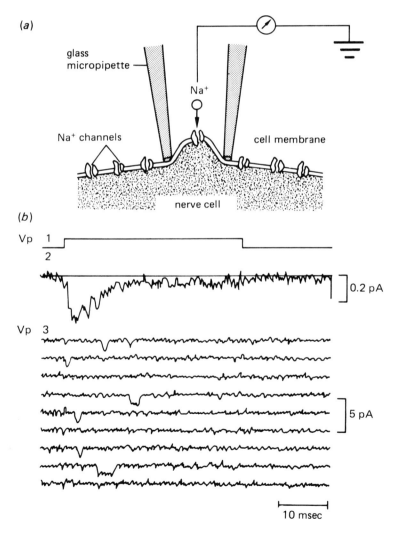

Research into the functional mechanism, the form and the chemical nature of the active ion channels as well as their number per unit of area is carried out mostly with *neurotoxins* that bond to the channel proteins and prevent them from opening and closing. Individual voltage-dependent channels located in minute localized regions of the membrane can be examined today by utilizing another specialized method known as the *patch clamp method*. It allows the researcher to observe how changes in potential effect individual voltage-dependent channels and, likewise, the manner in which chemical substances affect chemical-dependent channels (Fig. 6.7).

6.2.4 Subthreshold Potentials

The stronger the electrical stimulus, the less protracted it must be in duration. A strong electrical impulse is necessary to depolarize the membrane rapidly enough (within 0.2 to 0.5 msec) that an action potential follows. In any case, the impulse must evince a threshold of intensity and duration in order to trigger a reaction (Fig. 6.8).

The weaker the electrical stimulus, the longer it must be in duration. If, however, the stimulus is insufficient to generate an action potential, depolarization of the membrane will proceed

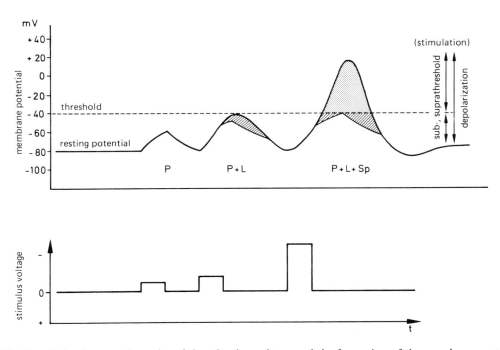

Fig. 6.8. Correlation between intensity of the stimulus voltage and the formation of the membrane potential. Only after the stimulus intensity exceeds a threshold—building upon a passive membrane response (P) and a local generator potential (L)—is an action or spike potential (Sp) triggered.

Fig. 6.7. Diagram of the "patch clamp method" for measuring current through individual voltage-dependent ion channels: (*a*) a glass micropipette draws in a minuscule area of the membrane including a Na^+ channel. The Na^+ current is measured with an ultrasensitive patch electrode; (*b*) Na^+ channel measurement in the muscle cell of a rat. (1) 10 mV impulse; (2) computer average of 300 separate tests of the Na^+ inflow through the Na^+ channel; (3) nine individual measurements of the 300 in which six opened Na^+ channels can be recognized (Reprinted with permission from Kandel and Schwartz, 1986).

slowly. The result is that the sodium system becomes inactivated (within 1 msec) as the Na^+-K^+ ion pump becomes activated. This occurs before the stimulus threshold can be reached.

Within the reaction spectrum of the nerve cell, such electrical stimuli that do not trigger an action potential, but rather remain *below a threshold*, are essential, too. Stimulation currents that increase very slowly can depolarize the membrane beyond the threshold point without it resulting in an impulse. The absence of impulses in the process of very slow depolarization is referred to as *creeping*. Subthreshold local stimuli play a significant role in the central nervous system because many synapses can be depolarized to the stimulus threshold and then reach the threshold very quickly from this state.

6.2.5 Impulse Generation and Conduction of the Action Potential Within the Nerve Cell

A nerve cell can be divided into four functional regions that differ primarily according to membrane properties (Fig. 6.9):

1. The dendrite and cell body regions (soma, perikaryon) in which an impulse triggering potential, the *generator potential,* accumulates;
2. The *transducer, trigger zone* in the axon

hillock in which the information flowing from the various inputs is coordinated;
3. The axon, or *conduction region* in which the nerve cell's reactions are conducted in the form of action potentials;
4. The *synaptic transmission region* of the nerve fiber end formations, the area where an impulse is transmitted to a receiving cell.

Diverse impulses are led to the nerve cell body via a ramified system of dendrites. Each of these impulses is triggered by a specific stimulus and each carries specific information. These information-bearing impulses are balanced out in the cell body and can result in the following conditions:

1. When information of a similar nature is being conducted, the individual impulses accumulate to generate, effectively, a single stimulus that is of sufficient intensity to enable the electrical stimulus of the nerve cell body to trigger an action potential in the area of origin. The action potential is then conducted further.
2. When information of dissimilar nature is being conducted, the collective electrical stimulus that results is very weak. It brings about only a subthreshold depolarization in the axon and at the synapse that is not conducted further.
3. When hyper- and hypopolarizing stimulus

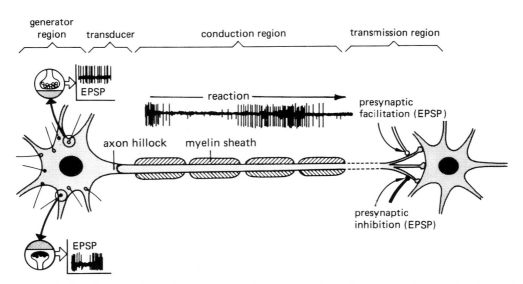

Fig. 6.9. The functional regions of a nerve cell with synaptic elementary processes (EPSP and IPSP).

impulses cancel one another, no electrical impulse is conducted to the axon.

4. The collective stimulus is hyperpolarizing in a subthreshold context and amplifes the negative membrane potential of the axon and synapse beyond the normal resting potential.

If the individual impulses are not conducted via the dendrites to the cell body as fluctuations in membrane potential, but are of a chemical, mechanical, or photic nature, as is the case with receptor neurons, then the nerve cell body must function as an energy *transducer*. These exceptional varieties of stimuli first must be converted into electrical stimuli in order to produce electrical currents so that a generator potential impulse may amass. This then spreads *electrotonically* within the membrane of the nerve cell body to the surrounding regions of the membrane, i.e., with exponentially decreasing amplitude. If it reaches the *base cone* of the axon with sufficient intensity, then it triggers a peak potential that is conducted further. Thus, peak, or spike potentials do not traverse dendrites and nerve cell bodies. "Creeper currents" do, however. Action potential is a phenomenon common to axons and synapses.

Two different types of neurons can be distinguished on the basis of their nerve fiber quality. They are either myelinated or unmyelinated (see Chapter 1). Unmyelinated fibers are older, ontogenetically speaking, than myelinated fibers.

6.2.6 Impulse Conduction in Unmyelinated Fibers

Unmyelinated fibers occur primarily in invertebrates. In vertebrates, especially mammals and birds, nerve fibers are predominantly of the more highly developed, myelinated variety. In the human, unmyelinated fibers are found mostly in the vegetative nervous system and in peripheral nerves which facilitate the perception of pressure, temperature, and pain.

A stimulus is propagated continuously in *unmyelinated fibers*. During this event, the electric current generated by an impulse flows passively (electrotonically) into the surrounding

area and depolarizes the membrane progressively through a chain, or cycle of spike potentials. In this manner, the stimulus can progress successively over the entire length of a fiber (Fig. 6.10).

One problem, however, is the relatively great waning of stimulus intensity as it transits the fiber, since a portion of the current is naturally lost to the environment *(decrement)*.

Another issue is the relatively slow conduction speed at which the stimulus can traverse the unmyelinated fiber. The fibers of nerve cells can attain quite extraordinarily long lengths. The *speed of stimulus conduction* depends largely on the longitudinal resistance of the extracellular fluid, membrane capacity, membrane resistance, and fiber diameter. In unmyelinated fibers of the human vegetative nervous systm, a conduction speed of 1 m/sec has been ascertained for a diameter of 0.5 μm.

The thicker the unmyelinated fibers are, the greater is the *conduction speed*. A rate of 20 m/sec has been measured at the giant axon of the cuttlefish, the diameter of which is 500 μm. Of course, fibers of this extreme thickness are rare.

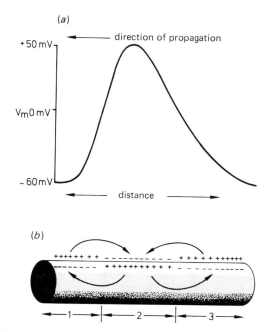

Fig. 6.10. Impulse conduction along an unmyelinated fiber.

By and large, the stimulus conduction speeds of unmyelinated fibers are too slow to facilitate rapid reactions. *Myelinated fibers* represent the specialized development that affords considerably greater speeds of stimulus conduction.

6.2.7 Impulse Conduction in Myelinated Fibers (Myelinated Axons)

The single, most characteristic feature of myelinated fibers is their specially developed insulation, the myelin sheath, which is composed of glial cells and enwraps the nerve fibers (see Chap. 1.2), effectively increasing the thickness of the membrane by a factor of 100. The electrical capacity of the fiber is greatly reduced whereas resistance is substantially increased.

The myelinated nerve fiber is divided into a chain of sections (*internodes*), 1 to 2 mm in length, each of which is enveloped in myelin. The links are separated by *Ranvier nodes* which are free of the insulating substance and are about 2 μm in length. A potential that spreads electrotonically sustains virtually no decrement and diminishes negligibly over distance. The action potential is conducted across the internodes at great speed. A specialized feature of the membrane of the Ranvier nodes facilitates this. It is replete with Na$^+$ ion channels that are subject to control.

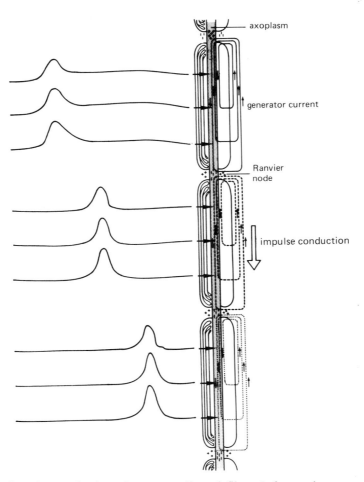

Fig. 6.11. Saltatory impulse conduction along a myelinated fiber. *Left*, membrane potential events measured at points along a myelinated axon designated by *arrows*. Conduction of the action potential (from *top* to *bottom*) is delayed at each Ranvier node.

TABLE 6.2. Correlation of impulse conduction speed and axon diameter of the nervus ischiadicus in various vertebrates (from various authors).

Genera	Number of fibers per nerve	Fiber diameter (μm)	Conduction speed (m/sec)
Frog	2,200	1.0–12.6	30
Mouse	3,700–5,200	1 –14	60
Rat	8,500–11,000	1 –15	70
Guinea pig	ca. 12,000	1 –17	80
Cat	22,285	2 –22	95
Human	—	10 –22	80–120

Increased depolarization can be achieved by means of a greater inward flow of Na^+. This means that the action potential is triggered periodically at the nodes thereby preventing the stimulus wave from falling below the stimulus threshold (Fig. 6.11). Thus, in myelinated fibers, the stimulus is conducted at great speed from node to node across the internodes, its progression along the fiber being retarded somewhat at the nodes. This manner of delayed continuity is referred to as a *saltatory*, or *discontinuous* (*jumping*) *stimulus conduction*. The rate of speed evidenced across the internodal sections is far greater in myelinated fibers than in unmyelinated fibers of the same diameter. In

vertebrates, all fibers that conduct at speeds in excess of 3 m/sec prove to be myelinated.

Table 6.2 shows examples of *conduction speeds* in the *nervus ischiadicus* (sciatic nerve) in various vertebrates relative to fiber diameter. The distance between the nodes is determined by the fiber type and can be as great as several millimeters. Accordingly, speeds as great as 80 to 120 m/sec can be attained. The greatest conduction speed to be observed in the human, 135 m/sec, has been observed in the tractus spinocerebellaris of the spinal cord.

In general, the spike, or peak potential of an individual nerve fiber is great enough to be measured by electrodes that are placed in proximity to the outside of a nerve fascicle. It is usually the case, however, that several fibers are active within a bundle simultaneously, such that cumulative action potentials are registered. Many fiber bundles contain various types of fibers that manifest dissimilar conduction speeds (Table 6.3). Additionally, many bundles also contain unmyelinated fibers of the vegetative nervous system. In actuality, a reciprocal impulse effect would have to take place during an impulse conduction event in light of the arrangement of individual nerve fibers and the fact that an action potential can be measured on the outside despite the considerable insulation. Just such a stimulus effect has been observed in unmyelinated fibers and might well take place in mixed bundles of

TABLE 6.3. Correlation between diameter and impulse conduction speed in various nerve fibers in the human. (Reprinted with permission from Schmidt, 1987.)

Fiber type[a]	Function	Fiber diameter	Conduction speed
Aα	Primary muscle spindle afferences	15 μm	100 m/sec
Aβ	Skin afferences for touch	8 μm	50 m/sec
Aγ	Motor to muscle spindle	5 μm	20 m/sec
Aδ	Skin afferences for temperature and pain	3 μm	15 m/s
B	Sympathetic preganglionic	3 μm	7 m/sec
C	Skin afferences for pain	0.5 μm	1 m/sec
	Sympathetic postganglionic	unmyelinated	—
I	Primary muscle spindle afferences (Ia) and tendon organ afferences (Ib)	13 μm	75 m/sec
II	Mechanicoreceptors of the skin	9 μm	55 m/sec
III	Deep muscle pressure sensibility	3 μm	11 m/sec
IV	Unmyelinated pain fibers	0.5 μm	1 m/sec

[a]Fiber types A, B and C are according to Erlanger/Gasser; types I, II, III, IV are according to Lloyd/Hunt.

unmyelinated and myelinated fibers. This phenomenon could cause a synchronization of the stimulus event in adjacent fibers that certainly could be of major consequence in higher neuronal processes in the CNS. However, there is no clear evidence of this to date relative to intact nerve tissue. This fiber-jumping phenomenon has been observed only in injuries to the myelin sheath, i.e., under pathological conditions, and is referred to as *ephapses,* or *pathological synapses.*

6.3. Transmission of Impulses in the Synapses

6.3.1 General Aspects of Synaptic Impulse Transmission

Chapter 1.1.3.4 detailed the fine composition of the end formations of nerve fibers, the synapses. The discussion in this chapter will center on those specialized transmission points in the nerve tissue where an electrical stimulating impulse is transmitted from a nerve cell to a receiving nerve cell or other effector cell (muscle or gland cell). The significance of the synapse in relation to the effective performance of the organism simply cannot be overstated. Not only does the organism rely on correctly functioning synapses for overall coordination, but for the processes

involved with learning and memory, as well. To these ends, synapses have special structure and specialized biochemical and electrical properties. The constituent elements that comprise a synapse are the specially configured end region of an axon, the *presynaptic region,* and the equally specialized *postsynaptic region* at a dendrite or body of a nerve cell or effector cell (a muscle or gland cell, for example). The section of the postsynaptic cell membrane that lies opposite the presynaptic ending is referred to as the *subsynaptic* area. Both have their own characteristic configuration (see Chap. 1).

Examination of the fine structure of synapses has revealed that stimulus transmission within the nervous system occurs only in specialized zones, i.e., where the pre- and postsynapse are already in particular proximity to one another. Two morphological and functional types can be distinguished in this regard: (a) synapses in which gap junctions exist between pre- and postsynapses, and (b) synapses in which the two specialized nerve fiber end formations are separated by a gap. These differing morphological types correspond in their functional modality to *electrical* and *chemical synapses* (Fig. 6.12a and b; Table 6.4).

6.3.2 Electrical Synapses

The contact area in electrical synapses is relatively large whereas the synaptic gap is very

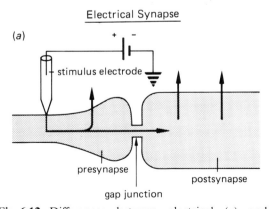

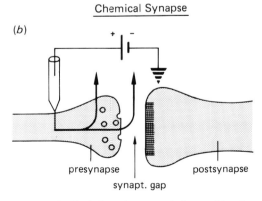

Fig. 6.12. Differences between electrical (*a*) and chemical (*b*) synapses; (*a*) following electrical stimulation of an electrical synapse, the current can flow across gap junctions directly to the postsynaptic cell and depolarize it; (*b*) in a chemical synapse, the

presynaptically induced current is incapable of reaching the postsynapse. Depolarization is dependent upon the release of chemical transmitters from the presynapse.

TABLE 6.4. Functional properties of electrical and chemical synapses.

Electrical synapses
1. reduced extracellular space (2 nm); cytoplasmic continuity between pre- and postsynaptic cells
2. mediator between the pre- and postsynaptic cell is an ion stream
3. negligible delay in transmission due to electronic impulse propagation across gap junction
4. bidirectional impulse propagation

Chemical synapses
1. increased extracellular space (20 to 50 nm); no cytoplasmic continuity
2. mediators between pre- and postsynapse are neurotransmitters ($\sim$ peptides)
3. significant delay (0.3 to 5 m/sec) in transmission is due to the opening of Ca^{2+} channels, the release, diffusion and reception of the transmitter and triggering of a synaptic potential
4. undirectional impulse propagation

slight, approximately 2 to 4 nm, or $\frac{1}{10}$ of the distance that normally exists between cells in tissue (see Fig. 1.15a). The gap junctions (see Fig. 1.16) in synapses are quite similar to the plasma connections that occur in many other body cells. A series of transcellular diffusion channels is located in these junctures that connects the cytoplasm of the sending and receiving nerve cells. Ions and smaller molecules (up to 1.5 nm in diameter) can pass through the connective channels from one cell to the other. The electrical stimulus transmission is effectuated via these connections. The action potential that arrives at the presynapse is conducted across the gap by means of the cytoplasmic bridge. The *postsynaptic* membrane is depolarized in less than 1 msec, whereby a spike potential is triggered and conducted further. In terms of the *all-or-nothing principle*, the postsynaptic membrane of the receiving cell in an electrical synapse reacts only if the threshold value is exceeded. Accordingly, the reaction modality of an electrical synapse is rather stereotypical, and very fast, as well.

6.3.3 Chemical Synapses

Chemical synapses predominate in more highly developed organisms. These function in a decidedly more differentiated and complex manner and do so, indeed, without plasmodesma. Chemical synaptic contact areas are normally rather small whereas the synaptic gap is quite wide at ca. 20 to 50 nm. A current, therefore, cannot flow from a presynaptic action potential directly to the postsynaptic side (Fig. 6.12b). The

electrical changes in field intensity that occur in the extracellular fluid are too weak to depolarize the postsynaptic membrane. Nonetheless, a transitory diminution of the membrane's resistance can be induced locally by a transmitter. This event is referred to as the postsynaptic potential (PSP).

The actual transmission of nerve impulses is effectuated by special transmitter molecules (see Chap. 7). Following the presynaptic action potential, transmitter substances are released in specific quantity from the presynaptic nerve fiber end region into the synaptic gap. Upon its release, the transmitter substance diffuses throughout the gap and establishes the vital connection with the corresponding receptor molecules at the postsynaptic membrane (see Fig. 7.7). This causes fluctuations in potential which result in the reduction of the membrane's resistance and, thus, permits the receiving cell to be stimulated (*excitatory synapse*). The converse may occur also, whereby membrane resistance is increased and a corresponding inhibition results (*inhibitory synapse*). The intensity of the fluctuation in potential varies, thus allowing the nerve tissue, through the events of inhibition and excitation, a degree of control in modualting stimulus transmission.

6.3.3.1 Chemical-Dependent Ion Channels

Chemical-dependent channels represent the most essential functional-morphological difference of the postsynapse. The electrical signals of the sending cell are transmitted chemically in the following manner: transmitter substances are released from vesicles located in the presynapse in response to an electrical signal at the presynapse (see Chap. 7.1.3). The transmitters diffuse throughout the synaptic gap in less than one one-millionth of a second and establish contact with the receptors of the postsynaptic membrane. For example, the transmitter acetylcholine will impinge upon acetylcholine receptors of the postsynaptic membrane. The receptors are channel proteins and bind 2 molecules of acetylcholine per channel, thereby opening the channel which, on average, will remain open for 1/1,000th of a second. The flow of Na^+ into the cell and the simultaneous flow of K^+ out of the cell transpires within this time period. Thereafter, the acetylcholine is broken down by acetylcholinesterase and the channel closes once again. The content of one synaptic vesicle opens approximately 2,000 ion channels at the postsynaptic membrane. Circa 20,000 Na^+ ions flow through each channel into the cell. The simultaneous outflow of K^+ ions is a lesser number and the resulting voltage at the membrane is altered by the uneven exchange of charges. The change in voltage, lasting 5/1,000ths of a second, generates the local *end-plate potential* (*EPP*) of a muscle cell or the correspondng local change in potential at the nerve cells. Potentials that originate via chemical-dependent channels have properties differing from those generated by voltage-dependent channels. Their duration is longer and their amplitude is not as great and, indeed, is dependent upon the quantity of the transmitter molecules released.

6.3.3.2 Excitatory Synapses as Evidenced in Neuromuscular Synapses

Research pertaining to the electrical and chemical events that occur during transmission has focused on synapses between a motoneuron and a muscle cell. These are readily accessible, simply configured and, above all, large enough that microelectrodes can be utilized with relative ease in taking electrical measurements and generating stimuli. Moreover, each muscle cell is innervated by a single axon that forms a specialized plate-like structure in its end region,

the so-called neuromuscular end plate (see Fig. 1.13). In the case of the excitatory synapse, the transmitter that is released into the synaptic gap (see Chap. 7.1.3) brings about a depolarization, the local *end-plate potential*, by its effect upon the receptors at the postsynaptic membrane. At sufficient intensity, this EPP infringes on surrounding membrane areas of the muscle cell by way of electronic currents and can trigger in those areas an *excitatory postsynaptic potential* (*EPSP*). In the case of a neuromuscular synapse, an EPSP causes a contraction of the muscle cell. Even if the EPP fails to attain threshold value, it does alter the membrane potential in the direction of the threshold value. It is possible, then, for subsequent EPPs, which arrive within a brief, prescribed period of time, to trigger the EPSP through their cumulative effect and thereby trigger the reaction, i.e., the contraction of the muscle cell (Fig. 6.13).

Slight, brief depolarizations of low amplitude, which are similar to an EPP, occur at irregular intervals in a resting nerve-muscle system due to the spontaneous release of minute amounts of the transmitter (acetylcholine). These occur in the absence of precursory presynaptic stimulus impulses.

Stimulation originates only on the postsynaptic side. These spontaneous depolarizations are referred to as *miniature end-plate potentials* (min. EPP). On average, they occur once every second. Apparently, miniature end-plate potentials are dependent upon the quantity of the transmitter found in a given vesicle. *Local end-plate potentials* are always produced by a multiple of the minimum EPP, i.e., the transmitter is activated in *quantum units*. After ca. 1 to 2 msec, the transmitter is broken down into its inactive constituent parts by a specific enzyme. This stage corresponds to the end of the time period when changes in potential at the postsynapse can occur.

Because the end-plate potential causes a depolarization of the membrane, it must be generated by an influx of positive ion charges or, in the case of chemical synapses, by chemical-dependent ion channels (see Chapter 7.1). Within this context, measurements have been taken by utilizing voltage clamps that clearly render the duration and the properties

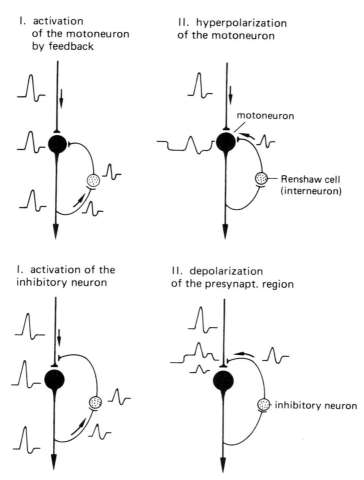

I. activation
of the motoneuron
by feedback

II. hyperpolarization
of the motoneuron

motoneuron

Renshaw cell
(interneuron)

I. activation of the
inhibitory neuron

II. depolarization
of the presynapt. region

inhibitory neuron

Fig. 6.13. Representation of the plastic properties of synapses: a single stimulation of the afferent fibers of a neuron induces a change in the membrane potential (MP) to a subliminal EPSP. A subsequent series of high frequency stimuli (100/sec, 5 seconds in duration) leads to increasing amplitude (frequency potentiation) resulting in the triggering of a spike potential. Subsequent to tetanic stimulation, EPSP amplitudes gradually diminish if individual stimuli are set (posttetanic potentiation, PTP).

of the end-plate potential. It has been shown that the EPP is greatest at its origination point and that it decreases in proportion to its distance from the end-plate region.

According to Ohm's law, the *intensity of the current of an end-plate potential* or of an excitatory postsynaptic potential (I_{EPSP}) is represented by the following formula:

$$(I_{EPSP} = g_{EPSP} \times (V_m - E_{EPSP})$$

whereby g_{EPSP} is the synaptic conductivity that is given by the conductivity of the ion channels activated by the transmitters; $V_m - E_{EPSP}$ indi-cates the electrochemical energy with which the ion current passes through the channels; V_m is the membrane potential; E_{EPSP} is the potential equilibrium for the post-synaptic excitatory potential.

In contrast to the equilibrium potentials of ca. $+55\,mV$ at axons and presynapses, the post-synaptic equilibrium membrane potential is $0\,mV$. Thus, it is not generated solely by a Na^+ equilibrium, i.e., several ions must be involved. Indeed, Na^+ and K^+ have proved to be the agents.

The postsynaptic stimulus differs from the

action potential of the axon in two essential ways: first, on the basis of the protein structure of the chemical ion channels, and second, on the basis of the mechanism by which the channels are opened and closed. In an action potential, the ion movement of Na^+ and K^+ occurs in succession (see Fig. 6.5); in a postsynaptic potential, however, the movement is simultaneous.

The chemical-dependent channels simultaneously allow the passage of even larger cations such as Ca^{2+}, NH_4^+, and even organic cations, but no negative ions such as Cl^-. This leads to the conclusion that the postsynaptic channel proteins repel anions. Postsynaptic channel pores have been calculated to measure at least 0.65 nm $\times$ 0.65 nm in width. By comparison, the Na^+ channel pores of the axon measure 0.31 nm $\times$ 0.51 nm whereas the K^+ pores measure 0.33 nm $\times$ 0.33 nm.

Another difference between an action potential and a postsynaptic potential lies in the fact that the ion conditions that are responsible for the action potential are voltage dependent and are opened by depolarization and closed by hyperpolarization. Opening of the channels at the postsynapse, however, is dependent upon the concentration of specific chemical transmitters. Thus, the number of opened channels is not increased by the process of membrane depolarization. This explains why postsynaptic potentials are relatively low and must function on the basis of cumulative activity.

6.3.3.3 Inhibitory Synapses

Although, under physiological conditions, an EPSP always results in a depolarization of the postsynapse, the consequence of an *inhibitory postsynaptic potential* (IPSP) is *hyperpolarization*, i.e., a heightening of the resting potential that prevents further conduction of the stimulus within the postsynaptic cell and within the nerve fiber to the synapse *(presynaptic inhibition)*. In an EPSP, the transient change in the permeability of the membrane leads to an increase in the permeability of monovalent cations, but the same is true of the EPSP only for K^+ and/or Cl^{-7E}.

The principle of *postsynaptic inhibition* was illustrated by the *Renshaw cells* (RC) in the *anterior horns* of the spinal cord. These cells form a functional unit with motoneurons of the spinal cord (see Fig. 5.4) whereby they are synaptically linked to the collaterals of motor neurites and are activated by the release of the transmitter *acetylcholine*. An RC dispatches neurites to the motoneuron and, through hyperpolarization, is capable of inhibiting the activity of the motoneuron by releasing an inhibitory transmitter substance (γ-aminobutyric acid, for example) via the IPSP mechanism. Thus, the arrangement of motoneuron-RC-motoneuron represents a negative feedback mechanism, or so-called feedback system.

In a manner analogous to the mechanism of

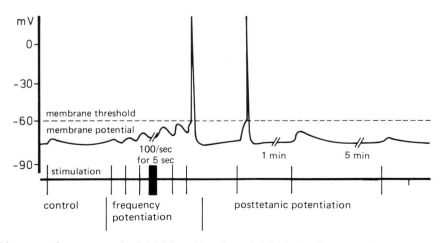

Fig. 6.14. Diagram of postsynaptic inhibition (Renshaw inhibition) effectuated by hyperpolarization of the motoneuron and presynaptic inhibition effectuated by depolarization of the presynapse.

postsynaptic inhibition, *presynaptic inhibition* (Fig. 6.14) prevents the further conduction of an action potential by means of depolarizing, *inhibitory synapses* that occur at the presynaptic membrane. These "predepolarizations" of the axon membrane diminish the incoming action potentials (Fig. 6.14) and result in a reduction in the amount of the transmitter substance released.

This form of inhibition predominates within the CNS and may serve to protect the higher associative centers from becoming overstimulated. Presynaptic inhibition may play a crucial role in the process of *facilitation,* as well (see Chap. 9.2.3).

6.3.4 Interneuronal Transmission

The study of synapses between two nerve cells presents considerably more difficulties than the examination of neuromuscular synapses because, in a receiving nerve cell, a profusion of simultaneously occuring synapses (as many as 10,000) of diverse impulse character can occur at dendrites and at the nerve cell body (see Fig. 6.9). The synapses are of the excitatory and inhibitory varieties. In any event, numerous stimuli arrive simultaneously that are similar to the EPP at the muscle fiber end plate. At the appropriate intensity, and by means of accumulating several PSPs, they can trigger an *excitatory postsynaptic potential* (EPSP). The intensity of the EPSPs, which are triggered simultaneously at one and the same nerve cell, accumulates. The EPSP and the action potential differ in this respect. Whereas the action potential always has the same amplitude, the excitatory postsynaptic potential exhibits variable amplitude in accordance with the stimulus input. If an EPSP, which is spreading over the nerve cell body, reaches the attachment point of the axon, the *axon hillock,* and has a value greater than the threshold value, it will trigger an action potential there. The action potential is then conducted along the axon touching off the same cycle once again at the axon end formation, i.e., it depolarizes the presynapse and conducts the stimulus to the next cell by means of the transmitter.

6.3.5 Plastic Electrical Response Behavior of Neurons

Increasingly, recent electrophysiological findings corroborate a hypothesis that was formulated by Ramon y Cajal in 1894 in which he postulated that neogenesis of synapses and changes in the conduction efficiency of the neuronal connections would have to form the basis of a memory imprint. Utilizing modern electrophysiological methods, synapses can be identified whose conductive properties remain constant despite prior activity. On the other hand, other synapses can be identified whose stimulus-response reactions change radically relative to their previous activity level. Thus, one speaks of *frequency potentiation* when, after repeated, uniform, electrical stimulation of an afferent fiber, the membrane potential of a receiving neuron i.e., the postsynaptic potential (PSP), is increased slightly with each stimulation, whereby the threshold of the action potential is ultimately exceeded (Fig. 6.13). Following stimulation, the postsynaptic amplitudes persist subsequent to individual stimuli. These improved conductive properties of the synapse, which can be affected by changes in concentrations of Ca^{2+}, are referred to as posttetanic potentiation (PTP) (see Chap. 9.2.3; Figs. 9.16 and 9.17).

Exceptionally plastic synaptic response reactions have been recorded for certain varieties of neurons, especially in the *hippocampus,* but also in the *striatum* and in the *cortex* of mammals. Here, low frequency stimuli of 10 impulses per second at defined afferent fibers are sufficient to increase the synaptic conductive properties within a few seconds. In this instance, the heightened amplitude of the post synaptic potentials remains elevated for as long as several hours or even days in the wake of subsequent individual stimuli (see Fig. 9.18). This nerve cell behavior is referred to a *long-term potentiation* (*LTP*). A certain reaction specificity can be ascribed the LTP phenomenon since it fails to manifest itself if other afferent fibers of the same neuron are stimulated. For this reason, LTP events of the type previously described are regarded as a bioelectric correlate to medium-term *memory* formation, as well (see Chap. 11).

In addition to these short- and medium-term adaptive changes in the bioelectrical activity of neurons, extraordinarily long-term adaptation of the electrical response characteristics (*postsynaptic potential amplitude*) has been observed that extends over a period of several weeks and is dependent upon changes in ecophysiological parameters such as temperature acclimatization (see Chap. 9.2.3). Since changes of these types can be paralleled by adaptive changes in learning behavior, the bioelectrical adaptation phenomena reviewed here underscore the fact that the plastic nature of nerve cells must be regarded in connection with the formation of both short- and long-term memory (see Chapter 10).

6.4 The Electroencephalogram (EEG) and Reaction Potential

6.4.1 The Electroencephalogram

An earlier discussion dealt with the fact that electrical potentials of even individual nerve cells are so great that they can be measured on the outside of their membrane and, in the case of nerve fiber bundles, through their insulation. Utilizing disk-shaped metal electrodes affixed to the cranium (or needles introduced beneath the skin) and a neutral electrode that is applied

outside of the cranial area (the ear lobe, for example), cumulative potentials, which are referred to as an *electroencephalogram* (EEG), (Fig. 6.15) can be measured at the cerebral cortex. The EEG records the total electrical activity of the brain's billions of nerve cells. EEG procedures have been standardized over time and it has developed into one of the most important clinical methods of examination used in the fields of neurology, neurosurgery, and psychiatry.

Two types of leads are employed. A *unipolar lead* means that the current is drawn from between one neutral electrode and one cranial electrode. A *bipolar lead* means that a differential in potential is drawn from each of two electrodes located on the cranium. An example of an EEG unipolar lead, representative of the internationally accepted format, is rendered in Fig. 6.15.

In contrast to the electrocardiogram (EKG of the heart), the EEG in the human contains no decidedly characteristic, regularly recurring groups of potential. Rather, it reveals a continuous production of voltage that fluctuates within limits. Analysis shows that, at any given moment, a different, specific fundamental rhythm predominates in the various regions of the brain. It can vary in frequency, amplitude, incidence, and localization of the prevailing waves, in its frequency-amplitude ratio, steepness of the

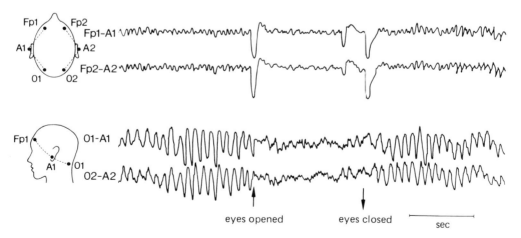

Fig. 6.15. Adult EEG; the cranial sketch shows the lead points for unipolar measurement at the earlobe. Inhibition of the fundamental rhythm is effected by opening the eyes (lead O → A). Artifacts appear during frontal-parietal leading (F_p → A) that are caused by eye muscle potentials.

potentials, the degree of correspondence, and the distribution of the various wave forms over the individual brain regions.

The predominant rhythm in an EEG, where the eyes are closed, consists of waves with frequencies of between 8 and 13 per second and amplitudes of 30 to 150 μV. These are most prominent above the occipital and parietal lobes (alpha waves, Fig. 6.15). More rapid waves of a lesser amplitude (15 to 30 per second, 10 to 30 μV) occur mostly frontocentrally (beta waves); however, these can occur occipitally, as well, superimposed on the alpha waves. The alpha rhythm is interrupted (desynchronized, active EEG) when the eyes are open (observing shapes) or somehow attentive (Fig. 6.15); as lower frequencies are generated once again, replacing the higher frequencies, amplitudes diminish. A comparison between the leads from the occipital and frontal lobes shows a very similar picture; however, distinct differences in the predominant frequencies are apparent.

In addition to the alpha and beta rhythms, gamma waves with frequencies of 4 to 7 per second are produced in the anterior central region of the cortex in juveniles and in people with increased vasovegetative lability. Otherwise, delta waves are quite rare in the normal EEG. They occur occipital-parietally at the cortex, have a frequency similar to that of alpha waves, and exhibit cortical response reactions to sensory stimulation.

The normal EEG in the human, which has become differentiated during the course of ontogenetic development into its own characteristic profile, speaks marginally to the physiological changes within the total organism. *Hyperventilation*, for example, during the course of which, among other events, blood flow to the brain is reduced, produces an increase in the amplitudes of the alpha waves that is followed by a reduction in alpha wave frequency. Ultimately, alpha and gamma waves appear that can be very large and slow (1.5 to 3 per second). Moreover, under pathological conditions, the EEG is altered, in part, in a characteristic manner. These deviations from the norm manifest themselves either in a generalized way as so-called *general changes* (following *barbiturate poisoning*, for example) or focal, i.e., local,

narrowly defined alterations of the EEG. The latter are designated as *foci findings* and can offer valuable diagonostic clues, for example, as to the location of brain tumors.

It has been shown experimentally that, essentially, only excitatory and inhibitory postsynaptic potentials are responsible for generating EEG waves. The stimuli that are conducted along the axons do not contribute in this regard. Since the electrodes are located at a considerable distance from the brain regions that are under examination and, to varying degrees, both bone and skin filter the potentials being measured, the measurements that result are between 100 and 1,000 times weaker than they actually are. Accordingly, this type of lead is not always sufficient. Especially in the field of neurosurgery, leads are taken from the open surface of the cortex (*electrocorticography, ECoG*) and, together with examinations from within the brain utilizing *subcortical electro-*

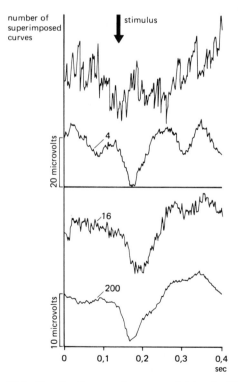

Fig. 6.16. Graph of a reaction potential in the EEG in response to a stimulus (optical or auditory) repeated at great intervals. Assessment of the superimposed measurement curves was computer-assisted.

graphy, epileptic foci or *tumors* can be localized. These latter methods, however, are being replaced increasingly by the process of *computer tomography*.

The normal EEG in the human, which has become differentiated during the course of ontogenetic development into its own general profile and the profile typical of the individual, speaks marginally to the physiological changes within the total organism.

Increasingly, a generalized expiration of the EEG (*baseline EEG*), brain death, is used as the criterion for the pronoucement of death in questionable situations since it is quite possible to rejuenate conditions of circulatory and respiratory arrest. Upon expiration of the EEG, however, the patient will regain neither consciousness nor spontaneous breathing.

6.4.2 Reaction Potential

Little can be gleaned from a cumulative EEG potential regarding individual reactions that are triggered in the brain in response to specific reactions such as those relative to sensory organs. Measurements of *reaction potential*, however, do yield information within this context. In contrast to EEG signals with amplitudes between 50 and 100 μV, individual reaction potentials have amplitudes of only ca. 0.5 to 1 μV. As such, individual reaction potentials do not lend themselves readily to measurement, although they are of great interest because these infinitesimal changes in voltage do provide information about brain activity.

At the present time, two different methods serve to filter out reaction potentials from the cumulative potential of the EEG:

1. An identical stimulus (optical or acoustic) is repeated at such great time intervals that the brain can return to the resting phase between stimulus events. By superimposing measurement curves with the help of a computer, the reaction potential becomes visible (Fig. 6.16). In time, with the great number of superimposed curves, all irregular changes in

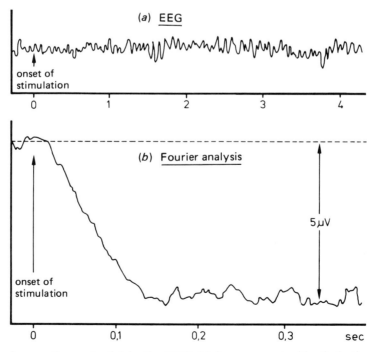

Fig. 6.17. Graph of a reaction potential in the EEG (*a*) in response to identical stimuli administered in rapid succession and overlapping, thus leading to (*b*) the formation of continuous potentials of low amplitude that were filtered out by computer-controlled Fourier analysis.

voltage that occur in the brain will appear in the collective curve.

2. In the second method, a stimulus is given in such rapid succession that a continuous stimulus is triggered in the corresponding active region of the brain. The successive stimuli overlap and build a continuous potential, correspondigly, of very low amplitude. It is this potential that is then filtered out of the conducted collective potential by means of Fourier analysis. The Fourier analyzer depicts only those changes in potential that are of the same frequecy as the stimulus, or a whole number multiple of this frequency. No other voltage changes are rendered (Fig. 6.17).

Reaction potential measurements will find application in myriad ways. Presently, however, they are still in the development stage and are used increasingly in medicine, for example, to analyze auditory and visual damage in early childhood. This type of research centering on reactions of the nervous system to target stimuli could offer the key to presently unexplained questions in the realm of functional analysis and, in conjunction with biochemical methods, could be instrumental in clarifying the nervous process.

7

Chemical Aspects of Neuronal Information Transmission in Synapses

7.1 Molecular Basis of Synaptic Information Transmission

The totality of mechanisms involved in the process of information transmission is immensely complex, comprised as it is of biochemical, biophysical, and ultrastructural elements. Moreover, these are not identical for all types of nerve terminals. Indeed, quite heterogeneous varieties have become apparent.

The principle of electrochemical impulse transmission from cell to cell was recognized as early as around 1900. While examining autonomic nerves, the English physiologist. Langley discovered that phenomena such as increased heart rate, which were triggered by the electrical stimulation of corresponding nerves, could be induced by administering adrenal cortical extracts.

T. R. Elliot then postulated in 1905 that electrical impulses caused a release of adrenaline-like substances from nerve terminals that, in turn, would cause gland cells to release secretions. Elliot felt that these gland cells contained receptive substances that were of an excitatory and inhibitory nature that, ultimately, determined the manner in which the organ would react.

This represents the first time that an interrelationship between chemical and electrical factors was presumed that incorporated the notion of molecular receptors. In 1921, Otto Loewi was able to establish that the *nervus vagus* reduced the heart rate by means of releasing acetylcholine. Henry Dale confirmed these findings around 1930 and expanded them with evidence to the effect that acetylcholine not only acts as a transmitter substance in the autonomic nervous system, but performs the same function at the nerve–muscle contact points, as well.

These somewhat speculative insights into the process of electrochemical transmission finally could be substantiated in the 1950s with the advent of microelectrode techniques and electron microscopy.

Communication between neurons as well as between neurons and nonneuronal effector cells occurs at highly specialized morphological structures, the synapses (see Chap. 1). Each synapse represents a specific link between a terminal of a presynaptic neuron and a receptive area of a postsynaptic cell. Because a neuron can send out as many as 1,000 nerve terminals and evidences just as many receptive areas, the human brain contains ca. 10^{14} to 10^{15} interneuronal contact points. The pre- and postsynaptic cells of most synapses are separated by a synaptic cleft of 30 nm that prevents direct electrical communication between cells. Thus, the *transmission* of electrical signals at these synapses is facilitated by molecules: the neurotransmitters. During the transmission process, an electrical signal at the presynapse triggers the release of transmitters that diffuse throughout the extracellular fluid of the synaptic cleft. The transmitters react with specific receptor points of the postsynaptic cell and effectuate a change there in the membrane potential (see Chap. 6).

A highly specialized mechanism has developed to facilitate the process of chemical communication between neurons. Neurons require special enzymes to synthesize and metabolize transmitters as well as mechanisms by which to store rather large quantities of transmitters in the nerve terminals. They also need specialized membrane structures that enable the release and recognition of transmitters. The question as to why, during evolution, the selection process favored chemical transmission over the more direct mechanism of electrical conduction (see Chap. 6.3.2) can be answered, perhaps, by understanding a few of the significant advantages offered by chemical transmission:

- Chemical transmission assures one-way transmission of information. This is not necessarily the case with electrical transmission.
- In manifesting a diverse mixture of variously composed transmitter substances that can function either in an inhibitory or excitatory manner upon their release from the presynapses of various neurons, a vast array of postsynaptic responses can be induced within a single, integrated neuronal process.
- Functional regionalization and differentiation of the brain relative to specific functional correlations is made possible by identifying individual brain regions where chemical transmitter substances predominate.

Due to their enormous significance relative to the functions of the nervous system (NS), synaptic contact points have been studied extensively from the perspectives of anatomy, electrophysiology, and pharmacology. A biochemical study of the underlying structures and molecular processes was begun in the last century and is still among the most relevant of research projects in neurobiology. Unlocking the molecular processes of synaptic transmission should lead to new insights into the mechanisms of neuronal processes and the causes of neurological disease. These insights, in turn, might lead to effective modalities of therapy.

Synapse research was thrust forward in 1962 when Whittaker and De Robertis discovered that the presynaptic nerve terminals avulse, or break off during the homogenization of nerve tissue, a process that is necessary for certain neurochemical studies. The products of this avulsion are discrete, self-contained structures referred to as *synaptosomes*. Since the synaptosomes apparently retain the morphology, chemical composition, and metabolic properties of the original nerve terminals, they offer ideal research opportunities in vitro for the study of various functional aspects of the synaptic and neuronal membrane.

7.1.1 Synaptic Membranes

When discussing the mechanism of impulse transmission in synapses, one is compelled to emphasize the functional importance of the composition of the synaptic membrane. Presently, it is possible to manufacture synaptosomal membrane fractions 80% in purity and to analyze them chemically. In the process, specific glycoproteins and polypeptides can be identified that correspond in their molecular weight to the purified *brain* Na^+-, K^+- and Ca^{2+}-*ATPase*, i.e., to those membrane-based enzymes that transport ions through the membrane by means of metabolic energy (ATP) and, thus, maintain membrane potential.

Of these enzymes, Ca^{2+}-*ATPase* is of particular importance in the synapse itself. This membrane-based enzyme, which occurs in two different configurations, is activated when calcium, having accumulated in the extracellular space, flows into the presynapse as a result of an action potential. It triggers secondary messenger effects here, but, subsequently, is pumped out immediately due to a concentration gradient based upon its cytoplasmic toxicity (for details, see Sect. 7.2).

In addition to studying membrane enzymes of this variety, the composition of the lipids in the synaptic membranes in vertebrate neurons has been pursued extensively, as well. Among the neuronal *lipids,* the *phospho-* and *glycolipids* are of extreme importance as is, of course, *cholesterol* since, as membrane producers, they establish specific membrane functions on the basis of their molecular polarity (hydrophilic-lipophilic) and, to a degree, their considerable negative charge. Because lipids (with the exception of cholesterol) have long-chained, aliphatic acids and, in some instances, even long-chained alcohols or a base, they are attributed common names and similar physical properties. However, the nature of the polar groups varies greatly, ranging from a single hydroxyl group (cholesterol) to six in the so-called *cerebrosides* and 20 to 40 in *gangliosides*.

Cholesterol (Fig. 7.1a) is found in the cell membrane almost exclusively in soluble form and not as cholesterol esters, as is the case in the membranes of peripheral organs. Most cholesterol (like most other brain lipids), is synthesized in situ. Because of a highly effective blood-brain barrier, passage of radioactively marked cholesterol from the blood into the brain is kept to a mere trace. The turnover rate of brain cholesterol is, likewise, extremely low. Even after one year, the content of marked

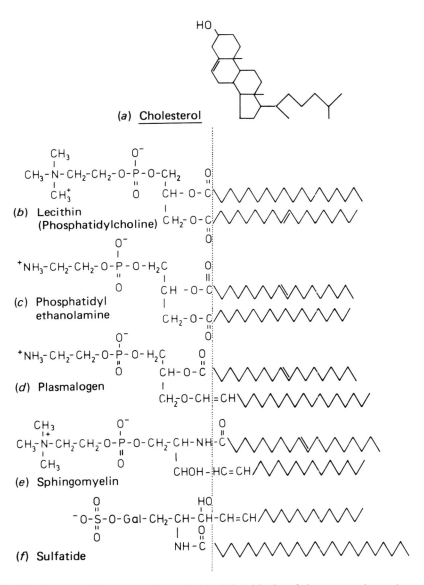

Fig. 7.1. Structural formulae of essential building blocks of the neuronal membrane.

cholesterol in the CNS was unchanged, whereas it disappeared from the blood and liver after 10 to 20 days.

The *phospholipids* of the CNS include *phosphatidylcholine,* cephalin, plasmalogens, and sphingomyelins which, due to their differing apportionment of fatty acids and other components, represent a highly heterogeneous molecule group. It is generally the case with phospholipids that phosphoric acid is esterized with glycerine and a second alcohol (choline, colamine, serine, inositol). In contrast to cholesterol, phosphatidylcholine *(lecithin)* (Fig. 7.1b) accumulates heavily in the synaptic membrane. To a certain degree, it represents the lipid matrix of the synaptic biomembrane in which proteins are submerged according to "fluid mosaic model" of the membrane (Singer and Nicolson, 1972). In addition to its dipalmitic acid constituent, lecithin can be identified in neuronal membranes by fatty acids between C_{13} and C_{22}, whereby palmitic acid and oleic acid account for more than 80% of all lecithin fatty acids. *Cephalines* such as *phosphatidylethanolamine* (Fig. 7.1c) and *plasmalogens* (Fig. 7.1d) are less prevalent in neuronal membranes and more prevalent in myelin, as is *sphingomyelin* (Fig. 7.1e), a nonglyceride.

The *glycolipids* have as their base not glycerine, but sphingosine. To this group belong the cerebrosides, sulfatides, and gangliosides, which are glycolipids by virtue of their carbohydrate residues. The hydroxyl groups of these carbohydrate chains cause the glycolipids to be relatively strongly polar compounds. The cerebrosides and sulfatides are predominantly fat soluble. The gangliosides, however, are water soluble due to their sugar residue.

The *cerebrosides* (Fig. 7.2) differ from other lipids by their sugar residue (galactose, as a rule). Otherwise, they also contain long fatty acid chains and are regarded as derivatives of the ceramides. They are prevalent especially in white matter (myelin). Disorders in cerebroside metabolism lead to neuronal disorders referred to as *cerebrosidoses.* The accumulation of glucocerebrosides, for instance, in which glucose replaces galactose, leads to *Gaucher's disease.*

The *sulfatides* (Fig. 7.1f) consist of galactose, sphingosine, and sulfate in equimolar portions.

They, too, accumulate in the white matter. They are reported to occur in rather high concentrations in the synaptic membranes of Arthropoda (insects) where they perform functions similar to those carried out by gangliosides in vertebrates. In a fashion suggestive of the cerebrosides, the sulfatides also contain long C_{24} fatty acid chains, partially hydrolyzed, partially unsaturated. Disorders in sulfatide metabolism cause *sulfatidelipidoses.*

Whereas the cerebrosides (Fig. 7.2) have a galactose molecule on the ceramide structure, the ceramide residue in *gangliosides* is bound to a complex carbohydrate structure *whose terminal molecule is always a sialic acid,* especially *N-acetylneuraminic acid* (NeuAc). The latter, in turn, plays a crucial part in the synthesis of *sialoglycoproteins.* The carbohydrate chain of brain gangliosides consists mostly of four hexoses in the sequence of glucose-galactose-galactosamine-galactose. The sialine residue group is always bound to the galactose molecule of the side chain and is of varying number. Due to the negative charge of the terminal NeuAc, sialoglycolipids of differing polar configurations (monohexasialogangliosides) come about, depending upon the number of gangliosides (1 to 6) the NeuAc molecules evidence. These differ in the degree to which they are hydrophilic (Fig. 7.3). According to Yu and Ando (1980), the bonding of differing numbers of sialic acid esters to the sugar side chain, facilitated by sialyltransferases, leads, by way of the three possible modalities of biosynthesis, to individual gangliosides whose polarities may differ greatly. Since the time this research was conducted, 40 or 50 various forms have been discovered.

On the cellular level, the primary synthesis of gangliosides occurs in the Golgi apparatus of the cell body (Fig. 7.4) from where they are sluiced via rapid neuronal transport (see Chap. 9.1.2) into the nerve fiber end formations. The presence of membrane-bound *sialyltransferases* as well as catabolic *neuraminidases* in synaptosomal fractions speaks to the fact that, to a certain extent, processes of synthesis and catabolism might be possible locally, i.e., in the synapses.

Gangliosides are anchored in the outer lipid layer of the cell membrane by their hydrophobic ceramide. From here, their hydrophilic carbo-

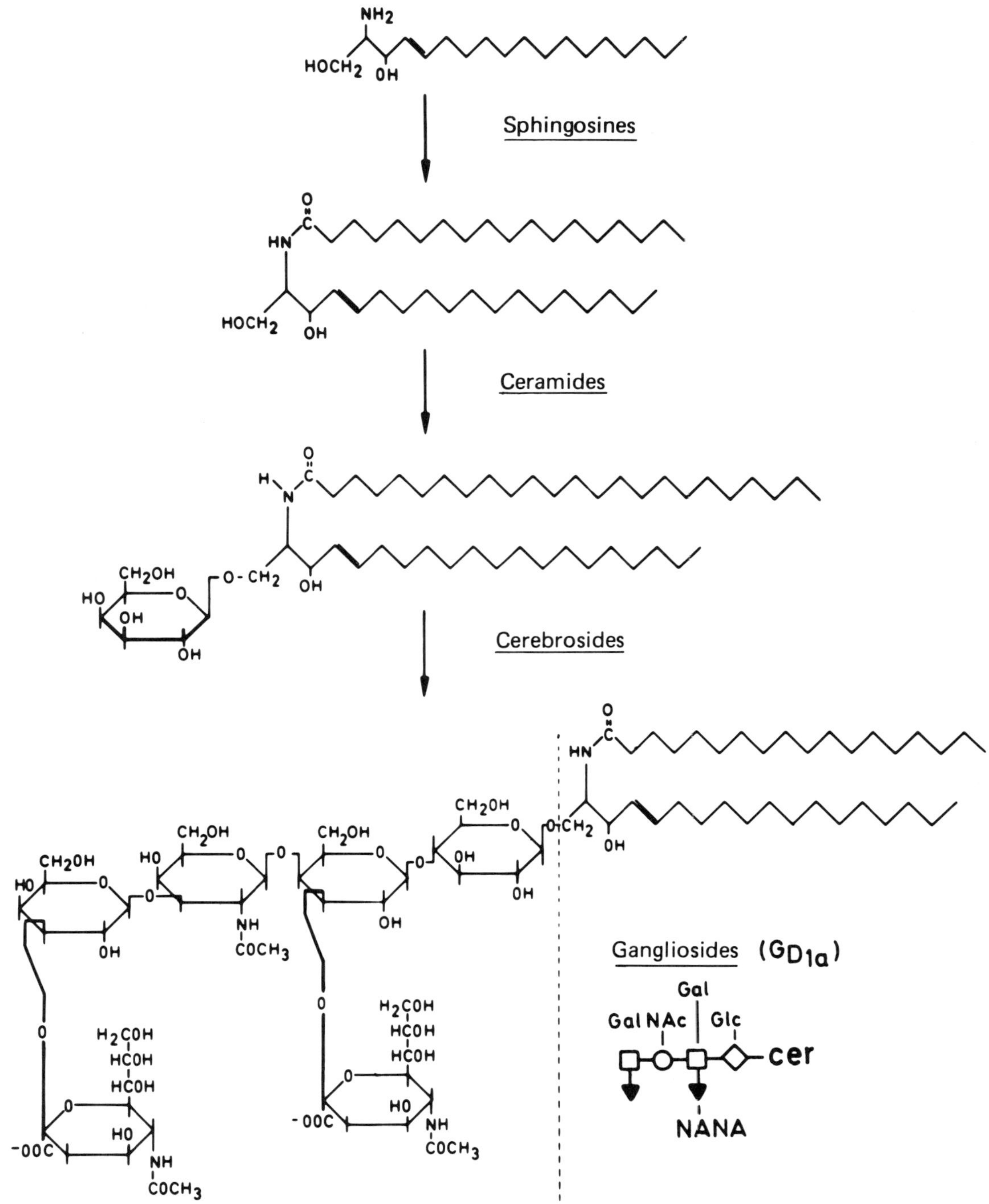

Fig. 7.2. Biosynthesis of the cerebrosides and gangliosides.

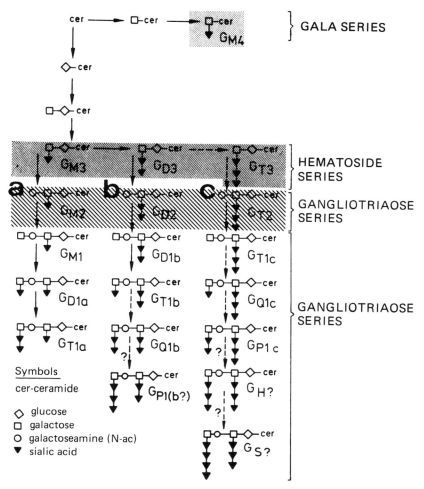

Fig. 7.3. Diagram of the three sequences of ganglio-side biosynthesis known to date (paths *a*, *b*, and *c* from Yu and Ando, 1980): the degree of polarity exhibited by the molecule is directly proportional to the number of negatively charged neuraminic acids per molecule (*black triangles*).

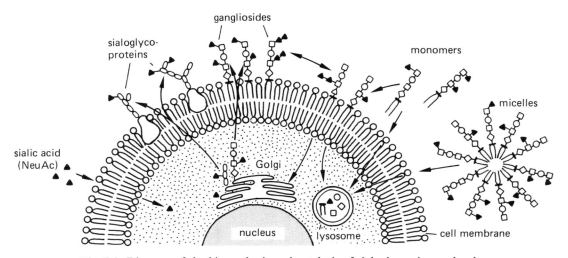

Fig. 7.4. Diagram of the biosynthesis and catalysis of sialoglycomicromolecules.

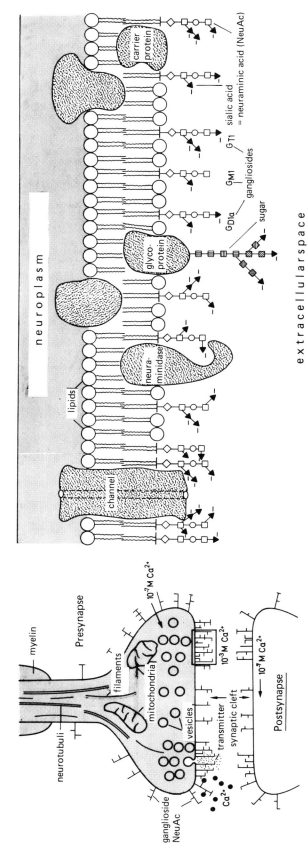

Fig. 7.5. *Left*: diagram of a synapse with ganglioside molecular fuzz (cluster) at the outer membrane. *Right*: diagram of a molecular membrane with gangliosides of differing polarity, integral glycoproteins, carrier and channel proteins, and the neuraminidase enzyme (sectional enlargement).

hydrate side chain, with the negatively charged sialic acid residues, extends into the extracellular space (Fig. 7.5). This enables interaction with cations (calcium, for example) or polyelectrolytes at the membrane surface. In this way, the gangliosides can become involved in receiving and passing on external signals and, functionally, can change the charge properties of the membrane surface, as well. The gangliosides are not distributed evenly within the membrane itself. Rather, they are arranged in molecule clusters. Presumably, these ganglioside clusters encircle channel, or carrier, proteins and, thus, modulate their functions. These glycolipids are attributed essential functional significance in the processes of impulse transmission and memory formation on the basis of the properties just described and due to the unusually high concentration of gangliosides in synaptic membranes (10 to 15% of all lipids) (Rahmann, 1976; see Chap. 11.2.4).

7.1.2 Synaptic Vesicles

Synaptic vesicles are the storage and transport receptacles of the transmitter substances which, based on studies carried out by Katz, are released at the motor end plates in quantum units. Details pertaining to the diverse chemical nature of transmitter substances and the manner in which they are released are discussed in Sect. 7.1.3 and 7.2. Not only is their content essential for the functioning of the vesicles, but their structure is, as well. Synaptosome preparations prove to be suited equally well for the examination of presynaptic vesicles and for the physical and chemical analysis of *synaptic membrane properties.* The protein and lipid composition of the vesicle membrane is very similar to that of the synaptic membrane as long as one recognizes that recently discovered vesicle gangliosides are located on the inside of the membrane. The similarity between the vesicles and the synaptic membrane enables their reciprocal fusion when the transmitters are released from the vesicles during synaptic transmission (see Fig. 8.15 and 8.16).

Just how the *synaptic vesicle is formed* is not fully understood. On the one hand, it might be

formed in the perikaryon of the neuron itself by the Golgi apparatus and filled with the transmitter substance. It would be sluiced from here into the presynaptic end formations by rapid neuronal transport. In the case of adrenergic vesicles, the latter facet probably could be established by fluorescence-histochemistry. On the other hand, vesicle origination is conceivable in the synapse itself through segmentation from the endoplasmic reticulum.

Another formation of synaptic vesicles occurs in the sense that the presynaptic plasma membrane is recycled by endocytotic constrictions of the plasma membrane as it joins in the transmission process. In this instance, reusable molecules, such as peptides, can be absorbed from the synaptic cleft into the newly formed vesicle. It is possible that certain neurotoxins, such as the tetanus and botulinum toxins, invade the neuron in this way.

7.1.3 Synaptic Transmitter Substances (Neurotransmitters and Neuropeptides)

Researching the chemistry of signal transmission in the nervous system is an immensely difficult undertaking largely due to the minuscule quantities in which the transmitter substances are present in the end regions of the nerve fibers. With the development of synaptosomal techniques, great strides were made relative to understanding the chemical nature of transmitter substances. Histochemical marking techniques utilizing fluorescent dyes and radioactive tracers have added greatly to our understanding of where transmitter substances are synthesized and where they remain in the nerve tissue. Finally, neurochemistry benefited enormously from the study of synaptic antibodies to these substances.

If one injects an enzyme, which has a key function in the synthesis of a transmitter substance, into another species, then that species will develop *antibodies* to the alien enzyme. These antibodies can be marked with a fluorescent dye or some other contrasting medium. Upon injecting the marked antibodies into the brain tissue of the original animal, they bind with all those neurons that contain the enzyme.

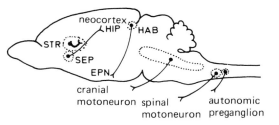

(a) ACETYLCHOLINE

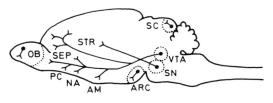

(b) DOPAMIN

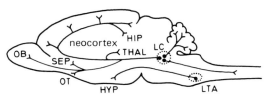

(c) NORADRENALINE

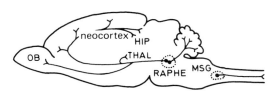

(d) SEROTONIN

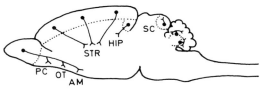

(e) GLUTAMATE/ASPARTATE

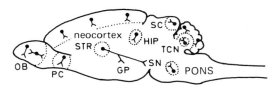

(f) GABA

Fig. 7.6. Diagrammatic overviews of the localization of neuron groups with varying neuron transmitters in the mammalian brain. AM, amygdala; ARC, nucleus arcuatus; DR, dorsal root; DRG, dorsal root ganglion; EPN, entopeduncular nucleus; GP, globus pallidus; HAB, corpora habenulae; HIP, hippocampus; HYP, hypophysis; LC, locus coeruleus; LTA, area tegmenti lateralis; MED, medulla; MSG, medullary serotonin group; NA, nucleus accumbens; OB, bulbus olfactorius; OT, tuberculum olfactorium; PC, pyriform cortex; PERI-V, periventricular nuclei; SC, colliculus superior; SEP, septum; SN, substantia nigra; STR, striatum; DCN, deeper cerebellum nuclei; THAL, thalamus; VTA, ventral tegmental areas. (Reprinted with permission from Shepherd, 1983. Copyright by Oxford University Press, 1983.).

These techniques have helped to determine that the various transmitter substances are not randomly distributed in the brain, but rather that neuron groups with identical transmitters are concentrated locally and that their nerve fibers extend to other brain regions (Fig. 7.6).

Two groups of transmitter substances can be differentiated; the *neurotransmitters* and the *neuropeptides*. Differentiating between the two groups and identifying them are among the most difficult tasks in the whole of neurobiology. Table 7.1 presents just such an attempt at differentiation. Neurotransmitters are small molecules with only two to ten carbon atoms whereas neuropeptides are medium-sized molecules with up to 100 carbons. The former exhibit a great binding affinity for their receptor compounds (see Sect. 7.1.3.1.2), whereas such

TABLE 7.1. Steps in the identification of neutrotransmitters and neuropeptides.

Neurotransmitters	Neuropeptides
Anatomical evidence of the substance in presynaptic processes	Evidence of the peptide nature of the neuroactive substance
Biochemical evidence of the occurrence and effect of enzymes of synthesis and breakdown in the presynaptic neuron	Development of a quantitative bioassay and corresponding procedures for extraction and separation
Physiological evidence that substances are released from the presynapse following stimulation and that direct application of the substance triggers a neuronal reaction	Ascertain presence of amino acid sequence
	Formation of peptide antibodies
Pharmacological evidence of the antagonistic effect of drugs on neuronal reactions	Development of an immunoassay for the immunocytochemical depiction of the peptide

binding affinity of the latter is minimal. The effective specificity of both groups, however, is equally great.

7.1.3.1 Neurotransmitters and Neuroreceptors

7.1.3.1.1 Neurotransmitters

Table 7.2 presents a summary of the most essential neurotransmitters, their effective modality (excitatory or inhibitory), and some of their pharmacologically effective antagonists. (Fig. 7.6 also offers an overview of the localization of neuron groups with various neurotransmitters in the mammalian brain.) It is apparent from Table 7.2 that some transmitters operate in both the excitatory and inhibitory modalities. This may be surprising since it has been assumed that one and the same transmitter can function only

TABLE 7.2. Neurotransmitters, their modality of effect, and their antagonists.

Transmitter	Modality	Structure	Antagonist
Acetylcholine (ACh)	exc., inh.[a]	$CH_3COOCH_2CH_2N^+(CH_3)_3$	Curare, atropine
Adrenaline (A)			Chlorpromazine, propranolol
Noradrenaline (NA)	inh. (exc.)		Several
Dopamine (DA)	exc., inh.		Haloperidol, spiroperidol
Serotonin (5-hydroxytryptamine; 5-HT)	inh. (exc.)		Methergoline
Glycine (Gly)		$HOOCCH_2NH_2$	Strychnine
γ-aminobutyric acid (GABA)	inh. (exc.)	$HOOCCH_2CH_2CH_2NH_2$	Bicuculline
Glutaminic acid	exc.	$HOOC-CHNH_2-CH_2-CH_2-COOH$	Tetrotoxin
Transmitter candidates: Histamine			
Taurine			
Others: ATP, aspartate, proline, octopamine, purine, carnosine			

[a]exc. = excitatory (stimulating); inh. = inhibitory.

in one direction. Recent data show that that is not necessarily the case. For example, acetylcholine (ACh) has an excitatory effect on muscle contraction when it is released at the motor end plate, although it has an inhibitory effect on heart musculature when it is released from terminals of the nervous vagus. Similar ambivalences can be found in *multiaction cells* of invertebrates (Kandel, 1976). Table 7.2 also reveals that various substances can effectuate one and the same function, for example, the triggering of an impulse, whereby the system achieves greater plasticity. However, it should be emphasized that, according to *Dale's principle*, a differentiated neuron at the most diverse presynaptic terminals can release only one variety of transmitter. Numerous presynapses of the most varied neuronal types do terminate, however, at a single neuron such that various transmitters act upon one cell. The most essential characteristic of neuronal information processing is the integration of all synaptic inputs and their conversion into a complex spatial and temporal pattern of membrane potentials, both localized and conducted further.

Elaborating upon *Dale's principle*, it was found a few years ago that one and the same neuron could synthesize two different transmitters in the course of its development, i.e., not simultaneously. It was also found that one presynapse can release classic neurotransmitters as well as neuropeptides. This might be a means of expressing *neuronal plasticity* (see Chap. 9), whereby the question remains as to which effects of induction ultimately establish the transmitter type of a neuron.

The functional principle of chemical synaptic impulse transmission is rooted in the fact that changes in the membrane potential of one cell (*presynapse*) influence ion channels and, thereby, the membrane potential in a second cell (*postsynapse*) through the release of a messenger substance (*transmitter*). Naturally, this process elapses in time.

In the case of the neuromuscular end plate, there is a delay in impulse conduction of about 0.5 msec. The transmitter requires one-tenth of this time (50 μsec) to diffuse throughout the synaptic cleft.

The most fundamental stages in the process of impulse transmission occur in an identical manner in all neurochemical synapses (Fig. 7.7):

- In an unstimulated synapse, a *resting potential* is maintained based upon the activity of ion channels and pumps in the membrane relative to various extra- and intracellular concentrations of mono- and divalent ions (stage 1)
- Upon arrival of an electrically encoded impulse signal (*action potential*, AP), and induced by the triggered changes in concentration of the mono- and divalent ions (2), Ca^{2+} leaves its extracellular binding area via specific channels, primarily to enter the presynapse (3) where one of its effects is to bring about a fusion of the synaptic vesicle (which is filled with transmitter substances) to the inside of the presynaptic membrane (4; see Chap. 8.2.4).
- The next stage consists of a *transmitter release* into the synaptic cleft brought on by exocytosis (5). It has been established through studies of motor end plates (Katz) that the transmitter release apparently occurs in very specific (quantum) units: the release of 100 to 200 *transmitter quanta* is required to trigger an AP, whereby each quantum, for example, in the case of acetylcholine, contains between 1,000 and 10,000 molecules. Since a vesicle contains about 2,000 molecules of acetylcholine, a quantum traditionally has been equated with the content of one vesicle. Recent studies would indicate, however, that the relative value of a quantum should be correlated with the quantity of transmitters released from several vesicles. (In addition to the transmitter substances, other compounds are present in the vesicles, as well, such as ATP or hydrolytic enzymes.) In addition to the vesicular release of transmitters, a direct, equally quantum release of the transmitter through the membrane is assumed, since free, non vesicle-bound transmitters have been found in the synaptoplasm of several neurons.
- Upon the release and diffusion of the transmitter through the synaptic cleft, the transmitter *binds to receptor molecules* (see Sect. 7.1.3.1.2) in the postsynaptic membrane (7). (In this context, it is still not known whether autoreceptor bonds with corresponding presynaptic molecules are possible.)

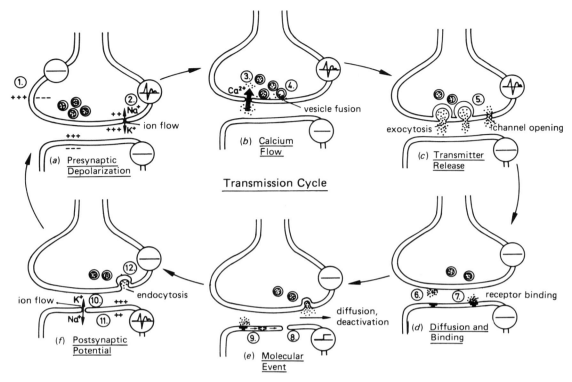

Fig. 7.7. General course of membrane-structural and biochemical processes during a transmission cycle in a chemical synapse (see text for details).

- In the simplest case, ion channels are opened directly in the postsynaptic membrane (8). It is possible, however, that conformation changes in membrane-based enzymes were triggered previously (9). Finally, an *ion current* of monovalent ions (10) results when the ion channels are opened. With it, a *generator* potential and a *progressive potential* are triggered in the postsynapse (11; see Chap. 6.2).

- While these presynaptic membrane events are in progress, the *transmitter is being broken down* in the synaptic cleft by inactivation or by hydrolysis; usable components are re-absorbed through the presynapse by means of *endocytosis* (12) and cleft products, which are no longer usable, pass by means of diffusion to adjacent glial cells. It is at this point that a synapse is free to perform its next transmission event.

However, such a *transmission cycle* must not be viewed necessarily in every instance as rigidly as it has been presented here. Indeed, the cycle is extraordinarily flexible and adaptive with regard to potential chemical or physical changes in the membrane environment. Within this context, *gangliosides* function as *neuromodulators* in vertebrates and, in conjunction with Ca^{2+}, ensure the optimal combination of physical membrane properties (viscosity, permeability, functioning of the ion channels) for each transmitter release (see Chap. 8.2.4).

Beyond the previously stated principles of information transmission by chemical means in synapses, there also exist considerable distinctions in biochemical detail that are relevant to the specific chemical substance utilized as a transmitter. These differences are of extreme importance in the field of pharmacology in as much as points of influence for exogenous agonistic and antagonistc compounds have been discovered, especially in the field of receptor chemistry of the postsynaptic membrane. The

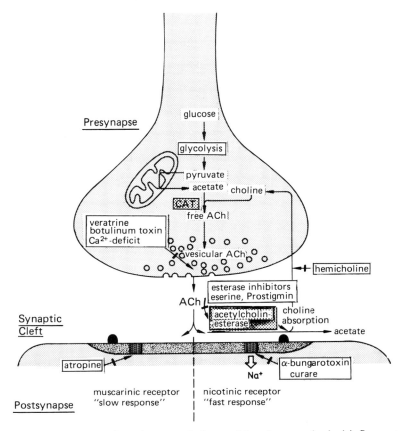

Fig. 7.8. Diagram of a cholinergic transmission and its pharmacological influenceability.

following section deals with the most important transmitters, their metabolism, and their pharmocology.

Acetylcholine (ACh). The first neurotransmitter, and the one best understood to date, is ACh. Its properties have been described in classic experiments at the motor end plate and more recently in research of the electro-organ in electric fish. Interest in ACh as a transmitter has lead to intensive studies of its biosynthesis. ACh is synthesized from *choline* and *acetyl coenzyme A* (acetyl-CoA) in a reaction that is catalyzed by the enzyme *choline acetylase* (also *choline acetate transferase;* CAT) (Fig. 7.8).

$$CH_3\text{-}C\overset{\displaystyle O}{\underset{\displaystyle SCoA}{<}} + HO\text{-}CH_2\text{-}CH_2\text{-}N^+(CH_3)_3 \rightarrow$$

$$CH_3\text{-}\overset{O}{\overset{\|}{C}}\text{-}O\text{-}CH_2\text{-}CH_2\text{-}N^+(CH_3)_3 + HS\text{-}CoA$$

$$acetyl\text{-}CoA + choline \xrightarrow{\text{choline acetylase}} acetylcholine.$$

The primary source of acetyl-CoA for the synthesis of ACh is the oxidative metabolism of glucose via pyruvic acid within the neuron.

Choline, on the other hand, is present from having been reabsorbed from the synaptic cleft following hydrolysis of ACh in the extracellular fluid, from the blood supply, and from phospholipids of brain tissue.

Choline acetylase, an enzyme with a molecular weight of about 65,000, is found primarily in the soluble fraction of nerve terminals and not in the vesicles. The molecular mechanism of acetylation is not yet fully understood. Accordingly, very little is known about the factors that enable regulation of enzyme activity in vivo. Nonetheless, it is assumed that feedback inhibition of ACh synthesis occurs at a certain concentration of ACh, whereas a decrease in concentration leads to stimulation of its synthesis.

The assumption is that there are two different ACh pools present in the nerve terminals:

- Unstable ACh (cytoplasmic pool) and
- Stable ACh (vesicular pool, ACh bound to vesicles).

It has been shown in marking experiments that newly synthesized ACh first appears in the cytoplasmic pool, later to appear in the vesicular pool. These findings agree with the cytoplasmic localization of choline acetylase from which it may be concluded that ACh is synthesized in the cytoplasm and then is stored in part in the vesicles. The mechanism by which ACh reaches the vesicles, and the nature of the dynamic equilibrium between the cytoplasmic and vesicular ACh, are questions that remain unanswered. The *vesicles*, each of which contain about 2,000 molecules of ACh, are clearly transmitter reservoirs in the nerve terminals that also may play a major role in transmitter release. Upon its release, ACh diffuses throughout the synaptic cleft and binds with specific receptors of the postsynaptic membranes (see Fig. 8.8). To date, two types of choline receptors are known: *muscarinergic receptors* at which a transmission either can be triggered by *muscarine* or inhibited by *atropine*, and *nicotinergic receptors* at which the transmission either can be triggered by *nicotine* or blocked by *curare* or *bungarotoxin*. ACh stimulates ganglion cells of the sympathetic nervous system very quickly via the nicotinergic receptors, whereas its effect is less rapid, but of longer duration, via the muscarinergic receptors. With the exception of the acetylcholine receptor (see Sect. 7.1.3.1.2), very little is known about the chemical nature of postsynaptic receptors. ACh is rendered inactive in the synaptic cleft by the *acetylcholinesterase* (AChE) of the postsynaptic membrane. AChE catalyzes the hydrolytic degradation of ACh. It is one of nature's most active enzymes. Hydrolysis can occur within microseconds.

$$CH_3\text{-}\overset{\displaystyle O}{\overset{\|}{C}}\text{-O-CH}_2\text{-CH}_2\text{-N}^+(CH_3)_3 \xrightarrow{\ \text{AChE}\ }$$

$$CH_3COO^- + HO\text{-}CH_2\text{-}CH_2\text{-}N^+(CH_3)_3$$

$$\text{acetylcholine} \xrightarrow[\text{esterase}]{\text{acetylcholine-}} \text{acetate} + \text{choline.}$$

The enzyme AChE and the molecular mechanism of enzymatic hydrolysis have been examined extensively. The active center of AChE has two important regions that are located about 5 Å from one another: an anionic region which attracts the positive charge of the quarternary N-atom in the choline portion, and an ester region which binds the carbonyl C-atom of the acetate portion and separates it from the choline. This leads to the temporary formation of an acetyl-enzyme compound that, ultimately, is hydrolyzed. Nothing specific is known about the location of the acetate residue of ACh; there is, however, a reabsorption mechanism for choline. At least some of the choline returns to the presynaptic nerve terminal where it is used to synthesize new ACh.

Meanwhile, a wealth of pharmacological findings is available that addresses a specific inhibition of the metabolism (ACh) and its effects (see Fig. 7.8):

- One group of styrylpyridine analogues specifically inhibits the activity of choline acetylase and, thus, prevents the synthesis of ACh.
- The reabsorption of choline from the synaptic cleft is inhibited by hemicholinium.
- The release of ACh is blocked completely by *botulinum toxin* (a lethal protein that is released by various strains of anaerobic organisms).
- The receptors can be blocked, or their interaction with ACh can be diminished, by
  (a) *curare* and *nicotine* at the nicotine receptors and
  (b) *atropine* and *muscarine* at the muscarinergic receptors.
- Inhibiting the esterase activity stimulates ACh transmitter activity. A potentiation of the ACh effect is caused by
  (a) *eserine* (*physostigmine*) and *prostigmine*, two inhibitory esterases, and
  (b) *alkylfluorphosphates*, by which the phosphate residue reacts with the ester region and blocks it.

The *occurence of ACh* is shown in Fig. 7.6. It accumulates quite noticeably in the prosencephalon, particularly in the projection pathways

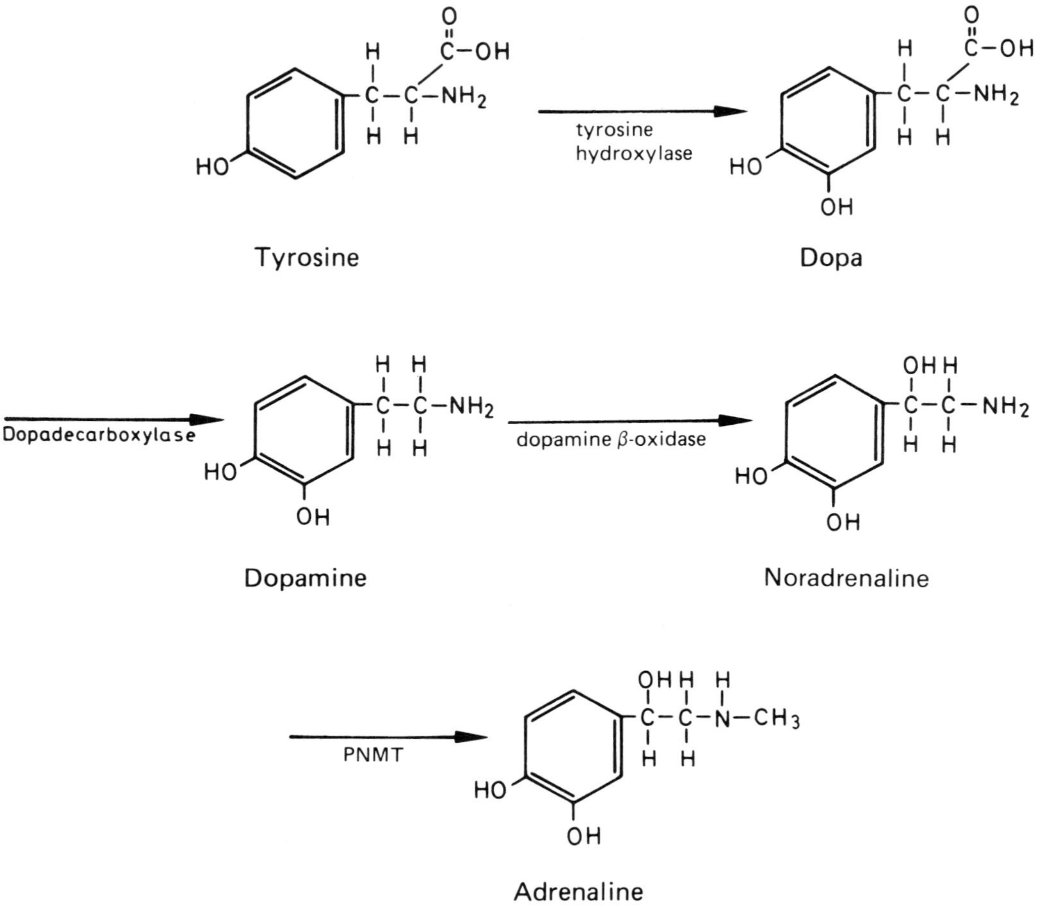

Fig. 7.9. Biosynthesis of the catecholamines.

between the *septum* and the *hippocampus* and between the *corpora habenulae* and the nucleus entopeduncularis in the hypothalamus. Especially in the hippocampus, ACh brings about both slow excitatory as well as inhibitory reactions in the pyramidal cells.

Catecholamines. The catecholamines *dopamine, noradrenaline,* and *adrenaline* (the transmitters of the sympathetic nervous system) occur in particularly high concentrations in brain stem areas and in the hypothalamus. These *biogenic amines* are formed from the amino acid tyrosine through reactions of decarboxylation and hydroxylation. The enzymes that are required for the process are synthesized in the perikarya of the sympathetic neurons. Three enzymes are

needed in order to produce noradrenaline from tyrosine (Fig. 7.9):

1. *Tyrosinhydroxylase* transforms tyrosine into *dopa* and is found in cell bodies as well as in axons and nerve terminals. Necessary cofactors are tetrahydropteridine, O_2, and Fe^{2+}. Tyrosinhydroxylase can be inhibited by noradrenaline. This constitutes a significant control mechanism in the synthesis of noradrenaline.
2. *Dopa-decarboxylase* transforms dopa into *dopamine*, catalyses the dissociation of CO_2, and is a relatively nonspecific decarboxylase. It also serves as co factor for pyridoxal phosphate and, as the so-called Schiff's base, is firmly bound to the apoenzyme.

3. *Dopamine-β-oxidase (DBH)* ultimately transforms dopamine into noradrenaline through oxidation of the side chains. This involves a relatively nonspecific calcium enzyme that requires for its synthesis both oxygen and ascorbic acid. Dopamine-β-oxidase is the only enzyme that is found within the vesicles and, thus, which is discharged into the synaptic cleft upon release of the transmitter.

4. *Phenylethanol-N-methyltransferase* (PNMT) forms *adrenaline* from noradrenaline by methylation, whereby S-adenosylmethionine functions as the methyl donator.

The catecholamines are broken down, in part, neuronally by deamination and subsequent oxidation that is catalyzed by *monamine oxidase (MAO)*. Extraneuronally, the breakdown of the

catecholamines is brought about by O-methylation, the reaction being catalyzed by *catechol-O-methyltransferase (COMT)*. Even as these two reactions play a vital role in the deactivation of the catecholamines, they also appear to be essential to the process whereby the products released are returned to the neuron (Fig. 7.10).

Various mechanisms of the sympathetic neurons control the formation and storage of the catecholamines. Regulation occurs, first of all, by rapid changes in tyrosine hydroxylase activity that results from a feedback mechanism in the nerve terminals. Further regulation takes place in the cell body itself inasmuch as the formation rate of enzymes is influenced by changes in nerve activity. This link results from the fact that the biosynthesis of the enzymes takes place in the cell body from where they are transported to the

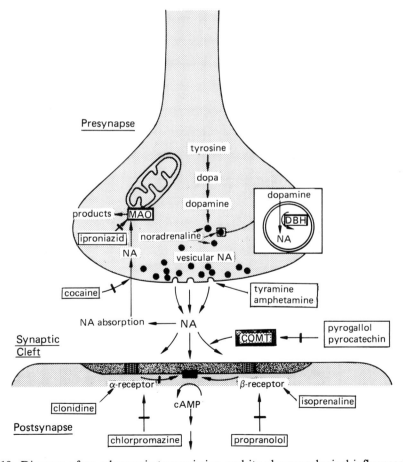

Fig. 7.10. Diagram of an adrenergic transmission and its pharmacological influenceability.

nerve terminals by means of *neuronal transport* (see Chap. 9.1).

To date, the following list represents the essential findings relative to the pharmocology of the adrenergic systems (Fig. 7.10):

- Dopa-decarboxylase activity is inhibited by α-methyldopa;
- *Reserpine, tyramine,* and *amphetamines* lead to the complete release of catecholines from the nerve terminals;
- *Cyclocholine* prevents the release of catecholamine;
- *Cocaine* inhibits reabsorption of the catecholamines through the presynapses;
- *Pyrogallol* and *pyrocatechin* inhibit COMT;
- *Hydrazine* derivatives such as iproniazid inhibit MAO;
- Chlorpromazine binds the α-receptor, and propranolol binds the β-receptor of the postsynaptic membrane;
- *Clonidine* and *isoproterenol* in contrast, have been recognized to be agonists of the α- and β-receptors.

Dopaminergic projection pathways are found in the mesencephalon from the *substantia nigra* to the *striatum.* Parkinson's disease is associated with a degeneration of these pathways. One concludes, then, that the striatum is the causal focus for manifestations of the disease.

Another dopaminergic neuron population is located in the ventral *tegmentum* of the mesencephalon, the region from which pathways radiate into various areas of the prosencephalon and which, in part, belong to the limbic system.

An inadequate dopamine supply to these structures results in emotional deficits and can lead to aggressive behavior. It can also lead to the breakdown of higher associative brain functions. *Schizophrenia* may be a result of an imbalance in the dopamine distribution to these regions.

Noradrenaline is concentrated only in a few neuron populations in the mesencephalon, primarily in the *nucleus coeruleus,* an extraordinary cell region in the brain. Projection pathways emanate from only a few hundred cells to virtually all regions of the brain. It appears that noradrenaline is secreted con-

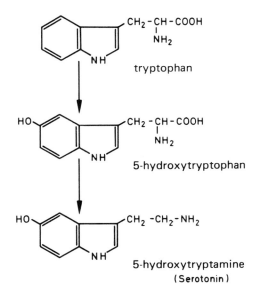

Fig. 7.11. Biosynthesis of 5-hydroxytryptamine (serotonin).

tinuously from the nerve terminals, a fact that suggests a neurohumoral control function.

5-Hydroxytryptamine (Serotonin, 5-HT). The biogenic amine 5-hydroxytryptamine (*serotonin*) is synthesized in a manner similar to that of the catecholamines, i.e., by hydroxylation and decarboxylation. In this case, synthesis is from tryptophan (Fig. 7.11). In like manner to the synthesis of tyrosine, oxygen, iron, and pteridine are essential cofactors. Serotonin is found primarily in the mesencephalon, the *raphe nuclei,* and the medulla. Serotonergic fibers project from these centers into the prosencephalon, cerebellum, and spinal cord (Fig. 7.6). As does noradrenaline, 5-HT appears to influence the higher associative functions, sensory perceptions, and emotions.

LSD and antidepressive drugs work on the serotonergic system both pre- and post synaptically. The level of 5-HT in depressive patients is lowered.

Glutaminic Acid, Aspartate. Glutaminic acid (*glutamate*) and its close relative *aspartate* are both definitive representatives of the family of amino acid neurotransmitters. Glutamate was first discovered to be capable of functioning as a transmitter in the neuromuscular synapses in crayfish. It is relatively difficult to localize in the brain of mammals, but it has been shown to occur in the granular cells of the cerebellum, those cells estimated to number in their vast abundance between 10 and 100 billion in the human. Other glutaminergic conduction pathways have been discovered in the prosencephalon between the olfactory bulb (bulbus olfactorius) and the olfactory cortex and in the fascia dentata of the hippocampus (Fig. 7.6).

Glutamate synthesis is facilitated by glutaminase from the *glutamine* which is supplied to the synapse by neighboring glial cells (Fig. 7.12). Prior to its storage in a vesicle, glutamine is still capable of being enhanced by glutamate that is derived from the mitochondrial Krebs cycle activated by α-ketoglutarate transaminase. Upon its release, the glutamate absorbed by the glial cells is broken down, a process facilitated by glutamine synthesis, to the diaminic acid

glutamine. The last, subsequently, is capable of accommodating a new transmission process.

γ-Aminobutyric Acid (GABA). Unlike glutaminic acid, which is found as an intermediary product of metabolism within the overall organism, GABA occurs only in the brain. It was discovered as an inhibitory transmitter in the extensor musculature in crayfish, whereas it was discovered in the vertebrate brain with the help of immunocytochemical tests that were intended to identify the enzyme of synthesis, glutaminic acid decarboxylase (GAD). In a process similar to that of glutaminic acid, GABA is absorbed by glial cells upon its vesicular release into the synaptic cleft where, facilitated by mitochondrial metabolic enzymes, it is transformed into *glutamate* and then into *glutamine*, the last able to accomodate the presynapse in processing a new impulse transmission (Fig. 7.13). Unlike the

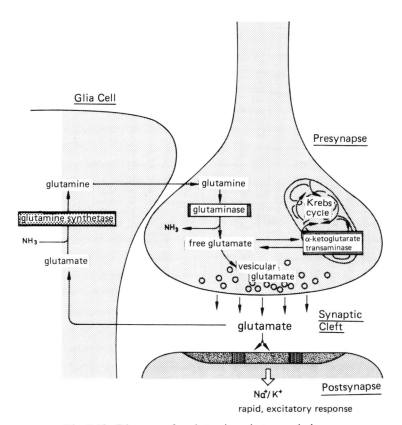

Fig. 7.12. Diagram of a glutaminergic transmission.

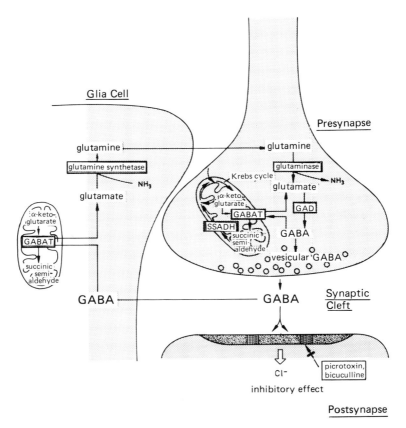

Fig. 7.13. Diagram of a GABAergic transmission and its pharmacological influenceability.

excitatory transmitters that cause a depolarization of the postsynapse by virtue of an exchange
of Na^+ and K^+ ions, GABA has an inhibitory
effect in that it induces a flow of chloride anions
into the postsynapse which, subsequently,
becomes hyperpolarized.

GABAergic neurons are located throughout
the brain, and occur in high concentrations
($\mu M/g$) in the cortex, olfactory bulb, hippocampus, cerebellum, and the retina. The
inhibitory effects primarily serve to control
sensory organ inputs in the sense of negative
feedback. Drugs such as *picrotoxin* and *bicuculline*
block GABA receptors at the postsynaptic
membrane and, thus, cause spasms. This leads
to the supposition that *epilepsy* might be linked
to disregulation of the metabolism of GABA in
interneurons of the cortex.

7.1.3.1.2 Neuroreceptors

Upon their presynaptic release and diffusion
throughout the synaptic cleft, the transmitter
substances are recognized by, and become bound
to specialized protein molecules at the postsynaptic membrane known as *neuroreceptors*.
The individual neuroreceptor molecules actually
represent the specific, molecular points of action
for the neurotransmitters. It is the task of the
neuroreceptor, subsequent to binding with the
transmitter in the postsynaptic membrane, to
trigger changes in potential (depolarization,
hyperpolarization) such that molecular channels
are altered to accomodate the passage of ions.
There is more than one type of receptor for each
transmitter. These receptors are not only
essential to the transmission of information from

neuron to neuron, they are also vitally important as the locations at which substances foreign to the body (pharmacological substances, drugs, and poisons) impart their effect. Where disorders in the mechanism of the neuroreceptors are evident, grave conditions emerge (*Parkinson's syndrome, myasthenia,* even *schizophrenia*). Moreover, discussion of these postsynaptic receptors also must include their connection to plastic neuronal changes, especially in regard to the phenomena of learning and memory (see Chap. 11).

Binding tests help to *identify neuroreceptors,* whereby not the actual transmitters but rather their antagonists are utilized. *Antagonists* are substances that are alien to the body and bind extremely strongly with the receptor when introduced in place of the transmitter (agonist). Several pharmacologically active antagonists for various types of transmitters are rendered in Fig. 7.8 and 7.10 through 7.13. Two different receptors have been discovered for cholinergic synapses: the so-called *muscarinergic receptor,*

for which *muscarine* is the agonist and *atropine* is the antagonist, and the *nicotinic, or nicotinergic receptor,* which is named after its agonist, *nicotine*. Antagonists to the nicotine receptor are the arrow poison *curare* and snake venom α-bungarotoxin (Fig. 7.8). One can depict the thickness of the binding areas between the toxin and the receptor on the postsynaptic side of the membrane by introducing α-bungarotoxin that has been labeled with iodine 125. Biochemical methods of separation then allow for the chemical characterization of the receptor.

Two *types of neuroreceptors* are known to date:

- *Receptor type* I allows for the ion permeability of neuronal membranes and, thus, regulates their capacity to depolarize as a result of its direct interaction with ion channels (Fig. 7.14). The *nicotinic acetylcholine receptor* of the neuromuscular end plate and of the electric organ in electric fish belong to this category. The type I receptor consists of two functional components: a "sluice" that can be opened by

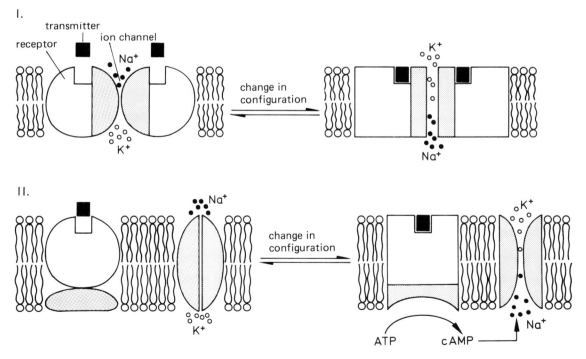

Fig. 7.14. Diagrammatic representation of receptor types I and II and their interaction with a neurotransmitter.

the incoming transmitter, and a "selection apparatus" that determines which ions may be transported through the membrane and which may not.

- *Receptor type II* does not directly influence the ion channel, but rather does directly affect membrane-bound enzymes such as an *adenyl cyclase* whose product, cyclic adenosine monophosphate (cAMP), causes the ion channel to open. It brings this about through intermediary stages that are, as yet, unknown but might involve a membrane-bound, cAMP-dependent kinase (Fig. 7.14). One known example of a type II receptor is the *β-adrenergic receptor*, one function of which is regulation of the heart rate. This receptor also has a regulating functional component that, in this particular instance, recognizes and binds with adrenaline, but, nonetheless, is not linked directly to the ion channel. Rather, this functional component is linked to adenyl cyclase which catalyzes the formation of the cell regulator cAMP from ATP. Finally, *cAMP* regulates membrane potential in a way that is still not understood. Nonetheless, in exerting this control, it also controls the triggering of the action potentials in the postsynaptic membrane.

Recent neurobiological research has revealed the great significance of these types of receptors, especially for neuropharmacology. An *opiate receptor*, for example, has been discovered to be linked to an enzyme and to be capable of functioning in two different circumstances: as an agonist with a great affinity for *naloxone* and *nalorphine*, and as an agonist for *morphine* and *heroin*. Opiate receptor research has produced a highly plausible explanation for the phenomenon of addiction (see Sect. 7.1.3.2). Of course, opiates do not occur naturally in the body although there are endogenous substances in the CNS that, as neurotransmitters and/or neuromodulators, regulate pain sensation and conduction. The neuropeptides (see Sect. 7.1.3.2) *enkephalin* and *endorphin* bind aggressively to the opiate receptor and act as powerful inhibitors. However, as do the exogenous opiates, they simultaneously give rise to manifestations of tolerance and addiction.

"*Diazepine receptors*" have been the focus of attention recently, as well, since many modern tranquilizers (*Valium*, *Tranxilium*, etc.) bind to these with great affinity. Drugs of this type are considered inhibitory in that they enhance the effect of GABA (see Sect. 7.1.3.1.1).

7.1.3.2 Neuropeptides

One of the most important developments in neurobiology in recent years was the recognition that *neuroactive peptides* are widespread throughout the nervous system. Of course, these types of compounds have been known for years relative to the peripheral organs: vasopressin and prostaglandin, for example, identify their primary areas of effect through their names. The names of the various releasing or inhibiting hormones, such as that of the luteinizing hormone releasing hormone (LH-RH) in the hypothalamus, very clearly identify the specific functions that these compounds perform. It is highly improbable that a decade ago one would have recognized the essential role that neuropeptides play in the nervous system as neuromodulators (see Chap. 8.1) of highly specific control functions. Their immense significance to the field of neuropharmacology is still not fully appreciated.

Some ways of distinguishing between neuropeptides and neurotransmitters, and identifying them, have been discussed in Sect. 7.1.3.1.

The history of discovery of the neuropeptides is closely related to the emergence of hormone physiology. *Neuroendocrinology* evolved as an independent field of research from the area of general endocrinology when cells with secretory properties, the *neurosecretory cells*, were discovered (E. and B. Scharrer, 1928) in the CNS of vertebrates and invertebrates (worms, snails, crayfish, insects). Generally speaking, we know today that neuroendocrine cells, which secrete neuropeptides, must have existed among the oldest forms of neurons because, phylogenetically, they were present even in the oldest of nervous systems. Since many neuropeptides occur in peripheral organs as well, one concludes that they derive embryologically from neuroectodermal origination cells.

By definition, a neurosecretory cell is a neuron that releases considerable quantities of an active substance into the bloodstream or into the intercellular space of the nerve tissue in order to exert an effect of considerable duration upon the function of other cells that are not in direct contact with that neuron. Thus, *neurosecretions* differ clearly from the *neurotransmitters* (see Sect. 7.1.3.1.1) which are released in the most minute quantities and have the short-term effect of so-called *diffusion activators* only in the immediate vicinity of their place of release.

Despite their particular histochemical signature (evinced by Gomori dye, immunofluorescence), neurosecretory cells are quite similar to normal neurons. Accordingly, their action potential is readily measurable. However, neurosecretory cells exhibit granulated axon swellings caused by their secretions. These bodies are referred to as "*Herring bodies*" and often give the fibers of this variety the appeareance of a string of pearls. On the ultra structural level, the neurosecretory neurons have uniquely characteristic nerve terminals that are designated "synaptoid configurations" from which the neuropeptides are released. It is likely that the neuropeptides are synthesized in the rough endoplasmic reticulum in the cell body of the neurons, become wrapped in the Golgi apparatus, and then are transported along the axon to the nerve terminals as membrane-encased granules. Electron microscopy reveals that the large neuropeptide granules become fragmented in the nerve terminals and that small *synaptoid vesicles* result from this fragmentation process (Fig. 7.15). These granules do not contain the actual neuropeptides (neurohormones) in free form; rather, they are bound to a carrier protein, the polypeptide *neurophysin*. The latter is not biologically active and is easily separated from

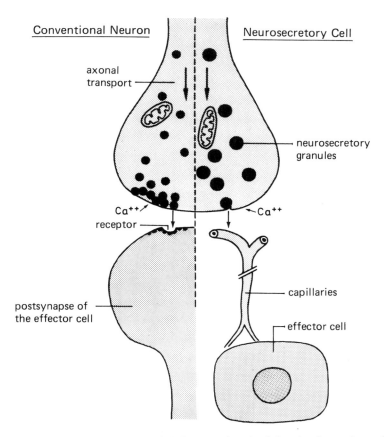

Fig. 7.15. Diagrammatic overview of the special features involved in the formation of a conventional neurotransmitter synapse and a neuropeptidergic nerve fiber terminal.

the active component, for example, using trichloroacetic acid.

Depolarization of the plasma membrane and the presence of calcium are two of the many conditions that are necessary for the neuropeptides to be released. Upon increasing the permeability of the membrane by transmitter induction, Ca^{2+} ions flow into the nerve terminal of the neuropeptidergic cell at a rate consistent with an electrochemical gradient. Ca^{2+} reduces the negative charges of the secretory granules intracellularly, a phenomenon that leads to the granules becoming attached to the inside of the membrane. At the same time, Ca^{2+} activates a phospholipase that opens the membrane and, thus, brings about an increased extrusion of the granular content.

The close relationship of all peptidergic neurons is substantiated by the finding that the neuropeptides that have been released in the course of phylogeny have been preserved resolutely such that substances of great similarity, or similar amino acid sequences, occur in the most diverse species of vertebrates and invertebrates.

Some of the most widely known neuropeptides are presented in Table 7.3 arranged according to the number of their carbon atoms. Many of these peptides contain two to ten carbons and, thus, correspond in size to the neurotransmitters. Others conform in size more closely with other hormones of the body. Several of the substances represented in Table 7.3 are examined more closely in the following sections.

Endorphin. Of all the neuropeptides, the enkephalins and, especially, the endorphins have been the focus of widespread attention in recent years. In 1975, Klosterlitz and Hughes were the first to extract endorphins from the bovine brain, whereby they ascertained that opiate antagonists (naloxone) inhibited their effect. Cells containing endorphins are found only in the hypothalamus from where projection pathways extend to the septum, periventricular regions, the locus coeruleus of the brain stem, and the raphe nuclei (Fig. 7.16).

Regarding their manner of effect, the endorphins seem to behave as *opiates* in that they inhibit the perception of pain (see also enkephalins and substance P). It is due to this similarity that research of these compounds has been greatly intensified in the hope of discovering morphine analogues that do not induce deleterious side effects such as addiction but do have the affect of alleviating pain.

Enkephalin. In their peptide composition, which consists of 31 amino acids, the endorphins contain one pentapeptide segment that corresponds to that of the enkephalins. It is for this reason that the term endorphin is applied to them, as well. Cells that contain enkephalin resemble GABAergic interneurons which function as modulators for local impulse circuits. Thus, in conjunction with substance P (see below), they are involved with the modulation of information in the region of the spinal cord that is associated with pain fibers leading to the

TABLE 7.3. Neuropeptides arranged by the number of their carbon atoms.

Neuropeptide	C-Atoms	Neuropeptide	C-Atoms
Carnosine	2	Luteinizing hormone releasing hormone	
Thyrotropin releasing hormone (TRH)	3	(LH-RH)	10
Met-enkephalin	5	Substance P	11
Leu-enkephalin	5	Neurotensin	13
Angiotensin II	8	Bombesin	14
Cholecystokinin-like peptide	8	Somatostatin	14
Oxytocin	9	Vasoactive intestinal polypeptide (VIP)	28
Vasopressin	9	β-Endorphin	31
Corticotropin releasing hormone (CRH)	10	Adrenocorticotropic hormone (ACTH)	39
Follicle-stimulating hormone releasing			
hormone (FSH-RH)	10		

CNS. Otherwise, the affect of the enkephalins is very similar to that of the endorphins (see Fig. 7.17). Neurons that contain enkephalin are widely distributed throughout the brain (Fig. 7.16). Clearly, they are not limited to the diencephalon as are the endorphin cells.

Substance P. Substance P was discovered in 1931 by von Euler in the mammalian brain and pharyngeal area. The first neuropeptide to be discovered, it has a stimulating effect upon smooth musculature. Since its discovery, substance P has been identified in various centers of the CNS with short projection pathways especially in the striatum, the raphe nuclei, the medulla oblongata, and in the dorsal roots of the spinal cord (see Fig. 7.16). In the cells of the last, substance P, administered externally (ionophoresis), is 200 times more powerful than glutaminic acid. Nonetheless, it is slower to take effect and, thus, is longer lasting in its effect. In this regard, it is clearly different from a neurotransmitter. In conjunction with the endorphins, especially with enkephalin, it is hypothesized that neuropeptide substance P would regulate the transmission of *pain sensation* originating in the periphery of the body and passing into the brain via the spinal cord (Fig. 7.17). According to this hypothesis, the cell that senses the pain would transmit its impulse by releasing substance P to a spinal cord cell that controls corresponding substance P receptors. Cells containing enkephalin would prevent the transmission of the pain signal by suppressing the release of substance P. This process would represent a type of presynaptic inhibition (see Chap. 6.3).

From the standpoint of neuropharmacology, administered opiates such as the pain killer morphine could combine with the endorphin (enkephalin) receptors such that the release of substance P would be prevented and further transmission of the pain signal would be blocked. Beyond this pharmacological aspect, it is possible that methods for the treatment of chronic pain (*hyponosis, electrotherapy* of the brain, *acupuncture*) might owe their effect to the fact to that they induce the release of enkephalins or endorphins in the brain or spinal cord and thereby prevent the release of substance P. One draws this conclusion from the fact that the methods cited above for the alleviation of pain are rendered virtually ineffective if the patient is being treated with *naloxone*, a drug that blocks morphine from binding to its opiate receptor.

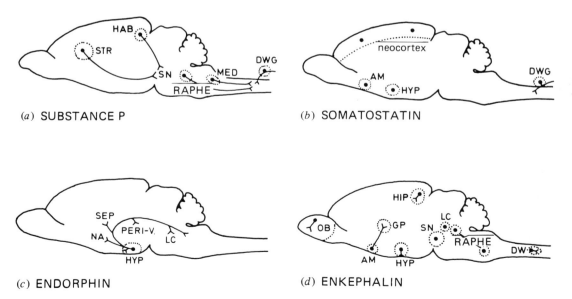

(a) SUBSTANCE P

(b) SOMATOSTATIN

(c) ENDORPHIN

(d) ENKEPHALIN

Fig. 7.16. Diagrammatic overview of the localization of neuron groups in the mammalian brain that contain neuropeptides (see Fig. 7.6 for brain region abbreviations).

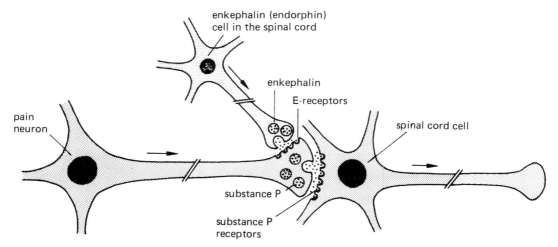

Fig. 7.17. Diagram of the interaction of the neuropeptide substance P with enkephalin/endorphin in the neuronal conduction of pain sensations.

Somatostatin. This neuropeptide has an inhibitory effect on the secretion of the growth hormone *somatotropin* of the adenohypophysis. In the periphery of the body, it occurs in the autonomic nerve fibers of the intestines. In the brain, it is found in the hypothalamus, the cortex cerebri, the corpora amygdaloidea, and in the dorsal root ganglion of the spinal cord (Fig. 7.16). When administered exogenously, it inhibits motor activity in the organism.

Hypothalamo-Hypophyseal Peptides. In addition to the previously listed neuropeptides that are formed widely throughout their corresponding peptidergic neuron regions of the brain, other neurosecretions increasingly have been found to be of significance. These are formed in the hypothalamus and then dispatched to the hypophysis in a number of ways. The *hypothalamus* is recognized to be the superordinate vegetative control system whose brain centers regulate not only the nervous activity of the organism's vegetative functions, but the humoral and neurosecretory mechanisms, as well. Coordination of the hypothalamic control mechanisms is mediated by the *hypophysis* (see Fig. 3.16), a detailed discussion of which can be found in Chap. 3.2.2.4.1. Linked to the polypeptide *neurophysin*, the neuropeptides that are formed in various areas of the hypothalamus can be sluiced into the neurohypophysis by axonal

transport as neurosecretions (*oxytocin* and *adiuretin: vasopressin*); or, various polypeptides known as *releasing factors* and *inhibiting factors* can be transported via short-circuited, brief portal circulation from the hypothalamus to the adenohypophysis where they regulate the release of numerous hormones that control the vegetative body functions. (See Chap. 3.2.2.4.1, Fig. 3.16, and Table 7.3 for details.) The selection mechanism of these hypothalamic neuropeptide hormones might well be regulated by humoral feedback, i.e., by the degree of concentration of the given peptides in the blood; or it might be regulated by neuronal steering mechanisms, whereby various neurotransmitters (noradrenaline, dopamine, or even serotonin) are integrally involved.

All neuropeptide factors (hormones) that are produced in the various regions of the hypothalamus and transpoted in the blood to the adenohypophysis take effect within only a few minutes, i.e., before they become inactive in the blood. They trigger a rapid release of hormones in the cells of the hypophysis and, simultaneously, induce an extended time period in which these hormones are synthesized anew. As is the case with other peripheral peptide hormones, the effect upon the cells of the hypophysis is brought about by the influence of membrane receptors in the presence of cyclic AMP.

These kinds of neuroendocrine systems are not limited to vertebrates. *Invertebrates*, especially insects, crayfish, and cuttlefish, evidence similar integrated neurohormonal systems, as well. In *insects*, for example, pupation as well as the reproductive processes are controlled by central *neuroendocrine cells* in conjunction with a kind of glandular appendage of the brain, the *corpora allata*. In *crayfish*, the process of molting is regulated by the interaction of neurosecretory cells in the so-called *X-organ* of the eye stalk and a *sinus gland*, a neurohemal organ. This eye stalk system controls a so-called *Y-organ* which is located in the antenna or maxilla segment of the body and has a mutual effect on molting. In *cuttlefish*, sex differentiation is a factor of incretory control: light stimulates the neuro-secretory cells of the so-called *subpeduncular lobes* the secretions of which normally inhibit an eye gland located at the eye stalk. Thus, the cells of the eye gland are active in darkness and secrete hormones that activate sex differentiation.

7.2 Calcium and Neuronal Functions

Section 7.1 treated the molecular basis of synaptic information transmission and the attendant complex of changes manifested in the membranes involved. Attention now should be directed toward the role of calcium, an essential component in all neuronal processes. Table 7.4 reveals in its overview that the functional role of calcium is essential to all short-term events, e.g., impulse conduction, impulse transmission, transport, enzyme activity, etc., and longer term adaptive processes such as learning, memory, and ecophysiological adaptation.

The differences in concentration in the individual mono- and divalent cations and anions in the extracellular fluid (blood, spinal fluid, tissue fluid) vis-à-vis the cytoplasm (Fig. 7.18) are sometimes extreme and give rise to the differences in potential between the inside and outside of the cytoplasmic membrane.

The concentration of free calcium ions in the extracellular fluid of the nerve tissue in animals amounts to a few mM/l. Considerable differences are evident, however, in the course of phylogeny: the extracellular calcium concentration in the nerve tissue in marine invertebrates (the cuttlefish, for example) amounts to 12mM and corresponds to that of sea water (isomolar). The concentration in cartilaginous fish is 5 mM, whereas it is 3 mM in bony fish and only 1 to 2 mM in mammals.

TABLE 7.4. Functional significance of calcium in the nerve cell.

Controls *excitation* of neuronal membranes through its effect upon the K^+/Na^+-pumps

Controls *conduction* along the nerve membrane by regulating the metabolic pumps and transmembranic ion exchange

Regulates the *conductivity* of electrical synapses

Releases transmitter by activating adenylate cyclase

Controls the reaction of the transmitter with the postsynaptic membrane, thus controlling *transmission* of the impulse

Regulates the *release of neurosecretions* and *neuropeptides*

Regulates *neuronal transport* through Ca^{2+}-neurotubuli-neurofilament interaction

Regulates nerve fiber *growth*

Activates and regulates intracellular, synaptic *metabolism* (change in pH, enzyme induction, cAMP, cGMP, phosphoinositol activation etc.)

Regulates intracellular *storage of Ca^{2+}* in the ER and in mitochondria

Regulates *macromolecular Ca^{2+} binding*, especially to calmodulin

Regulates *long-term potentiation* (LTP) as possible electrophysiological correlate to learning processes

Controls long-term neuronal processes of adaptation
  a) in ecophysiological *adaption*
  b) in processes of *learning and memory*

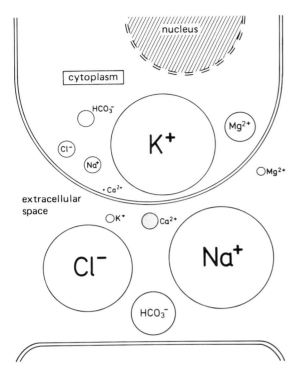

Fig. 7.18. Diagram of the distribution of the most essential physiologically active ions inside and outside the cell membrane in animals.

In contrast to the extracellular concentration of calcium, however, the intracellular concentration is considerably less, only about 1/10,000, i.e., less than 1 $\mu M = 10^{-7} M$. A continuing increase in the intracellular calcium content has a cytotoxic effect in that it induces the activation of proteases that break down the cytoskeleton. On the other hand, a rise of brief duration in intracellular calcium is necessary for many of the neuronal events listed in Table 7.4 relative to a secondary messenger.

Two processes allow for an increase in the intracellular concentration of free calcium ions: the inflow of calcium from the extracellular space and the release of calcium from intracellular reservoirs.

Inflow of calcium from extracellular space to intracellular space: In principle, cell membranes are not calcium permeable unless specific ion channels are present and open, in which case calcium ions flow in the direction of their electrochemical potential gradient into the cell.

Various types of calcium channels have been described. These differ according to their electrical properties (duration of opening, opening probability) and the extent to which they are subject to the action of ligands, i.e., substances that bind to them.

Calcium channels are controlled by various mechanisms from both outside and within the neuronal membrane. Calcium channels can be activated from the extracellular space by depolarization. Some channel types in the area of the synapse are voltage dependent, i.e., they open in response to changes in the electrical fields in individual regions of the synaptic membrane. Other calcium channels open in response to specific substances from the extracellular space, for example, ATP or certain transmitter substances.

One particular type of calcium channel has been studied extensively. It is linked to a specific glutamate receptor (NMDA receptor, see Sect. 7.1.3.2) and causes long-term potentiation postsynaptically. The channel opens for Ca^{2+} only if glutamate is bound to the postsynaptic receptor as the membrane becomes depolarized.

A change in the calcium channels is also possible from the side of the cytoplasmic membrane, for example, by processes of phosphorylation that are caused by protein kinases or cAMP (cyclic adenosine monophosphate). Since the activity of many protein kinases is dependent upon the intracellular concentration of calcium, there is clear evidence here of a feedback mechanism (Fig. 7.19). Activation of the calcium channels leads to the inflow of Ca^{2+} ions and, thus, to an increase in the intracellular concentration of Ca^{2+}. This activates protein kinases that, in turn, further activate (positive feedback) or deactivate (negative feedback) the Ca^{2+} channel.

The exact sequence of events in such reactions is dependent upon the concentration of all the substances in the cell that play a part. The sequence can result in different effects depending on the condition of the cell.

Release of calcium from intracellular reservoirs: The endoplasmic reticulum (ER) is the cell compartment from which Ca^{2+} ions can be released very rapidly within the cell. As a channel system, the ER runs through the entire cell

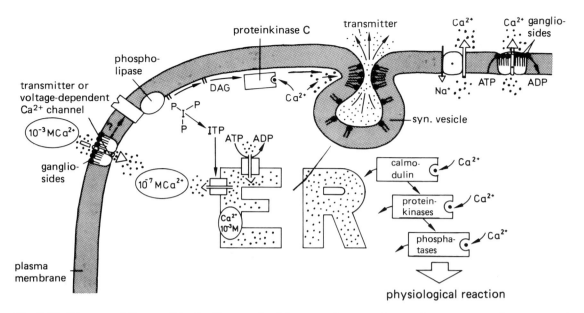

Fig. 7.19. Diagram of the possible significance of calcium as primary and secondary messenger for extra- and intracellular processes at the synapse.

(Fig.7.19). If an external stimulus (transmitter or voltage change) is to cause calcium ions to be released from the ER, then the information from the plasma membrane must reach the ER through the cytoplasm. A second messenger, 1,4,5-inositol triphosphate (ITP), serves this purpose. The cytoplasmic side of the cell membrane contains phosphatidylinositol-4, 5-biphosphate (PIP_2) which is anchored to the membrane by two fatty acid chains. Activation of the cell by various mechanisms (the binding of an agonist to its receptor, for example) causes PIP_2 to split into diacylglycerine (DAG), which remains bound in the membrane, and ITP which moves about freely in the cytoplasm. The ITP diffuses to the ER, binds there to a specific receptor and triggers the release of Ca^{2+} ions. DAG, the second product of the splitting of PIP_2, is relatively short-lived; however, it binds only to one highly specific protein kinase, protein kinase C. This increases its affinity for calcium so greatly that the calcium becomes bound, thereby activating the kinase and allowing the various proteins to phosphorylize. In this way, the binding of an agonist induces the onset of numerous processes that are associated with calcium ions at various junctures.

Because of calcium's cytotoxic effect, the temporarily increased intracellular concentration of Ca^{2+} must be returned to its lower, normal level. A portion of the Ca^{2+} ions can bind for a term of moderate duration to proteins such as parvalbumin, calbindin, and calmodulin which function as calcium buffers since they reduce the concentration of free Ca^{2+} ions. Calcium binding alters their properties, whereby various other mechanisms of regulation can be triggered. For example, the Ca^{2+}-calmodulin complex activates protein kinases and calcium pumps. However, to assure long-lasting cell function, the Ca^{2+} ions must be pumped out of the cytoplasm back to the extracellular space, a process directed against the extremely high concentration gradient of 10^{-7} to $10^{-3} M$ Ca^{2+}. As such, it is quite energy intensive.

The process could involve various mechanisms (Fig. 7.19):

- *Ion exchange*: The most quantitatively significant transport of Ca^{2+} is facilitated by a sodium-calcium exchange system in the cell membrane (see Fig. 8.18). Sodium ions flow in the direction of their concentration gradient from outside to inside. The counterflow transports calcium ions to the outside. Ultimately, energy is derived from the Na^+-K^+ pump which builds up the sodium potential by consuming ATP (see Chap. 6). The extremely low concentration of calcium, however, cannot be achieved solely by this

mechanism. Additional pump systems must be activated.

- *Calcium pumps*: Ca^{2+} ions can be transported by Ca^{2+} ATPases through the membrane channels directly by consuming ATP. Such Ca^{2+} pumps are located both in the plasma

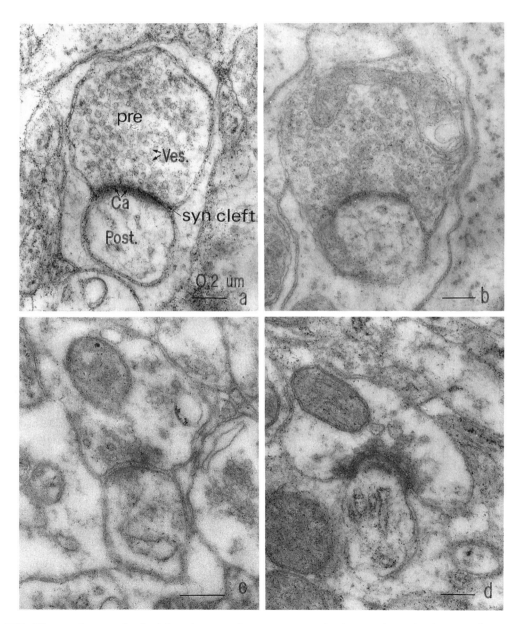

Fig. 7.20. Electromicrograph of calcium in synaptic contact zones in the vertebrate brain: (*a*) active summer period vis-à-vis (*b*) winter dormant period (carp); (*c*) birds (chicks) and (*d*) mammals (mouse). Note the great species-specific and function-dependent differences in calcium content of the synaptic extracellular space.

membrane and in the membrane of the ER. Their activity may possibly determine the concentration of free Ca^{2+} ions in the resting state of the cell.

- *Vesicles and mitochondria as supplemental Ca^{2+} reservoirs to the ER*: In addition to being stored in the ER, calcium ions can be stored in synaptic vesicles and within mitochondria. The importance of mitochondria as Ca^{2+} reservoirs in the nervous system is currently under debate. If the cells are damaged and the intracellular Ca^{2+} concentration rises sharply, then the mitochondria absorb the Ca^{2+} ions that are precipitated within the cell as Ca^{2+} phosphate. As such, great quantities of Ca^{2+} can be stored and, thus, rendered harmless. It is unclear if such processes occur under normal conditions.

In general, it can be said that there is a multitude of mechanisms that can affect the intracellular concentration of calcium and, in turn, control numerous cell functions. Accordingly, calcium plays a central role in regulating the events of the cell.

There are molecular structures in both the intra- and extracellular space of the synapse that bind calcium to a greater or lesser degree.

Until now, strongly *bound calcium* has been very difficult to localize and quantify. Electron microscopy has made it possible in recent times to identify calcium in precipitates of calcium osmiate phosphate, in intracellular reservoirs (ER, synaptic vesicles), in the extracellular space of the nerve tissue and, of course, in the area of the synaptic contact zones where it is especially concentrated (Fig. 7.20). As has been

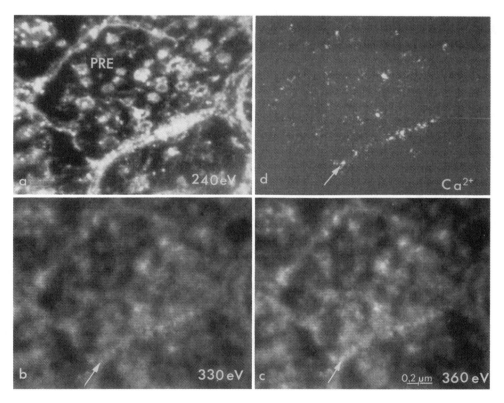

Fig. 7.21. Evidence of calcium in the region of the synaptic contact zone of a nerve cell from the tectum opticum of a carp (enhanced by computer-assisted electron microscopic spectroscopy): (a) dark field-like overall view of the synapse; (b) and (c) computer-enhanced image of the upper half (360 eV) and lower half (330 eV) of the calcium-specific energy absorption range of the overall view (a); (d) computer-enhanced differential image of (b) vis-à-vis (c) which covers a definitive calcium event in the synaptic cleft, in the smooth ER and in the region of the synaptic vesicle.

shown, calcium occurs in considerably greater concentration in the synaptic contact region in lower vertebrates (fish) than in birds or mammals. Functionally dependent differences can be shown in the strongly bound synaptic calcium fraction between the active summer period and dormant winter period in certain species of fish; calcium is packed considerably more densely in the former than in the latter. Recently, computer enhanced electron microscopic spectroscopy has furnished the ultimate proof that the electron-dense precipitants of Fig. 7.20 actually are a calcium issue (Fig. 7.21).

Of special interest, however, are the macromolecular structures of the cell membrane and from within the cell that can weakly bind calcium, relinquish it readily, and reabsorb it. *Free calcium* is immensely important to the most diverse sequences of the neuronal event. In its *extracellular form,* it performs the function of a *primary messenger* in the processing (internalization) of electrically encoded data. In its *intracellular form,* it fulfills the role of a *secondary messenger* in that it accommodates neuronal metabolism by performing a stimulatory function.

The question as to how extracellular calcium is able to perform its function of primary messenger is still unanswered in most of its facets. In pursuit of answers, scientific research is directed toward specific macromolecules that are present in the outer layer of the neuronal membrane, and that can bind calcium weakly (maintaining functional efficacy) through changes in electrical field or through interaction with transmitters. Within this framework, great significance is attributed to the strongly polar *gangliosides* that belong to the glycolipids and contain negatively charged neuraminic (sialic) acids (see Chap. 8.2). In invertebrates, however, which have no neuraminic acid and, thus, no gangliosides, just such a function could be fulfilled by other negtively charged glycolipids, such as those containing *glucuronic* or *phosphonic acid.*

8

Modulation of Neuronal Information Transmission

8.1 General Aspects of Neuromodulation

In light of the myriad details presented thus far relative to the molecular structure of the synaptic membrane, the vesicles, the synaptic transmitters (neurotransmitters, neuropeptides), and their neuroreceptors, one might well conclude that the relatively strict sequencing of events in the transmission process of electrical information signals from one neuron to the next requires that the component processes be *modulated* in order to attain transmission. Of course, a neuronal impulse triggers the release of transmitter substances. Nonetheless, the quantity of the released substance as well as the strength, duration, and modality of the postsynaptic effects (inhibitory or excitatory) must be coordinated precisely.

Extended and repeated stimulation quickly depletes the released quantity of the transmitter since only a relatively small supply of it is available at any given moment. The amount of available transmitter is largely a factor of its neosynthesis within the body of the neuron, its neuronal transport (see Chap. 9.1.2), and the extent of its reabsorption in the synapse itself (see Chap. 9.1.3).

Numerous presynapses (several thousand per cell) from various sending cells terminate at the rather large postsynaptic region of a receiving cell. Consistent with their diverse nature is the likelihood that dissimilar transmitter substances will comingle. Simultaneous stimulation of several synapses requires that the nature of the various transmitter mixtures be recognized adequately with regard to their overall content of information. An additional factor to be considered is *autoinhibition* of the stimulated synapse. This occurs when the quantity of the transmitter released into the synaptic cleft reaches a specific concentration.

From the example in Chap. 7.1.3.2 regarding the neuronal conduction of pain sensation, we have learned that the synaptic terminals control mechanisms that can help to regualte the effect of a released transmitter upon the postsynaptic receptor region by eliciting the emergence of another substance. This indicates that there are, apparently, systems of deactivation and absorption that can affect synaptic transmission by controlling the concentration of the released transmitters and the duration of their effect at the postsynaptic membrane. All substances are significant in the process of *synaptic modulation*, which effects changes in the neuroplasm or in the neuronal membrane in a manner that parallels the interaction of a neurotransmitter and a neuroreceptor, or which effects the release, the binding, or any other action of a neurotransmitter.

Presently, those substances that bring about short-term activity in a synapse are referred to as *neurotransmitters*, whereas those whose effects are long-term are called *neuromodulators*. By these definitions, the *neuropeptides,* which were addressed in Chap. 7.1.3.2, are regarded often as modulator substances whose effect is based upon the activation of cyclic AMP (adenosinemonophosphate) in the neuronal membrane of the receiver cell. Such activation alters postsynaptic cellular metabolism by degree of stimulation.

If one chooses to conceptualize the notion of *synaptic transmission* primarily as frequency-dependent amplitude change of transmitter release, then the idea of *neuromodualtion* ought to encompass each step that affects impulse transmission adaptively in synapses. All synaptic processes, processes such as enzyme activation, transmitter storage and release mechanisms, transmitter interaction with corresponding receptor molecules, ion channel regulation, as well as the kinetics involved in deactivating the released transmitter and in its reabsorption by the presynapse, serve to modulate impulse transmission and, thus, the transmission of information from one neuron to the next.

In all instances where a synaptic mechanism of modulation ensues from adaptive bioelectrical phenomena such as *PSP, PTP*, and especially *LTP* (see Chap. 9.2.3), the neuromodualtor, upon a presynaptic action potential, brings about a change in the number of Ca^{2+}-ions that flow into the nerve terminals. In such a process, it is characteristic of a neuromodualtor that it does not cause the ion channels to open or close in and of itself, but rather that it induces specific ion channels to respond to another stimulus. In this way, they increase or diminish the flow passing through a specific number of channels that become activated as a result of depolarization.

In contrast to the neuromodulator, the classic synaptic transmitter substances behave as chemical stimuli, usually acting to open the membrane channels. A modulator substance, on the other hand, controls the sensitivity of channel proteins for the depolarization of the membrane.

The degree and speed of the neuromodulatory event in synaptic transmission is probably different for each type of synapse. In each instance, however, it is likely that the effects of modulation can be affected by changes in external factors such as physical parameters (pressure, temperature) or chemical parameters (ion milieu, enzyme activities, exogenously administered substances).

8.2 Significance of Gangliosides as Neuromodulators

Chapter 7.1.3.2 addressed the significance of the neuropeptides as neuromodulator compounds. At this juncture, it is timely to consider those mechanisms and processes that ensure a functional synaptic impulse transmission, the transmission event that is adapted to the prevailing background conditions of the total organism. Thus, viewed in the long-term, the storage of information (memory) ultimately is at issue.

Two particular paths ought to be followed when discussing mechanisms of synaptic modulation:

- The search for a *primary messenger system* that ensures the reception of an electrically encoded impulse pattern from the presynaptic outer membrane to the inner membrane such that, from there, secondary messenger systems can continue conducting a stimulus-dependent change in neuronal metabolism,
- Based on numerous electrophysiological and biochemical findings, *calcium* (Ca^{2+}) appears to be of extreme functional importance in neuromodulation (see Sect. 8.2.4).

A controlled Ca^{2+} exchange between the extra- and intracellular space of the synapse is considered to be an essential requirement for the conversion of the impulse. Accordingly, research is directed increasingly toward those membrane components that could be involved to a greater degree with the process of synaptic transmission on the basis of their physical properties relative to calcium.

Against this backdrop, we proffered a functional hypothesis as early as 1976 (Rahmann, 1976, 1983, 1987; Rahmann et al. 1976, 1984) in which we suggested that, particularly in vertebrates, glycosphingolipids that contain sialic acid (N-acetylneuraminic acid), especially *gangliosides* (see Chap. 7.1.1), could function as *neuromodulators* in both the short-term process of synaptic transmission of information and in the long-term phenomenon of memory formation. Our hypothesis was based on the knowledge that brain gangliosides in particular, as opposed to other neuronal membrane substances, exhibit numerous phenomenological (physiological), bioelectrical and, especially physicochemical properties that could explain the extraordinary adaptive capacity of synaptic processes.

8.2.1 Physiological Adaptive Capacity of Brain Gangliosides

Numerous findings have been obtained in the last decade relative to gangliosides in the vertebrate brain that verify their essential involvement in brain function (Table 8.1). The findings most vital to the question of memory (see Chap. 11) are delineated below, especially as they concern the great physiological adaptive capacity of the brain gangliosides.

- In the development of vertebrate *phylogeny*, the concentration of brain gangliosides increases in more highly evolved groups. The ganglioside concentration in cold-blooded, lower vertebrates (fish, amphibians, reptiles) varies between 110 and 800 μg per gram of

brain weight, whereas in warm-blooded birds and mammals the range is from 400 to 1,200 μg. Parallel to the increase in concentration, the complexity in the organization of the brain gangliosides is greatly reduced. In lower vertebrates, there occur, at times, considerably more single gangliosides than in mammals and birds (Fig. 8.1). Associated with this is a substantial shift in the preference for the three modalities of biosynthesis (a, b, or c path) illustrated in Fig. 7.3. The *polarity of the brain ganglioside pattern* is dependent upon the negatively charged neuraminic acids (NeuAc). In the bird and mammalian brain, the polarity is relatively slight due to only two or three neuraminic acids; it is considerably greater in fish and amphibians, however, due to gangliosides with as few as four and as many as seven neuraminic acids. Additionally, the organization of the brain gangliosides differs in the brain from region to region. Since the individual sections of the brain have different functions to perform, and contain neurons

TABLE 8.1. Brain gangliosides as neuromodulators: Overview of the physiological, bioelectrical, and physicochemical properties of brain gangliosides as compared to other membrane lipids.

A. Physiological properties
   1. Change in composition ($\hat{=}$ polarity) in gangliosides dependent upon:
      —phylogeny: level of organization
      —ontogeny: degree of differentiation
      —disturbances in ganglioside metabolism (gangliosidoses)
      —ecophysiological adaptation (environmental adaptation)
      —administering of seizure-inducing agents
      —sensorial stimulation
   2. Stimulation of membrane-bound enzymes (protein kinases) by Ca^{2+}-ganglioside complexes

B. Bioelectrical properties
   —Restoration of electrical excitability subsequent to the exogenous addition of gangliosides
   —Significant effect upon postsynaptic excitability following degradation of neuronal gangliosides, e.g., by neuraminidases
   —Cell-specific triggering of depolarization, desensitization, and change in conductivity subsequent to exogenous application of ganglioside

C. Physicochemical properties of Ca^{2+}-ganglioside interactions
   —Change from hydrophilic to lipophilic properties in the two-phase system (chloroform/water) subsequent to addition of Ca^{2+}
   —Ca^{2+} released from Ca^{2+}-ganglioside complexes by metallic ions (K^+, Na^+, Li^+, Mg^{2+}, Ca^{2+}), neurotransmitters (ACh, serotonin), neurotoxins (curare), temperature changes
   —Discontinuous Ca^{2+}-binding with two binding points
   —Micellar aggregation capacity in aqueous solutions
   —Peculiarities in membrane behavior of monolayers (space requirement of molecules, surface potential) attendant to change in Ca^{2+} concentration, temperature, or addition of protein
   —Voltage-dependent ad- or desorption of Ca^{2+}-ganglioside complexes at mercury electrodes

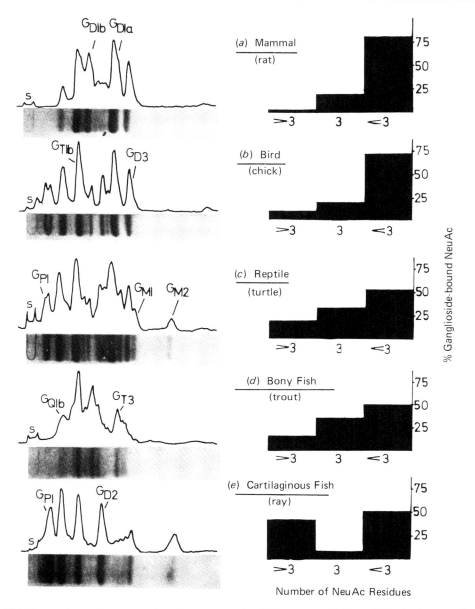

Fig. 8.1. Thin-layer chromatograms, densitograms, and relative composition of the brain gangliosides in various vertebrates organized according to their polarity, which is based on three or fewer, or more than three neuraminic acids per molecule.

with differing transmitters (see Chap. 7.1.3.1.1), altered concentrations of gangliosides indicate functional connections.

- During *early ontogeny*, especially during critical and progressive phases of development such as birth, hatching, opening of the eyes, there occur in all vertebrates conspicuous shifts in the organization of brain gangliosides in addition to marked changes in concentration. In mammals and birds, these shifts follow Haeckel's rule, i.e., in the early stages of development, the ganglioside pattern is similar to that in fish; only later does it take on the characteristics of the higher vertebrates.

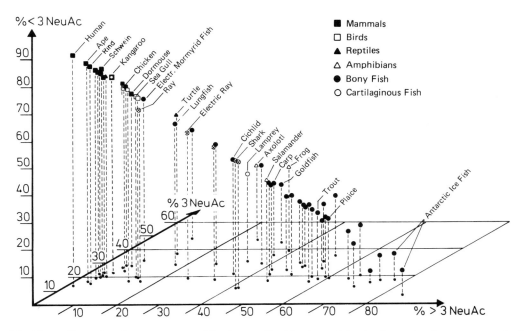

Fig. 8.2. Three-dimensional arrangement of brain gangliosides in vertebrates from six classes, some of which represent extremes in habitat temperature.

- In the development of an individual neuron (*neurogenesis*), individual gangliosides assume a *marker function* for the existing state of differentiation (see Fig. 2.14). Thus, increased synthesis of the GD_3 ganglioside can be correlated to the phase of cell division and migration from the central neural tube; polar gangliosides with 4 or 5 neuraminic acids (GQ_{1b} and GP_{1c}) characterize fiber growth and ramification; increased synthesis of GD_{1a} or GT_b is indicative of the phase of synapse formation whereas GM_1 and GM_4 serve as markers for the formation of myelin.

- In the course of neuronal differentiation of primary or secondary in vitro cultures of neurons or degenerated neuroblastoma cells, exogenously administered gangliosides increase the life span (*neurotrotrophic effect*) and enhance fiber growth (*neuronotropic* or *neuritogenic effect*). Indeed, the question remains open as to whether these effects can be traced back to the gangliosides since several other substances evince similar effects indirectly, for example by influencing the ions of the

neuronal membrane (see Chap. 2.2.3.4).

- In the course of *senescence*, i.e., the phase of aging, the ganglioside pattern in mammals shifts in individual brain regions (the cortex, for example) to favor polar fractions, a phenomenon that can be traced back to a reduction in the GD_{1a} gangliosides associated with synapses (see Fig. 2.13) and that usually manifests itself in diminished membrane viscosity and, hence, permeability.

- Genetic disorders in the activity of enzymes that break down gangliosides (hydrolases) usually result in serious, often lethal neurological diseases, the *gangliosidoses Tay–Sachs* and *Sandhoff* diseases, for example. These *ganglioside storage diseases* manifest themselves in the form of serious mental retardation. Additionally, antiganglioside immune activity can be shown in the blood of patients with other neuronal deficiencies (*schizophrenia, brain tumor, multiple sclerosis*). Such activity indicates that gangliosides are of immense importance for maintaining proper neuronal activity.

- In the course of *ecophysiological temperature*

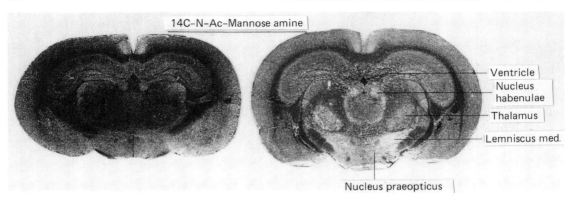

Fig. 8.3. Contact autoradiograms (negatives) of cross-sections of the forebrain of a summer-active and a hibernating dormouse (*Glis glis*) subsequent to intracranial administering of 14C-*N*-acetylmannose amine to depict regional variations in ganglioside metabolism in the brain (*light areas* = regions of strong radioactivity that represent the areas with increased brain activity).

adaptation, the brain ganglioside pattern in vertebrates has adapted variously in accordance with the rule: "The lower the environmental (body) temperature, the greater the polarity of the brain gangliosides." Evidence has accrued from the examination of brain gangliosides in vertebrates (a) with differing requirements of habitat temperatures (extremes: tropical fish vis-à-vis antarctic ice fish; Fig. 8.2); (b) with seasonally or experimentally stimulated temperature adaptation; and (c) with the markedness toward hibernating in the winter (Fig. 8.3). Evidence has also accrued from birds and mammals during their heterothermally early phase of development (Fig. 8.4).

- Parallel to the temperature-dependent change in polarity of the brain gangliosides, there occurs a change relative to their sensitivity to the membrane-bound enzyme *neuraminidase* which breaks down gangliosides. This further substantiates the above cited findings of ecophysiological temperature adaptation of these membrane components of the nervous system.
- *Seizure-inducing agents* (tetrazolium, bemegride, cobalt) alter the ganglioside composition in the brain as well as their organization. Local applications of *antiganglioside sera* cause long-lasting spike changes in the electro-encephalogram.
- In general, *stimulation by light* brings about

an increase in the molecular precursors that are incorporated into the gangliosides of the retina and, subsequent to their synthesis there, are transported axonally to the optic centers of the brain. In similar fashion, ganglioside metabolism in electric fish is affected in the brain centers that are involved in electrocommunication when the fish are subjected to artificial electrical stimulation by an implanted electrode (Fig. 8.5).

8.2.2 Brain Gangliosides and Bioelectrical Activity of the Nervous System

The phenomenological findings relative to functional adaptation in brain gangliosides show clearly that a strong selection took place with regard to the enrichment in concentration and differentiated composition in these lipids in the course of vertebrate phylogeny, especially in the nerve tissue. From the standpoint that brain gangliosides might perform as neuromodulators in the process of synaptic impulse transmission, the question is relevant as to what influence might be exerted upon the *bioelectrical activity* of the nervous system through changes in gangliosides. In recent years, studies in this area have yielded promising findings.

- *Electrical excitability* can be prolonged considerably in brain sections by administering gangliosides exogenously. These findings could

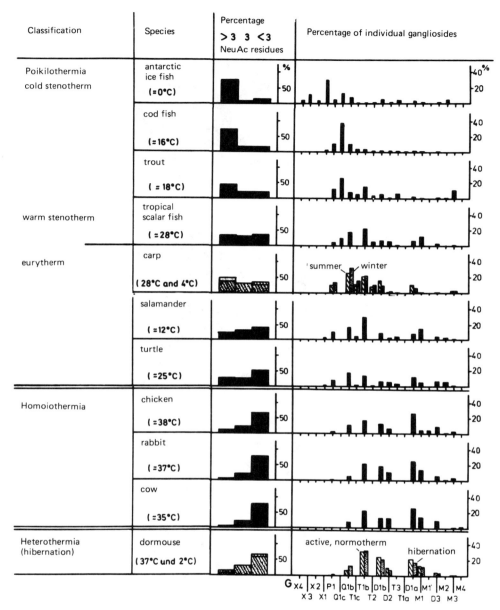

Fig. 8.4. Relative polarities of brain gangliosides in vertebrates with differing strategies for surviving the cold. The percentages of individual gangliosides with 1 (mono-), 2 (di-), 3 (tri-), and 4 (quadri-) neuraminic acids ($\hat{=}$ polarity) are given here.

be explained by the fact that exogenously administered gangliosides have been shown to become incorporated into the membrane of cells in vitro and, indeed, can be transported from there into the cell (see corresponding model, Fig. 8.6).

• It has been shown that gangliosides apparently do not play a significant role in impulse conduction subsequent to *neuraminidase* (the enzyme that breaks down gangliosides) taking effect on spontaneously active, electrically or sensorially stimulated neurons. However,

Control

(individual from
school grouping)

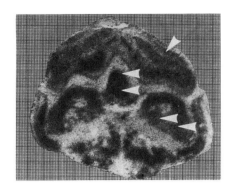

Stimulated

(40 Hz, 48 hours)

Deprived

(2-month isolation)

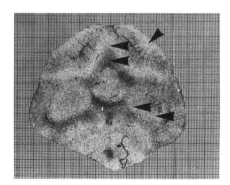

Fig. 8.5. Autoradiograms of cross sections of the midbrain of the electric mormyrid fish (*Gnathonemus petersii*) subsequent to removal of stimulus, artificial electrical stimulation, or control, and following cranially administered 3H-*N*-acetylmannose amine, a specific ganglioside precursor (*dark areas* = regions of strong radioactivity that represent areas of increased brain activity).

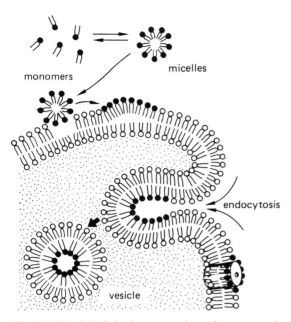

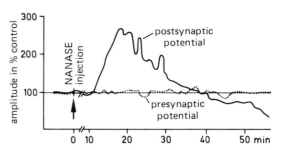

Fig. 8.7. Effect of neuraminidase on electrically triggered (7.3 V; 0.5 sec) cumulative potentials in the optic tectum of the carp: only the postsynaptic potential event is changed, indicating an effect solely upon impulse transmission in the synapses, and not upon impulse conduction along the nerve fibers.

Fig. 8.6. Model of the incorporation of exogenously administered gangliosides into a cell membrane and vesicular transport into the cell.

gangliosides do seem to play a major role in impulse transmission in the synapses (Fig. 8.7). Section preparations through the hippocampus and the cortex have revealed the probability that individual gangliosides (GM_1) are involved in modulating glutaminergic

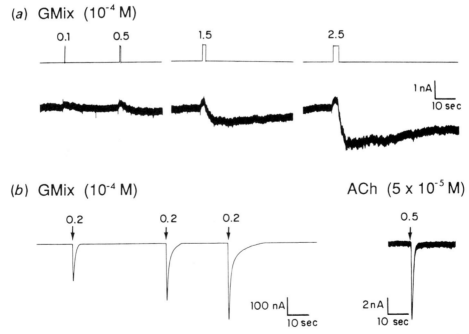

Fig. 8.8. Voltage-clamp analysis for the electrophysiological characterization of the effect of exogenously administered (pressure application) ganglioside mixtures (GMix) vis-à-vis acetylcholine (ACh) on a neuron

of the marine snail Aplysia. *A*: Increase of the inward flow with increasing application time. *B*: Comparable increase (sensitization) via gangliosides and after application of acetylcholine.

synapses whereas other gangliosides take part in cholinergic synapses.

• *Voltage clamp analyses* conducted to characterize electrophysiological properties have shown, relative to neurons of the sea snail *Aplysia sp.* which by nature has no gangliosides, that exogenously administered gangliosides are able to cause cell-specific depolarization, desensitization, and increased ion conductibility within milliseconds, even in these cell membranes (Fig. 8.8).

8.2.3 Physicochemical Adaptive Capacity of Ca^{2+}-Ganglioside Interactions for the Simulation of Membrane Events

In addition to the physiological–phenomenological features of brain gangliosides discussed above, as well as the evidence of their functional involvement in bioelectrical processes in nerve tissue (which is of great significance for future research), it is important now to determine which of the *physicochemical properties that gangliosides*

have might function as neuromodulators during the synaptic transmission of information. The answer to this question should be approached from the perspective of examining possible physicochemical interactions between gangliosides and calcium (Fig. 8.9), since all functional neuronal events are dependent upon the presence of extracellular calcium (see Chapter 7.2).

Gangliosides are both lipid soluble and water soluble. (Their molecular structure, biosynthesis, and incorporation in the neuronal membrane have been discussed in Chapters 2.2.3.4 and 7.1.1.) Due to their amphiphatic nature, they react with extraordinary sensitivity to each change in a given solution and to the addition of complex-forming molecules such as albumin. In organic solutions, gangliosides occur as monomers, i.e., as individual molecules, whereas they occur in a micellar aggregation configuration in aqueous solutions [see Fig. 7.4; critical micellar concentration (CMC) ca. 10^{-8} mol/l]. The aggregation properties of the gangliosides can be affected by changes in pH, the addition of metallic ions, and, especially important with

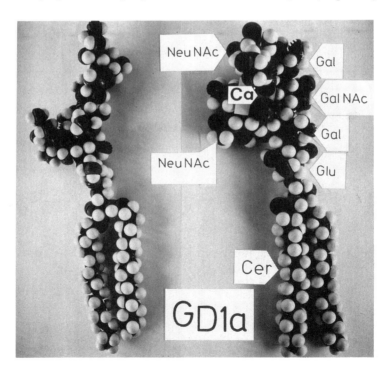

Fig. 8.9. Molecular model of the disialoganglioside GD_{1a} and of a Ca^{2+} complex. Ca^{2+} causes a change in conformation that might be of extreme neuromodulatory importance withing the neuroplasm membrane.

regard to impulse transmission in the synapses, by changes in the electrical field.

Some basic physicochemical properties of gangliosides, as opposed to other membrane lipids, lend themselves quite well to examination using in vitro model experiments. *Liposome* experiments, for example, are well suited to this task, whereby artificial lipid vesicles are manufactured that consist primarily of basic lipids of the biomembrane, lecithin (phosphatidylcholine) and cholesterol, to which gangliosides of varying concentration and composition are added. With such liposomes, i.e., those containing gangliosides, binding studies can be carried out with cations using radioactively labeled $^{(45)}Ca^{2+}$. The strength of the Ca^{2+} binding can be tested and it can be determined if the binding can be released by adding other substances such as mono- and divalent cations (Na^+, K^+, Li^+, Mg^{2+}) or transmitters (acetylcholine, serotonin), and if micellar conformational changes occur in the gangliosides subsequent to additions of Ca^{2+}. Additionally, examining the molecular arrangement of lipids in so-called *monolayers* is particularly germane to the analysis of the physicochemical properties of lipids. The procedure prescribes that one float the lipids to be

tested on the surface of water contained in a special trough such that a monomolecular lipid film results. The individual molecules first arrange themselves in a loose and expanded manner within the film. However, if the surface film is compressed with a barrier, then the molecules orient themselves one upon the other under the increasing lateral pressure. A conformative molecular change (transition phase) between a liquid-expanded and a liquid-condensed state can arise (Fig. 8.10). The characteristics (surface pressure isotherms) of such surface pressure curves and, especially, of the point of collapse, (i.e., the point, under increasing pressure, at which the lipid film collapses) are lipid-specific and, thus, are ideally suited to characterize their physicochemical properties. The knowledge obtained in this way about the behavior of lipids in a monolayer has been extremely helpful in extrapolating the behavior of lipids that might be found in a biological membrane. Because of the particular importance that such physicochemical model research has had in clarifying cellular events, some of the in vitro findings relative to *ganglioside-calcium interaction* are reviewed below. These findings might be of great help in

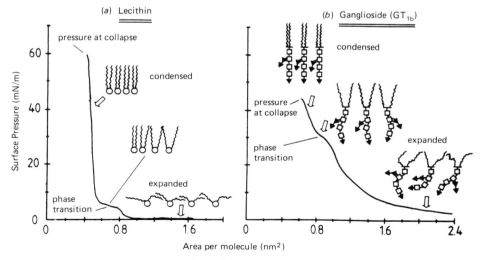

Fig. 8.10. Phase behavior of lipids (*a*: lecithin and *b*: ganglioside GT_{1b}) in a monomolecular film at the water/air interface. The surface requirement of the molecules decreases with increasing surface pressure. Depending upon the nature of the lipid, surface-pressure–area isotherms exhibit a phase transition from the expanded to the condensed liquid state, and pronounced differences in the collapse of their film. Pressure conditions in the biological membrane are between 20 and 30 mN/m (milliNewton/m).

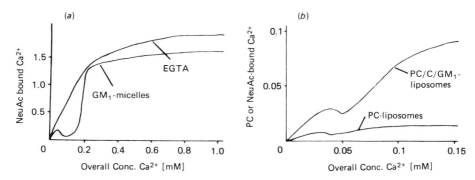

Fig. 8.11. Discontinuity in the binding of Ca^{2+} to gangliosides: (*a*) ganglioside micelles in comparison to Ca^{2+}-chelator EGTA (*β*-amino ethyl ether-*N*,*N*,*N*′,*N*′-tetraacetic acid); (*b*) liposomes created from pure lecithin (PC), from lecithin-cholesterol (C) and from the GM_1 ganglioside.

interpreting processes within synaptic membranes since they bring to light special functional qualities of the gangliosides as opposed to the other lipids that are present in the neuronal membrane, the phospholipids, in particular (see Table 8.1):

• Individual gangliosides (see Fig. 7.3), as well as natural gangliosides mixtures from the brain of various vertebrate species, exhibit extremely *unstable binding capacities* with regard to calcium (complex chelate binding, no ion binding). In contrast to magnesium, gangliosides

have two binding points with calcium; Ca^{2+} binding is about 20 times stronger than magnesium binding.

• In contrast to binding with other membrane lipids, Ca^{2+} binding to gangliosides occurs discontinuously, i.e., in certain ranges of concentration, bound Ca^{2+} is released before further binding of additional Ca^{2+} occurs (Fig. 8.11).

• In contrast to binding with other lipids, Ca^{2+} binding to gangliosides is highly sensitive to the addition of cations (Na^+, K^+, Li^+, Ca^{2+}, Mg^{2+}), transmitter substances (acetylcholine,

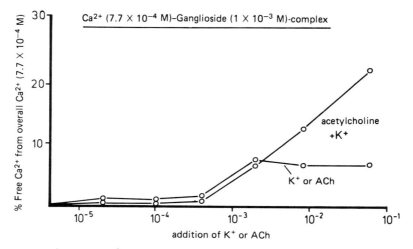

Fig. 8.12. Release of Ca^{2+} from Ca^{2+}-ganglioside complexes by K^+ or acetylcholine, and by simultaneous administering of acetylcholine and K^+.

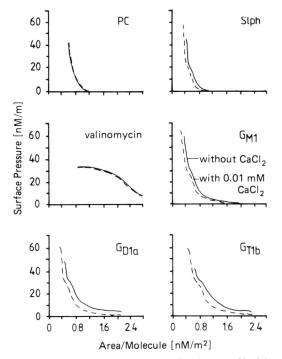

Fig. 8.13. Surface pressure–area isotherms of lecithin, sulfatide, the peptide valinomycin, and the variously polar gangliosides GM_1, GD_{1a}, and GT_{1b} with and without the addition of Ca^{2+} (0.01 mM).

serotonin), and changes in temperature and pressure (Fig. 8.12).

- In artificial membrane systems, the gangliosides act in concert via hydrogen bridges in the presence of Ca^{2+} in the form of molecular clusters.
- Presumably, the structure of these ganglioside clusters is affected, both in solution and in the membrane, by the degree of the Ca^{2+}-ganglioside formation complex.
- The addition of Ca^{2+} causes characteristic changes in the surface behavior (Fig. 8.13) in pure *monolayers* of gangliosides with increasing polarity ($GM_1 < GD_{1a} < GT_{1b}$) in contrast to other lipids (lecithin, sulfatide) and, particularly, peptides (valinomycin): reducing the space required by the molecules and distinctly increasing stability of the lipid film (shift in the point of collapse).

- In *mixed monolayers*, gangliosides, in their interaction with other building blocks of the membrane such as peptides (here, the ion carrier *valinomycin*) or *cholesterol*, cause the condensation of the common molecular surface requirement by 10 to 20% of the calculated surface requirement (Fig. 8.14). The addition of Ca^{2+} further reduces the required surface area. However, surprisingly, a separation of the components occurs here, either through the formation of smaller, unmixed partial surfaces or, less probably, through complete phase separation. Furthermore, the stability of the combined peptide–ganglioside film is considerably greater than the film stability of the individual components.
- Recently, leading potentiometric experiments have shown that the addition of Ca^{2+} to a "hanging mercury electrode" causes a radical shift in the so-called desorption potentials of gangliosides in the sense that considerably higher voltages are required to desorb ganglioside films from the surface of the mercury electrode when Ca^{2+} is present than when it is not. In these desorption and adsorption experiments, the polar gangliosides GD_{1a} and especially GT_{1b} prove to be highly effective as opposed to the merely negatively charged GM_1. Magnesium proved to have a far less stabilizing effect in this model.

On the whole, the in vitro findings up to now have uncovered extraordinary characteristic properties of gangliosides in conjunction with calcium. Similar properties still remain to be discovered relative to other membrane lipids under scrutiny. These findings serve as a functional model for biological membrane properties inasmuch as they show how complex compounds of gangliosides and calcium might exert lasting, modulatory effects on membrane events.

In particular, the findings last mentioned, those relative to the interaction of calcium, gangliosides, and peptides, as well as the references to the unstable, voltage-dependent behavior of ganglioside-calcium complexes at an electrode surface, justify referring to these compounds as modulator substances in the regulation of the synaptic transmission.

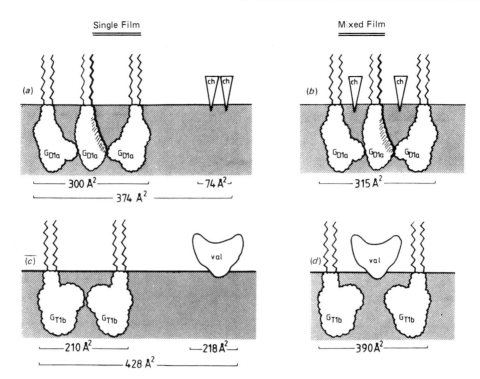

Fig. 8.14. Diagram of the molecular area requirement of gangliosides (GD_{1a} and GT_{1b}) and cholesterol and valinomycin at the water/air interface. Gangliosides cause a condensation of the common area requirement by 18% and 10%. Addition of Ca^{2+} causes further condensation of the molecules.

8.2.4 Functional Model of the Neuro-modulatory Effect of Ca^{2+}-Ganglioside Interactions in Synaptic Transmission

Given the need for flexible neuromodulator substances to regulate impulse transmissions adaptively (see Sect. 8.1) and current experimental findings as to the pronounced physiological and physicochemical adaptive capacity of brain gangliosides in vertebrates (see Sects. 8.2.1 and 8.2.3), this section offers a functional model relative to the functional involvement of gangliosides in the transmission of neuronal information in synapses. The model is based on the notion that stimulation-dependent *changes in conformation* occur in the gangliosides that are localized on the outside of the synaptic membrane and that these changes in conformation, in turn, bring about a change in the structure and/or function of membrane-bound synaptic channel or enzyme proteins, thereby

effectuating *neuromodulation*. The functional changes in local membrane sections brought about in this way lead to a stimulation-dependent release of transmitter substances that is regulated by Ca^{2+}-ganglioside complexes and, thereby, lead to the transmission of electrically encoded information from one neuron to the next. According to this model, membrane-bound, ganglioside-associated Ca^{2+} functions as *primary messenger* in internalizing an electrical signal. It does so based on the instability of its binding as opposed to changes in environmental factors such as in electrical fields. Subsequent to this, calcium that has been released from its extracellular reservoirs penetrates the synapse and triggers in the synaptoplasm a torrent of *secondary messenger reactions* that modify the metabolism of all cells involved relative to the altered state of stimulation.

The most important steps in the *modulation of the synaptic transmission* process are summarized below relative to conformative changes

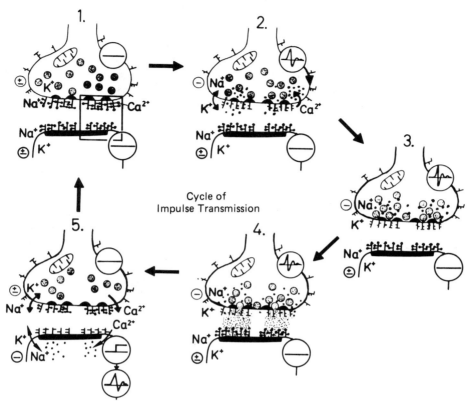

Fig. 8.15. Supplemental diagrammatic overview to Fig. 7.7 relative to the function of calcium and gangliosides in synaptic impulse transmission: (1) unstimulated state, tightening of the synapse membrane by Ca^{2+}-ganglioside complexes; (2) stimulation of the presynapse; transmembranic exchange of Na^+ for K^+, and/or voltage-dependent, allows Ca^{2+} to flow into the synapse; (3) Ca^{2+} binds transmitter vesicles to the inner synapse membrane; (4) transmitter release; (5) triggering of a generator potential and a progressive potential in the postsynapse and return transport of the ions; binding of Ca^{2+} to the gangliosides. (Section enlarged and detailed in Fig. 8.16).

in the membrane-bound *Ca^{2+}-ganglioside complexes*. For purposes of clarification, reference is made to the overview model of a synaptic transmission cycle rendered in Fig. 8.15 and its corresponding detailed model (Fig. 8.16). These representations complete the general plan, already discussed, of the other biochemical processes involved in transmission (see Fig. 7.7) with regard to the roles of calcium and gangliosides:

• In vertebrates, the membranes of a nonstimulated nerve terminal in the region of the actual synaptic contact zone might be tightened by the collective effect of Ca^{2+}-ganglioside complexes (Fig. 8.15.1). In this regard, Ca^{2+}-ganglioside clusters could affect integral membrane

proteins and/or ion channel proteins (Ca^{2+} channels, for example) on the basis of their diverse conditions of conformation (more or less rigid) in that they either reduce the Ca^{2+} flows (blocking) or activate them. The conformative changes of the modulator ganglioside in the absence or in the presence of Ca^{2+} would be the trigger for this, as described earlier (Fig. 8.16a).

• Upon the arrival of an electrically encoded impulse signal (*action potential*), Ca^{2+} is released voltage-dependently from its ganglioside binding points outside of the synapse membranes. This is brought about by the change in the electrical field and/or the transmembranic exchange of the monovalent

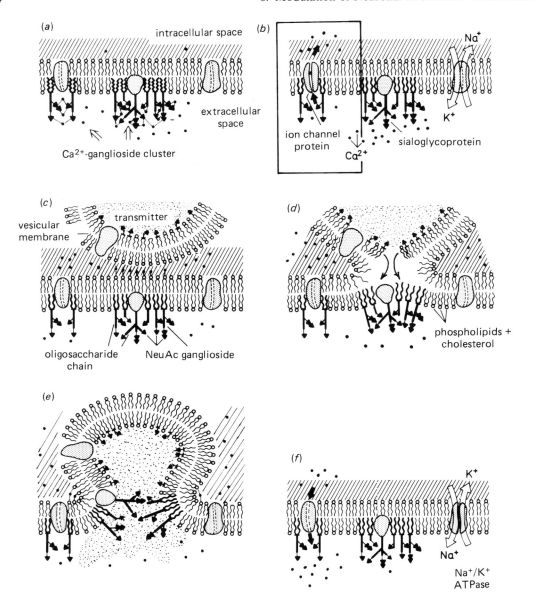

Fig. 8.16. Detailed molecular model (see Fig. 8.15) of the role of Ca²⁺-ganglioside complexes in synaptic transmission (see text for details).

cations Na⁺ and K⁺ (Fig. 8.15.2). As a result of this, the gangliosides experience the *change in conformation* mentioned above: the corresponding membrane region becomes more fluid, a condition that, in the case of ganglioside clusters, could effectuate the opening of the calcium channel (Fig. 8.16b). Released calcium flows through the opened channels

into the presynapse and allows Ca²⁺, which is present in abundance in the extracellular space, passively to follow behind in a flow created in the draft of its displacement. The voltage-dependent change in conformation in the gangliosides subsequent to their release of Ca²⁺ represents a *primary messenger mechanism* for receiving the signal.

TABLE 8.2. Amount of brain ganglioside and Ca^{2+} concentrations in the brain of warm- and cold-blooded vertebrates that are relevant to the discussion of a functional model of the neuromodulatory effect of Ca^{2+}-ganglioside complexes in synaptic transmission.

	Cold-blooded vertebrates (fish, amphibians, reptiles)	Warm-blooded vertebrates (birds, mammals)
Concentration of brain gangliosides ($\mu g/g$ net wt. of tissue)	110–800	400–1,100
Ganglioside composition (= polarity)	highly polar: 4–7 neuraminic acids per ganglioside	less polar: 2–3 neuraminic acids per ganglioside
Extracellular Ca^{2+} concentration (mM)	3–5	1–2
Number of Ca^{2+} ions per transmitter quantum release	4–5	2–3

- Ca^{2+} in the synaptoplasm, which has flowed from the extracellular reservoirs, forms *trans*-complexes between the membrane of the synaptic vesicles and the inside of the membrane of the presynapse (Fig. 8.15.3, 8.16c). The precise localization of endogenous calcium at points mentioned earlier can be identified through electron microscopy (see Fig. 7.20 and 7.21).

- Due to the short-term fluidization (reduction of the lateral membrane tension) of the plasma membrane, the transmitter vesicles, which were led to the synapse membrane assisted by Ca^{2+}–trans-complexes, fuse with the membrane and the transmitter is released into the synaptic cleft (Figs. 8.15.4, 8.16d and e). The transmitter interacts with specific *receptor molecules* on the surface of the postsynaptic membrane, whereupon *ion channels* located there are opened, either directly or indirectly, by the effect of membrane-bound enzymes (*adenyl cyclases*). Due to the resulting flow of Na^+ into the postsynaptic cell, impulse conduction occurs at the membrane of the adjacent neuron. Parallel to this, the transmitter released from the Ca^{2+}-ganglioside complexes of the postsynaptic membrane could release *calcium*. In the draft of its displacement, extracellular calcium can penetrate the postsynaptic membrane into the postsynapse in a manner analogous to the presynaptic phenomenon and induce secondary messenger functions there. (A release of calcium from Ca^{2+}-ganglioside complexes can be achieved in vitro both by various cations and by transmitters; see Sect. 8.2.3.)

- Two sets of conditions lead to repolarization at both synaptic membranes: (a) when the mono- or divalent ions are pumped back through the membranes to their origination points upon activation of ion pumps (Na^+, K^+-*ATPase*, Ca^{2+}-*ATPase*) (calcium into the extracellular synaptic cleft, into the intracellular reservoirs of the *endoplasmic reticulum*, or into *reservoir vesicles* known as *calcisomes*; see Fig. 7.19); or (b) upon the rapid breakdown by enzymes of the released transmitter. In both instances, the *calcium*, which has been returned to the extracellular space, can accumulate loosely at the negatively charged residues of neuraminic acid of the gangliosides. This event causes a *change in conformation* that ultimately leads to the closing of the ion channels and, thus, to the sealing of the membrane (Figs. 8.15.5, 8.16f).

Thus, according to this functional model, Ca^{2+}-*ganglioside complexes* bring about an optimal physicochemical constellation (viscosity, permeability, opening and closing of the ion channels, etc.) of the synaptic membranes in each individual, a quite short-lived *transmission* event whereby a carefully determined, or quantum *release of the transmitter* is ensured, as is a defined transmission of an electrical signal from neuron to neuron.

Furthermore, an important *neuromodulatory effect* can be expected of the *brain gangliosides* insofar as these glycolipids, more so than almost any other class of substance, are capable of releasing measured quantities of transmitter substances. They can do this because their concentration and composition in diverse vertebrates can vary greatly in individual fractions of differing polar nature when adapting to ecophysiological conditions.

Accordingly, it can be seen in Table 8.2 that, in cold- and warm-blooded vertebrates, the concentration and composition of brain gangliosides are inversely proportional to the extracellular concentration of *calcium*. Future research will have to determine the degree to which these ratios relate to presently inexplicable differences in the supply of varying amounts of Ca^{2+} ions for the measured release of a *transmitter quantum*.

The functional model of the neuromodulatory effect of Ca^{2+}-ganglioside complexes in short-term *transmission* as presented here represents the basis for the hypothesis of long-term memory formation via molecular facilitation in synapses (see Chap. 11).

9

Neuronal Plasticity

Early research in the fields of morphology and physiology held that the neuron was essentially a rigid, static component of the nervous system where the phenomenon of impulse conduction and transmission took place and, thus, where the transfer of information from cell to cell occurred. The study of membrane properties, the phenomena of de- and repolarization, and related electrical events were almost exclusively the domain of neuro- and electrophysiology.

The trophic functional component of the neuron went largely unnoticed until about the middle of this century, as did the pronounced flexibility of neuronal fibers in adapting their growth to prevailing external conditions, both during early ontogeny (when the neurons first begin to function) and during regeneration. Only gradually did neurobiology begin to assess the scope of *neuronal adaptivity*, otherwise referred to as *neuronal plasticity*. Previous research into the effects of this phenomenon were carried out with an eye toward the potential of nerves to regenerate in the neuronal assemblies of the peripheral and central nervous systems subsequent to injury. Generally speaking, the phenomenon of neuronal plasticity might play an important, if not the most decisive role, in the processing of impulse stimuli and both the storing and reactivation of memory content in the normal functioning of the brain.

The phenomenon of neuronal plasticity will be examined here from the perspectives of *neuronal transport* (in various forms), *synaptic plasticity*, and *de- and regeneration*.

9.1 Neuronal Transport

In the second century A.D., the Roman physician Galenus taught that the brain controls the body's functions and movements through its production of "pneuma," a substance that reached its intended destination by means of minute, invisible channels in the nerves. Moreover, the notion that prevailed into the 18th century was that the brain was a sort of gland that secreted a special liquid that was transported through the nerves by a system of tubes to those sections of the body in need of management. Today, we recognize that these curious theories were not all that far from the facts as we know them. Causal analysis of the phenomena underlying the events of de- and repolarization of neuronal fibers has been advanced considerably in recent years, in large part due to the long-standing recognition of those dynamic processes.

From the middle of this century, neuroscientists have endeavored to explore both the structure and function of nerve tissue. It is only in this recent time period that the neuron has been regarded as a basic structural, genetic, functional, and, above all, trophic unit of the nervous system. The last characteristic is especially significant in that the neuron is unique among all other cell types in an organism: extremely long (1 meter or more) branches (one axon and numerous dendrites) can extend from a relatively small (50 μm) cell body. The metabolic needs of these branches and ramifications must be attended to by the perikaryon, the only region of the cell able to form new proteins and other essential cell components (see Chap. 1.1.3). The ribosomal RNA required for the synthesis of proteins is found only in the perikaryon and the dendritic points of exit; the distal fiber regions and the synapses do not contain ribosomal RNA. Neither is direct coding-dependent biosynthesis to be found in these locations. Active absorption of metabolic compounds from the extracellular space must be ruled out, as well (with the exception of certain pinocytosis processes in the synaptic region, see Sect. 9.1.3).

The task of neurobiology, in particular that of *neurodynamics*, is to study these trophic components of neuronal function. Primary among the considerations is the phenomenon relating to the distribution of metabolic products within the nerve fibers, i.e., *neuronal transport*. Nature has developed a complicated transport system for supplying the nerve fiber terminals. The system supplies products (membrane components, vesicles, transmitter substances) to the synapses and, from there, catabolites are transported back to the cell body where some of them are broken down completely. Neuronal transport is one facet of the extremely complex, plastic nature of the nervous system; it manifests itself in various ways (Fig. 9.1). Products of synthesis reach the periphery of the cell by way of *anterograde transport*, i.e., they proceed from the cell body to the fibers. Most substances that are translocated in this way move at a basic rate of 1 to 3 mm/day. This *slow neuronal transport* exists side by side with a faster delivery system (*rapid neuronal transport*) that hurries a small percentage of supply substances at a rate 100 times that of its slower counterpart. Some substances expedited through the rapid system exit the axon terminals and are absorbed by postsynaptic receiving cells (*transneuronal transport*). However, small quantities of substances (smaller proteins, lipid components, toxins) can be absorbed in the region of the nerve terminals and can be transported rapidly from there intracellularly back into the perikaryon (retrograde transport). Finally, ions and smaller molecules can be exchanged directly between cells through channel pores in the membranes (transmembranic transport). Substances, therefore, are transported constantly for purposes of supply and communication from segment to segment within the neuron as well as from cell to adjacent cell (see Fig. 9.1).

9.1.1 Slow Neuronal Transport

The first, clear-cut evidence of the existence of slow neuronal transport can be gleaned from merely observing the growth of a neuron. It progresses from a largely spherically shaped neuroblast cell. As it matures, the cell body sends out a variety of processes through an amoebic-like outflow of cytoplasm that thickens to

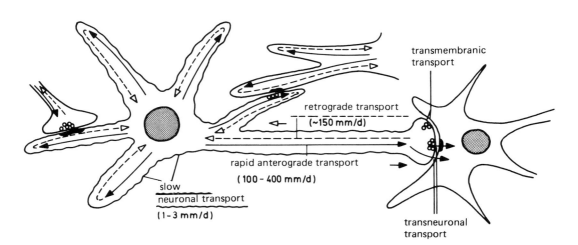

Fig. 9.1. Diagram of the various types of neuronal transport.

Fig. 9.2. Diagram of neuronal fiber growth during early neurogenesis.

become fiber processes (Fig. 9.2) as a membrane takes shape. The same principle occurs in the regeneration of a peripheral nerve fiber (see Fig. 9.22). The speed with which a severed nerve fiber regenerates and ultimately reaches its destination is about 1 to 3 mm/day, depending on the type of nerve, its age, its overall organization, and, in the case of poikilothermic animals, environmental temperature.

The first experiments carried out to investigate the phenomenon of slow neuronal transport, and which are now considered "classic," were published by Paul Weiss et al. in 1948. They completely tied off nerve fibers of the sciatic nerve in rats, chickens, and monkeys. In time, substances collected above the point of constriction. These swellings represented an accumulation of nerve fiber cytoplasm. In contrast, signs of degeneration became evident below the constriction that were similar in nature to conditions observed in severed nerve fibers. Upon removal

of the ligation, the accumulated neuroplasm flowed into the degenerating section, and did so at a rate of 1 to 3 mm/day, a value that coincided with the speed of regeneration in severed nerve fibers (Fig. 9.3). These observations lead to the first definitive, dynamic functional model of the neuron according to which a constant flow of substances occurs through the fibers (dendrites, axons) to the nerve fiber terminals (synapses). This is one of the most important concepts to be considered when discussing the phenomenon of memory (see Chap. 11.2.4).

The process of slow neuronal transport was analyzed first by using microscopic and microcinematographic techniques in experiments involving severance and ligation. Procedures borrowed from fluorescence histochemistry lead to the first evidence of the transport of *acetylcholinesterase* within axons.

Beginning in the early 1960s, radioactive tracer techniques were utilized increasingly to try to

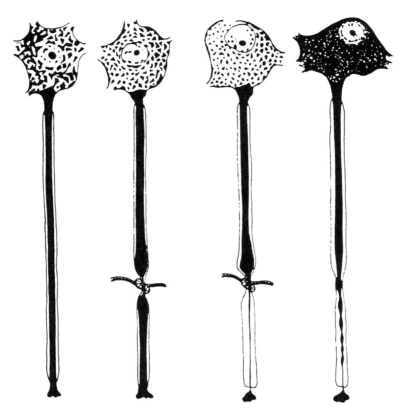

Fig. 9.3. Evidence of slow neuronal transport using fiber ligations: material accumulates on the cell body side of the ligation; upon removal of the ligation, backed-up material flows again to the fiber terminals.

identify within a neuron the points of synthesis and, especially, the whereabouts of newly synthesized neuronal compounds. The principle in effect here is that a preliminary metabolic stage in the synthesis of macromolecular compounds, having been labeled radioactively, reaches the neuron via the bloodstream or by direct administration. Incorporation of the tracer, its modification, or its whereabouts within the cell can be determined by means of radiochemical or autoradiographic techniques.

The location of the tracer within the various regions of the neuron (cell body, fibers, synapses) is extremely difficult to ascertain when the tracer is administered intravenously. Suitable larger neuronal structures were sought, therefore, that were clearly distinguishable in their separation of cell body regions as opposed to fiber and synaptic end terminals. Such structures would facilitate differentiated analyses. The *optic system,*

especially in lower vertebrates, proved to be the ideal target structure. It contains, in the retina of the eye, the perikarya of the optic nerve cells. The fibers run from here in bundles (forming the optic nerve) to the primary optic center in the brain (in lower vertebrates to the *optic tectum*), which can be located at a considerable distance from the retina, and where they establish synaptic contact with cell processes of the more deeply situated layers of neurons.

Upon its administration to the vitreous body of the eye, absorption of a radioactively labeled metabolic compound by the retina ganglion cell bodies can be traced. Depending on the type of substance administered, transport of the labeled compounds can be followed in time through the optic *nerve* via the *optic chiasma* to the optic nerve terminals in the contralaterally situated *optic tectum* (Fig. 9.4).

Autoradiography is helpful in depicting the

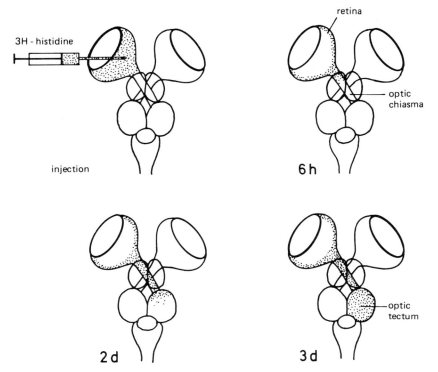

Fig. 9.4. Diagram of a radioactively labeled substance administered intraocularly (here, the amino acid 3H-histidine) and of the time-dependent transport of compounds thusly labeled in the optic tract in fish.

intra-axonal transport of proteins in the optic system that were labeled with a radioactive amino acid ([3H]histidine, for example) during their synthesis in the cell body of the retina ganglion cells. In this process, appropriate tissue sections are covered with a radiation-sensitive photoemulsion that is "exposed" in the dark by the radioactively labeled tissue proteins just as normal film would be exposed by light. Developing the film produces black images in areas where the silver granules were exposed to radiation (conversely, white areas represent nonexposure) and these can be studied under the microscope. In this way, information is obtained about the location and concentration of the radioactively labeled compounds.

The spreading of the labeled proteins can be traced in the optic system at various intervals of progression. The bulk of the labeled substances (ca. 80 to 90%) is transported very slowly in the *optic tract* (Fig. 9.5). It requires several weeks to reach the nerve fiber terminals in the *optic tectum*

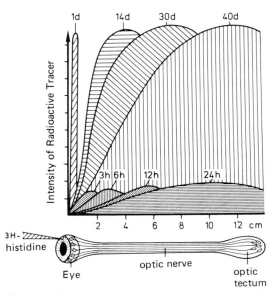

Fig. 9.5. Diagram of the slow and rapid axonal transport of radioactively labeled compounds in the optic nerve-subsequent to intraoculor injection of a radioactively labeled amino acid (3H-histidine).

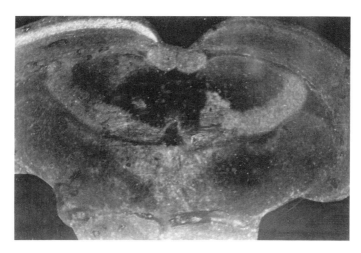

Fig. 9.6. Dark-field photograph of the autoradiography of an histological section through the midbrain of a fish (carp) subsequent to one-sided intraocular injection of 3H-histidine. The projection pathways of the optic nerve are labeled selectively only in the optic tectum which is associated with the injected eye.

(depending on the length of the system), a fact that correlates to a rate of transport of about 1 to 3 mm/day. The projection pathways are marked selectively in the optic tectum corresponding to the injected eye. This can be depicted readily utilizing dark-field exposure of the autoradiogram (Fig. 9.6).

It was shown also at the spinal ganglion cells, which run dorsally along the backbone, that proteins that are synthesized in the cell body of a neuron can be translocated over great distances to the nerve fiber terminals by means of slow neuronal transport. The *spinal ganglion cells* send out central fibers that lead to the spinal cord. Peripheral branches from the cell body also run to the various body parts, as in the case of the spinal ganglion cells located in the region of the lumbar vertebrae. Their branches lead in a bundle, the sciatic nerve, to the leg. Subsequent to the injection of radioactively labeled amino acids into the spinal ganglion cells, central and peripheral nerve filaments were removed at one week intervals, cut into sections 6 mm in length and analyzed for their content of protein radioactivity. Differential analysis showed that the proteins of the neurotubules and neuro- and microfilaments (see Chap. 1.1.3), which comp-

rise 80% of all proteins, migrate by means of slow neuronal transport. Transport in the fibers leading from the spinal ganglion cells to the leg proved to be two to three times faster than in the central fibers that lead to the spinal cord.

The rate of slow transport is apparently dependent upon the requirements of the nerve fibers in question. In mature plaice (*Pleuronectes platessa*), in which the optic nerve of one eye is substantially longer than that of the other due to body rotation in early ontogeny, the proteins synthesized radioactivey in the retina of each eye reached the nerve terminals in both primary optic centers of the brain at the same time.

The primary factor in slow neuronal transport, which corresponds in speed to the rate of nerve regeneration, pertains to the translocation of all neuroplasm columns, including the organelles contained within them, such as the mitochondria, neurotubules, neurofilaments, and smooth neuroplasmic reticulum. This form of transport continuously serves to renew the plasma and the membrane of the distal nerve terminals in the mature neuron by supplying necessary structural components. Moreover, it enables damaged fibers to grow out once again, resulting in their regeneration.

The *mechanism* by which *slow transport* is maintained is still unknown. The formation of pseudopods in amoebas seems to be the only applicable analogy, whereby the cell membrane extends outwardly for the purpose of forward motion and cytoplasm flows into the void that is created by the membranic extension. The behavior of the *growth cones* as they sprout from the nerve fibers seems to lend support to the amoeba theory; radical changes in the composition of the developing membrane have been recorded. In any case, a sure point of reference is the fact that the neuroplasm does not flow within the fibers as water would through a garden hose; rather, transport might occur in a simultaneous pushing forward of the net-like web of neurotubules, neurofilaments, microfilaments, and neuroplasmic reticulum. Paul Weiss was able to detect minuscule peristaltic waves in nerve fibers in situ utilizing time-lapse microfilm analyses. These waves roll over the fibers at regular time intervals and, in this way, transport the neuroplasmic columns from the cell body toward the fiber terminals. The question remains open as to whether, and to what extent, this endogenous neuronal rhythm is supported by rhythmic pulsations of the glial cells surrounding the neurons and their fibers.

9.1.2 Rapid Neuronal Transport

Research as early as the mid-1960s was directed toward the point of synthesis and the whereabouts of newly synthesized substances in the nerve tissue. It became apparent that a small portion of the compounds in the nerve fibers was being transported considerably more rapidly than the phenomenon of slow neuronal transport would explain. Radioactively labeled compounds appeared in the *optic tectum* within only a few hours following injection of radioactive metabolites into an eye of carp (*Carassius vulgaris*) (Fig. 9.5), whereas the bulk of labeled proteins required several days to reach the tectum of the midbrain (see Sect. 9.1.1). At first, the speed of the rapid transport wave could not be ascertained, since autoradiography was unsuitable for measuring accurately the parameters in question. Nevertheless, a speed was derived from the first

peak value of radioactivity to appear in the optic nerve. It amounted to at least 10 mm/day and, as such, was substantially greater than that of slow neuronal transport.

Ultimately, the *speed of rapid neuronal transport* was determined more accurately when radiochemical experiments were carried on the sciatic nerve in cats. Values of 100 to 410 mm/day were observed. These high values appeared to apply to sensory and motor nerves alike in warm-blooded animals, regardless of whether they were at rest or stimulated. Rapid neuronal transport is sustained for a period of time in nerve fibers even when contact with the perikaryon is interrupted. That means that the substances newly synthesized in the cell body are translocated to the fiber terminals spontaneously and actively. Of course, the nerve fiber must be supplied with sufficient oxygen and energy (ATP) for this to occur. Metabolic poisons such as *cyanide* and *dinitrophenol*, which block phosphorylation, bring rapid transport to a halt immediately upon being administered. Moreover, rapid transport has proved to be highly sensitive to temperature. The rate of transport in poikilothermic fish at $10\,°C$ is about 50 mm/day, and it jumps to 250 mm/day at $25\,°C$. Extrapolating these values to correspond to mammalian conditions ($37\,°C$), one arrives at a transport rate of more than 400 mm/day. It is apparent from this that the mechanisms that underlie neuronal transport in warm- and cold-blooded animals are quite similar.

Although structure-bound proteins make up the bulk of the substances translocated by slow neuronal transport (see Sect. 9.1.1), 10 to 20% (i.e., a far smaller apportionment) of the rapid neuronal transport system serves to deliver soluble and particle-bound *proteins, glycoproteins, mucopolysaccharides, phospholipids, sialoglycolipids (gangliosides)*, metabolites of low molecular weight (*amino acids, glucose, N-Ac-glucose amine, uridine, uridine diphosphate glucose (UDPG), cyclic nucleotides*), and, especially, *neurotransmitters* (*noradrenaline* and other *biogenic amines*, such as *putrescine*) to the nerve fiber terminals (Fig. 9.7).

Various models have been suggested for the *mechanism of rapid neuronal transport*. Although slow transport is inhibited by antimetabolites that block perikaryal protein synthesis (*cyclo-*

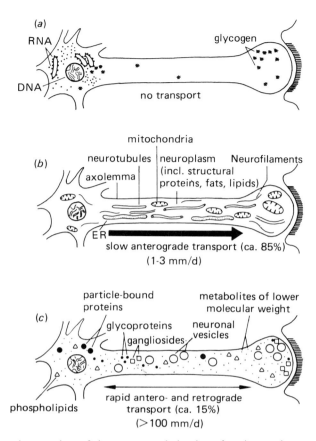

Fig. 9.7. Diagrammatic overview of the transport behavior of various substances in the neuron.

hexamide), rapid transport remains unaffected by these. It is, however, apparently dependent upon the presence of intact *neurotubule* structures, a collective grouping comprised of essential structural elements of the neuroplasm, axons, and dendrites (see Chap. 1.1.3). *Colchicine*, a substance that blocks the function of the neurotubules in a specific way (by dissociating the tubular protein *tubulin* into its inactive component parts), immediately arrests the process of rapid transport, as does *vincaleukoblastine* (vinblastine). One means of demonstrating this is by the fact that the concentration of soluble proteins, *acetylcholinesterase*, for example, is cut off above the neuronal ligation when colchicine is administered. This colchicine test, however, does not resolve all questions. Colchiceine (*O*-dimethylcolchicine), a colchicine derivative, causes rapid neuronal transport to be blocked far more effect-

ively even though this substance does not exhibit any active binding properties with regard to tubulin.

Opinions vary greatly about the internal mechanism of rapid neuronal transport. Although some propose great specificity in the interaction between neurotubules and transported substances, others posit a nonspecific, unified system for all transported compounds. Unanimity of opinion prevails, however, in regard to the notion that all substances that are translocated rapidly are transported from the perikaryon to the neuronal terminals in particle-bound from. The concept has emerged that *neuronal vesicles* act as transport vehicles in the process and interact with the *neurotubules*:

- The "*transport filament*," or "transport vesicle" model holds that the vesicles, which potentially would be bound to neurofilaments, would slide along the neurotubules as a result of their similar chemical affinities, much as they are known to do in the myosin–filament system of muscle contraction.

- A more *"mechanical-chemical"* model presumes that anionic polyelectrolytes (acid polysaccharides) surround the neurotubules which can contract in response to ion flows (Ca^{2+}) in such a way that the various compounds that are bound and charged at the polyelectrolyte mantle are pushed along the tubules.

- The *"microflow hypothesis"* asserts that neuronal components are moved along in mild, less viscous subcurrents of the neuroplasm columns. This hypothesis is based on the concept that regions of varying viscosity exist in the neuroplasm (as there are in the cytoplasm of all cells) from which diverse subcurrents originate. In the special instance of the nerve fiber, regions of lesser viscosity could surround the neurotubules like a tube. The surface of the tubules would have to be of a nature such that the effects of shearing force would be projected from that surface into its environment. These force effects could be directed by the structure of the neurotubules and by the concentrated release of electrostatic energy of the counteraction. The energy released by ATPase reactions at the surface of the tubules through the breaking down of hydrogen and

ion bindings would account for the formation of material of lesser viscosity. A variety of ranges in viscosity results in various flow rates.

- Recent views, based on videomicroscopic analyses of the transport movements of neuroplasmic organelles as well as supplemental biochemical studies, maintain that the rapid transport of organelles is directly related to their interaction with the neurotubules. Anterograde transport seems to be dependent exclusively upon the charge-related interaction of a *kinesin* (a protein with a molecular weight of 500,000 that is isolated from the axoplasm), the neurotubules, and the organelles. Retrograde transport, therefore, would be dependent upon another translocator protein. Here, too, conclusive evidence remains to be found as to the correctness of this hypothesis.

- These theories can be pursued quite effectively utilizing computer-enhanced contrast amplification, a new, powerful method of evaluating filmed recordings that was developed for light microscopy. Continuous, bidirectional organelle movement has been confirmed both within the giant axon in cuttlefish as well as in the free axoplasm derived from that axon. The movements are integrally related to the neurotubules that are present in copious abundance within the axoplasm. The model of *"active microtubules"* (Fig. 9.8) suggests that the sliding movement of the microtubules

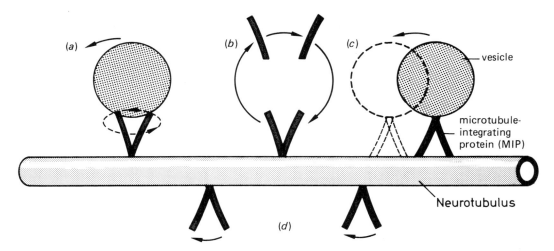

Fig. 9.8. Various hypotheses of the functional interaction among energy-producing ATPase molecular arms, microtubules, and transport vesicles that serve as the basis for rapid, bidirectional neuronal transport.

and the movement of the organelles are rooted in the same energy-producing enzyme (ATPase) that either is located atop the neurotubules (Fig. 9.8a, d), binds the tubules to the vesicles (Fig. 9.8c), or integrates into the substrate (Fig. 9.8b). In any event, the arm-like, lateral transverse bridges can be depicted by means of high resolution electron microscopy. According to the hypothesis, the small, lateral molecular arms of the neurotubules render a cycle of changes in configuration between an activated state and a resting state. Depending on the cycle, the tubule itself can be moved, or organelles (mitochondria, neuronal vesicles) can be advanced continuously. Anterograde and retrograde transport of the organelles at the same tubule might be explained by the ellipsoidal swinging motions of the arms, a phenomenon that actually can be observed under the light microscope.

In spite of the admirable results compiled by microphotographic detail analysis, conclusive evidence of the actual functional mechanism underlying rapid bidirectional neuronal transport remains elusive. However, a technique has been devised recently by which the movement of the neurotubules and other neuronal organelles can be observed directly in vivo. This new offering in perspective is essential for clarifying the process of neuronal transport.

9.1.3 Retrograde Transport

The counterpart to anterograde, rapid substance flow, i.e., transport from the cell body to the nerve fiber terminals, is referred to as *retrograde transport*. Scientists became aware of this phenomenon when substances (noradrenaline, for example) began to accumulate on the distal side of an incompletely ligated nerve fiber, albeit in quantities far below those appearing above the ligature. Studies utilizing radioactive tracers proved that labeled substances in nerve fibers from the periphery reached their corresponding cell bodies in the CNS.

Meanwhile, other researchers proved the existence of retrograde transport more quickly by using different marker substances. If *horseradish peroxidase* (an enzyme found in the horseradish and used as a marker compound in electron microscopy because of its electron density) were injected into the *musculus gastrocnemius*, its retrograde transport could be traced within the *sciatic nerve* to the motoneuron cell bodies in the spinal cord.

Detailed studies have shown that the indicator substance, first appearing extracellularly in the region of the synaptic contact zones of the neuronal terminals, is incorporated by means of pinocytosis into synaptic vesicles immediately after they release their transmitter content into the synaptic cleft. The vesicles that become enriched with peroxidase in this way eventually are attracted by *lysosomes* and transported to the cell body for the purpose of lysosomal catabolism (Fig. 9.9).

It has been shown by means of fluorescence-labeled *noradrenaline* that even components of transmitter substances are reabsorbed into the presynapse subsequent to their release into the synaptic cleft. They are then translocated via retrograde transport to the cell body for the purposes of catabolism or resynthesis. Even the so-called *nerve growth factor* (NGF), a protein that promotes both growth and differentiation of the peripheral sympathetic nervous system, is absorbed by adrenergic neuronal terminals and transported retrogressively to the cell bodies.

The retrograde transport of the *rabies* and *herpes* viruses as well as of the pathogens of *tetanus* might well be regarded as an atypical phenomenon in the realm of normal neuronal function. It is possible that these pathogenic organisms have used the ecological niche of synaptic pinocytosis and retrograde transport in a coevolutionary manner in order to replicate themselves within the neuron. The period of time, for example, between an injury and the outbreak of *tetanus* corresponds approximately to the time needed by the toxin to reach the cell body, whose metabolism it impairs, via retrograde transport. (The fact that membrane-bound *gangliosides* are involved in the incorporation of the toxic components of the tetanus toxin might represent a coevolutionary adaptation on the molecular level.)

The *rate of retrograde transport* is only about $\frac{1}{2}$ to $\frac{2}{3}$ that of rapid anterograde transport, i.e., about 100 to 200 mm/day. Its mechanism,

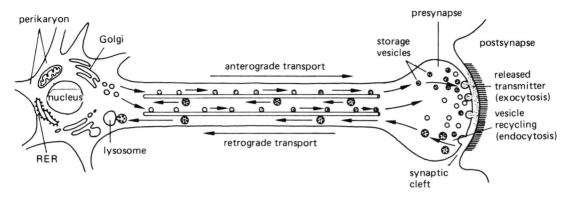

Fig. 9.9. Diagram of rapid anterograde vesicle transport and of the retrograde transport of more complex vesicles from the nerve terminal to the perikaryon subsequent to recycling.

however, is equally temperature-dependent and dependent upon the presence of intact neurotubule structures; retrograde transport, too, is halted after *colchicine* has been administered.

9.1.4 Transneuronal Transport

In discussions of the importance of the phenomenon of neuronal transport, the notion has prevailed for quite some time that the compounds newly synthesized in the cell body of a neuron are broken down to a great degree (to a small portion of compounds that flow retrogressively) after their journey to the nerve fiber terminals is completed. An elevation of catabolic enzymes in the fiber terminals lends credence to this theory. Meanwhile, evidence continues to mount indicating that neuronally transported compounds, and, indeed, not only transmitter substances (!), are transported from the terminals to receiver cells or even to the cerebrospinal fluid.

Transneuronal transport of this sort had been established for labeled amino acid derivatives, transmitter substances, and even larger molecules, such as nucleosides and neuropeptides. Apparently, not all of these compounds are broken down completely subsequent to their synthesis in the perikaryon, their neuronal transport, and their release from the fiber terminals (acetylcholine by acetylcholinesterase, for example); rather, they can be utilized in the postsynaptic ganglion cell by enzyme inductions

(synthesis of tyrosine hydroxylase, activation of secondary messenger systems, for example).

Against this backdrop, the field of neurobiology has attached great importance to the study of neuroactive peptides (*neuropeptides*) in recent years. These compounds, commonly named after the organ that they affect (vasopressin, prostaglandin, enkephalin, for example), induce neuroactive effects much as *neuromodulators* do. As peripheral hormones (insulin, glucagon, LH) or biogenic amines, neuropeptides, upon their release from nerve terminals, cause *adenylate cyclases* to be activated transneuronally in the membranes of receiver cells. These, in turn, activate the secondary messenger system intracellularly which ultimately causes the changeover in the output of protein synthesis (see Chap. 8.1 for details on distinguishing neurotransmitters from neuromodulators).

Recent discoveries of *ectoenzymes*, especially *ectonucleotidases* and *ectoglycosyltransferases*, are of great importance in assessing transneuronal transport phenomena, as described above, in relation to the overall neuronal event. The action of these subtances, which are located at the outer membrane of neurons, is greatly increased during maturation of the neuron and during formation of synapses. Of these neuronal ectoenzymes, perhaps the most important are the *ectosialyltransferases* inasmuch as their presence induces the synthesis of membrane-bound *gangliosides* in the synapse. In turn, the gangliosides play a crucial role in conjunction with extracellular Ca^{2+} in the short-term process of synaptic transmission as well as in long-term neuronal adaptations such as adaptation to temperature (see Chap. 8.2.4).

Regarding the secretion of neuronal compounds that are synthesized in the perikaryon and, subsequent to rapid transport, released from the fiber terminals to the *cerebrospinal fluid* (CSF), the mechanism in effect might be one of *cerebrospinal neurocriny*, i.e., the release of neuronal products for the purpose of humoral self-regulation within the CNS, i.e., regulation following the fluid pathway. In addition to secretions from *circumventricular organs* to the CSF, there appears to be a secretion of proteins from brain regions to the CSF that do not belong to the circumventricular system, such as from the terminals of the optic nerve subsequent to their transport from the retina. That means that CSF proteins would not have to be exclusively serogenous in nature (on the part of the plexus choroidei). They could derive from nerve tissue, as well.

Presently, it seems that the phenomenon of transneuronal transport in the nervous system is crucial to interneuronal communication and, thus, might be essential to the materialization of memory tracks in the region of the synaptic contact zones. In any case, the phenomenon of transneuronal transport is of great import when considered from the perspective of neuronal plasticity.

9.1.5 Transmembranic Transport

In essence, the term *transmembranic transport* refers to the molecular events attendant to the conduction of an impulse along nerve fibers, and to the transmission of the impulse (and, thus, information) in the region of the synapses. It centers on the processes of ion exchange as well as the exchange of compounds of higher molecular weight between the extracellular space and the neuroplasm as facilitated by complicated

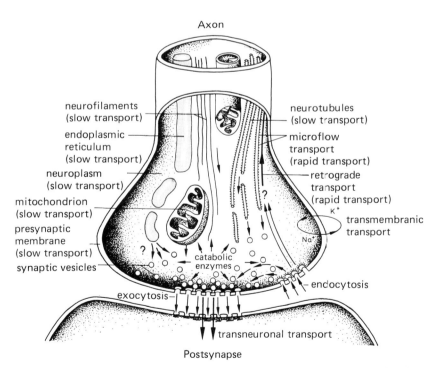

Fig. 9.10. Diagrammatic overview of the various processes of neuronal transport in the region of the nerve fiber terminal.

ion pump mechanisms and membrane-bound molecular channel systems. Chap. 7 addresses these processes in detail.

9.1.6 Significance of Neuronal Transport

In connection with the foregoing presentation of the various types of neuronal transport and their mechanisms, Fig. 9.10 depicts the most essential neuronal processes in the presynaptic fiber terminal.

Slow, anterograde neuronal transport (1 to 3 mm/day) apparently serves to maintain and renew those substances that are indispensable to the continued existence of the nerve fiber terminals, such as soluble proteins and membrane components. The emergence of such a system in the course of phylogeny in neuronal development was certainly inevitable, since, for one, it was desirable, from the perspective of impulse conduction, to develop the longest possible structures of conduction (axons and dendrites). Secondly, it was not feasible, from the perspective of molecular biology, to establish the entire code-dependent process of protein synthesis (translation) at each and every imaginable cell location and at nerve processes that are in a constant state of flux. The distances are too great between the nerve fiber terminals and the cell nucleus, with its store of DNA and molecular mechanisms necessary to form the various RNA types, to synthesize protein. Accordingly, synthesis of the most important substances to be received by structural components in the neuronal perikaryon (such as proteins) occurs in differentiated neurons under the regulating control of the cell nucleus after the necessary metabolites (primarily, amino acids) have been absorbed by the perikaryon under the selective control of the cell membranes. In addition to the synthesis of soluble proteins that comprise the neuronal hyaloplasm, there is synthesis of the proteins that serve as subunits in the formation of membrane elements, neurotubules, neurofilaments, endoplasmic reticulum, and mitochondria. The bulk of the hyaloplasm, as well as the organelles contained within, are transported continuously to the periphery subsequent to their synthesis in the perikaryon.

In contrast, *rapid, anterograde transport* (> 100 mm/day), which takes place within the neuroplasmic columns of the fibers (the neurotubules acting as structures of conduction), is apparently necessary for the renewal of those materials that are involved with impulse transmission in the synaptic terminals. Among these substances are those plasma components that are necessary for the formation of membranic structures and vesicles, those responsible for their regeneration and the regulation of their growth processes, and material for manufacturing transmitter substances.

As a rule, the notion of rapid neuronal transport is applied only to the most rapidly transported substances. Recently, however, several ranges of speed have been recognized for various substances. Due to the rather complex nature of transport mechanisms, the conveyance of all kinds of complete cell organelles is included now in the definition, and, indeed, in both directions. Rapid transport is apparently involved in the renewal of all sorts of substances needed for impulse transmission. Its immediate significance for synaptic transmission becomes especially clear: blocking it, for example, by administering colchicine, has no effect on impulse conduction; impulse transmission, however, ceases. Moreover, rapid transport might perform crucial tasks in connection with the trophic functions of the neuron, and, indeed, for all substances that underlie the particularly rapid consumption, or are required for the restructuring of specific other components. That applies, especially, to enzymes that are necessary for the synthesis or breakdown of transmitter substances, for the replacement of spent vesicles, and the control of membranic changes attendant to the vesicles that are released from the presynapse into the synaptic celft. The fact that the rapid transport of compounds of lower molecular weight is associated with the completion and restructuring of substances of higher molecular weight in the region of the synapses was exemplified by the sialoglycolipids (gangliosides) characteristic of membranes (see Chap. 7.1.1 and 8.2).

Retrograde neuronal transport, contrary to anterograde transport, moves from the nerve terminals toward the cell body. It apparently serves the process of molecular feedback and probably serves to transport substances from the synapses that ultimately are to be broken down by lysosomes in the cell body.

Transmemrbanic transport, especially of ions (Na^+, K^+, Cl^-, Ca^{2+}), constitutes the actual premise of impulse conductivity in neuronal membranes.

Transneuronal transport serves the interaction between a neuron and its receiver structures on a trophic level. Neuroendocrine interactions (those triggered by hormones) are introduced primarily in this realm. Neurons and other cells react directly to these substances that are released by neurons. This is the manner in which transmitter substances such as noradrenaline effect specific receptor molecules that are present at the postsynaptic membrane when such substances are released from the presynapse and diffuse through the synaptic celft. However, transmitter substances are not unique in the fact that they are released and dispersed transneuronally; numerous other substances, such as many neuropeptides and even nucleosides, advance transneuronally from one cell to the next where they induce changes in metabolism.

9.2 Synaptic Plasticity

The ability of individuals to change their behavior in accordance with past experience speaks to the considerable *plasticity* of the nervous system. This plasticity manifests itself not only in the phenomena of neuronal transport and the great capacity of nerve fibers to regenerate, but it is presumed that experience-dependent changes in neuronal processes also cause changes in synaptic functions. Within this context, neurobiological research has directed its attention increasingly over the last decades to the study of (a) stimulation-induced morphological changes in the region of the nerve fiber terminals and (b) functional adaptation.

Explication of the fundamentals of genetic coding in living organisms, and of the molecular-genetic background of their metabolism, was approached on a track mostly independent of developments in neurobiology. On the basis of the information obtained from such endeavors,

efforts in molecular neurobiology were directed toward examining the question of the extent to which the formation of neuronal circuiting assemblies were determined genetically, or to what degree they could develop epigenetically as a result of functional demand. In the case of the latter, the central question pertains to that point in time when experience-conditioned neuronal structures are initiated, either during early perinatal development or later, during the adult phase. This question is especially pertient inasmuch as the ability to learn and to form memories lasts throughout one's lifetime.

In light of the uncertainties touched upon here, the following will concern itself with the way in which neuronal networks are formed in early embryonic development. Then, discussion will center on the importance of adequate stimulation for the emergence of neuronal assemblies during postnatal development and adulthood. Finally, the possible mechanisms that underlie synaptic linkages and their significance for memory formation will be addressed.

9.2.1 Selective Stabilization of Synapses as Mechanisms for the Specialized Formation of Neuronal Networks During Early Development

The DNA present in the zygote is known to be capable of encoding only a few million types of proteins in the adult organism, of which an extraordinarily great percentage is common to various species. More than 60% of these proteins are involved solely with processes controlling metabolism and, thus, do not serve as structural components, as they are required, for example, in the conjoining of neuronal networks. One asks, then, how it is that the immense complexity of neuronal networks can be controlled by a relatively limited number of genes. (The number of synaptic connections in the human brain is in the area of 10^{14} to 10^{15}.) The answer to this question must be sought in the mechanism that is responsible for the manner in which the nervous system develops in its embryonic stage. The neuronal assemblies in the brain of an adult vertebrate are so complex that the assumption of a *"principle of gene economy"* is a foregone conclusion. The conceptual models offered to date take this into account. They assume that the fine wiring of neurons into neuronal assemblies must supersede a relatively gross, genetically determined pattern for development, and must do so independently of functional demand. On the one hand, one no longer assumes that the wiring of the neuronal network is predetermined to the last detail, i.e., determined prior to the onset of functional demand. Neither is it, however, fashionable to assume that the manner in which they are wired occurs in a purely accidental fashion among the participatory neurons.

In the spirit of compromise between the preformative and empirical schools to attempt an explanation of neuronal wiring, one assumes today that a fundamental genetic program directs the actual interrelationships among larger categories of neurons assisted by the mechanisms of *cell differentiation, chemoaffinity,* etc. as described in Chap. 2.2. Moreover, there exists within a genetically determined group of neurons an extremely high degree of limited

redundancy, or flexibility. This is due to the various types of neuronal contact that can occur at one and the same location. Ultimately, that means that early activation of neuronal assemblies, be it spontaneous during embryogenesis or induced externally after birth, substantially increases the specific formation and the efficacy of a neuronal group, whereby its redundancy inevitably is reduced.

In accordance with a "*selective stabilization hypothesis*" of neuronal wiring that developed from these concepts, three different stages must emerge in the formation of synaptic contacts in order that synaptic plasticity be achieved: an *unstable state, a stable state*, and a *regressive state.* In this context, nerve fiber outgrowth determines the formation of the unstable state. Neuronal contacts are capable of transmitting impulses in both the unstable and stable state. Understandably, this is not the case in the regressive state. In their unstable state, nerve fiber terminals either become stabilized or they are broken down, in which case the course of *regression* is

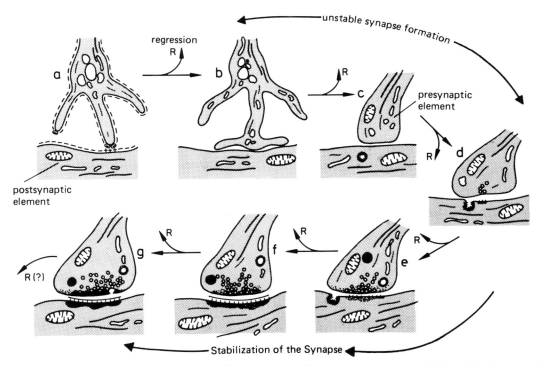

Fig. 9.11. Diagram of possible developmental sequences during synaptogenesis. *a–d*: Unstable phase in the formation of a synapse; *e–g*: stabilization of the synapse, R: possible regression.

irreversible (Fig. 9.11). A subsequent reemergence of nerve fiber outgrowth at an adjacent cell is entirely possible, however. It represents the basis for life-long, continuous nerve fiber regeneration (see Sect. 9.4).

An essential point in the emergence of long-term neuronal assemblies is that the transition in the formation of neuronal contacts from the unstable to the stable state depends, for one, on its point in time; secondly, upon the activity of the excitatory (presynaptic) fibers; and, finally, upon the state of activity of the postsynaptic cell to be stimulated. This means that incipient neuronal activity within a developing network selectively stabilizes only those emerging contact points (synapses) whose postsynaptic (receiving) cell is receptive to the process at a given, decisive moment. It follows, then, that there may be a critical phase in the development of each nerve terminal within which a specific, external pattern of stimulation can cause the hitherto unstable synaptic contact point to become stabilized.

The hypothesis of selective stabilization in developing synapses (and of the neuronal networks that result) has the advantage of the associated principle of "*economy of gene activity.*" It does not require the activation of a specific gene for connecting a specific, individual synaptic contact. Of course, *more general gene-dependent activations* are necessary for the general outgrowing of nerve fibers, supplying diverse transmitter substances to their corresponding categories of neurons, the maturation of postsynaptic neurons, and for providing substances that stabilize the synapse and maintain stabilization in the synapse for the duration of its life. Nonetheless, the concept promotes the notion that the fine structuring of the actual synaptic contacts follows epigenetically upon the foundation of a relatively less specific, basic genetic constellation of less finely structured neuronal networks, and is implemented with a great degree of flexibility. It can vary from one region of the brain to the next. It can even be so flexible that, in the case of conceivable losses of function in a given brain region (due to developmental disorders, for example), a given function can be taken over by another region. This epigenetic fine-linking of neuronal assemblies ensures the original complexity of an animal organism. *Synaptic plasticity,*

i.e., the capacity to form unstable, stable, or regressive synaptic contacts, underlies this phenomenon. Although the phenomenon of synaptic plasticity occurs only rarely in invertebrates, it is especially well established in vertebrates.

Some precedential findings are cited below from the wealth of experimental research that supports the above-mentioned *hypothesis of selective stabilization in neuronal networks*:

• Assuming that the linking of neuronal networks is determined solely by genetic factors, the anatomy of the nervous system in adult, geneticaly identical individuals of a species (e.g., identical twins) would have to be absolutely alike. In *cloned organisms*, or parthenogenically derived (and, thus, genetically uniform) water fleas and fish (guppies), it has been shown that the gross anatomy of the nervous system, including the cell count, is identical, although the ramification pattern of the nerve fiber terminals and, thus, their system of wiring, can vary greatly. One concludes from these findings that genetically determined formation does not extend to the level of rigid fiber linkage; rather, a considerable degree of flexibility is evidenced in the wiring system.

• It can be shown in neuronal tissue cultures through a variety of techniques including direct observation using time-lapse photography, structural analysis, biochemical markers, and electrophysiological leads, that the *growth cones of nerve fiber terminals* can be active functionally far in advance of the ultimate formation of a differentiated synapse. It appears that only a few growth cones on the developing end of the neurite are chosen selectively for defined synapses from the original surplus of nerve terminals. Selection is carried out by trial and error and is function-dependent. The remaining nerve terminals undrgo regression.

• During early embryonic development, there occurs, in accordance with the basic genetic program for the organization of the nervous system, a considerable *overproduction* not only of neurons, but also of unstable synaptic contacts. Regression of the surplus neuronal elements occurs largely as the nervous system matures. For example, *Purkinje cells* in the cerebellum of 8- to 9-day-old rats become innervated by at least two climbing fibers. This process of innervation is reduced to a ratio of 1:1 in 3-week-old rats and in adult animals. In tadpoles of the clawed toad *Xenopus sp.*, the ventral anterior horn regions of the spinal cord are each originally comprised of 5,000 to 6,000 cells. Subsequent to metamorphosis, however, the cell count is reduced to about 1,200. Similar regressions can be observed in the early embryonic phase in chicks and mice. A considerable regression then occurs in a later phase that centers on surplus axon collaterals and dendrites.

- A loss of target cells to be innervated during early embryogenesis (through the experimental removal of extremity formations in chick embryos or amphibian larvae, for example) results in a dramatic increase in normal *early ontogenic cell death*, especially of motor neurons and spinal ganglion cells. This leads to the conclusion that a functional interaction between the periphery and the developing nervous system is essential to the normal formation of neurons.

- *Hydrocephalus internus* (see Fig. 3.34) in humans can result from disorders in the flow of cerebrospinal fluid from the brain ventricles during very early embryonic development. Hyperpressure of the water in the brain chamber results, in varying degree, in the nonformation of the cerebrum (forebrain). Although most infants afflicted with this disorder die within several weeks after birth, numerous cases have been documented recently, based on examination by computer tomography (Fig. 9.12) and ultrasonic encephalography, that indicate that a large number of hydrocephalic children have grown to become adults and have developed normal intelligence. An intelligence quotient (IQ) of over 100 has been shown even in humans in whom more than 95% of the forebrain was dispersed by water. One of the cases was a mathematician with an IQ of 126. His skullcap was lined in the region of the forebrain with no more than 1 to 2 mm of cortex; the rest was water. For obvious reasons, autopsies cannot be used to support such studies of living patients. It is

apparent, however, that the brain is able to transfer functions of injured or undeveloped areas to other regions, and do so far more efficaciously than was once believed. As such, data pertaining to *equipotentiality* (i.e., reciprocal representation) that were obtained in the 1950s in related experimental studies on rats, have found their parallel in the human. The one prerequisite for this phenomenon is that enough brain mass be present in the mature state ("*mass action*," Lashley) so that a sufficiently great degree of wiring options can be realized.

- Contrary to the earlier assumption that there must be absolute genetic specificity in neuronal cell-to-cell recognition, numerous experiments have shown that functional synapses can be formed in the presence of the actual target cells or even subsequent to the *artificial displacement of nerve fibers* to another area. In order for this to occur, however, the receptor molecules of the postsynapses that are present in the new target area must be consistent with the transmitter of the outgrowing presynaptic fibers.

9.2.2 Function-Dependent Structural Formation of Neuronal Networks During Postnatal Development and in the Differentiated Nervous System

One would not have to go to great lengths to extend the list of experimental findings cited above that substantiate the phenomenon of selective stabilization in neuronal networks during early development. It is, perhaps, more important to draw attention to the fact that such phenomena occur not only during embryonic and early postnatal development, but that they can occur throughout the life span of a vertebrate. It bears emphasizing, too, that neuronal assemblies are formed with the onset of neuronal function. Accordingly, the formation of neuronal assemblies at the onset of neuronal functioning is doubtless the most important *morphogenic basis* for long-term development of *memory* (see Chap. 11).

Against this backdrop, it is conceivable that neuronal circuitry networks could emerge either

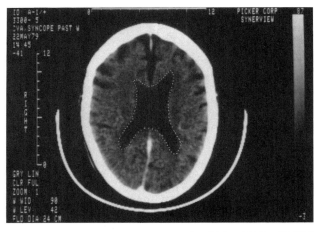

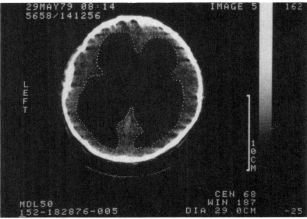

Fig. 9.12. Layer x-ray image (computer tomogram) through the forebrain of a normal human male (*a*) vis-à-vis one suffering from hydrocephalus (*b*). In the latter, about 70% of the cerebrum failed to emerge due to the buildup of cerebrospinal fluid during early ontogeny. The higher associative functions of the patient, a bank director, are taken over by the brain stem or the cerebellum. *Dotted line* is the boundary of the brain ventricles.

through long-term changes in the functional properties of genetically predetermined neuronal connections, through the selective, function-dependent formation and stabilization of new contacts, and/or through the elimination of the loosely established, unused, unstable connections (regression, see Sect. 9.2.1). The last proposal represents the viewpoint of the majority of researchers today.

Due to recent techniques in ultrastructural research that can be administered quantitatively, there has been a substantial increase in published works that have reported findings about selective and function-dependent changes in the synapse formation not only in developing nervous systems, but in differentiated, mature nervous systems, as well. The sequence of structural changes in the nerve terminals, as reported in these publications, is discussed ultimately in terms of learning and memory formation. Discussion applies particularly to the number of synapses and their manner of linkage in functionally accessed regions of the central nervous system (Fig. 9.13) and to their efficiency relative to

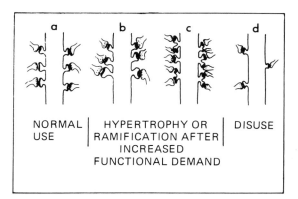

Fig. 9.13. Diagram of the structural plasticity of synapses with regard to their normal functioning, nonfunctioning, and increased functional demand.

impulse transmission from neuron to neuron, a phenomenon that can be measured electrophysiologically.

A few recent, relevant findings pertaining to the function-dependent formation of neuronal wiring systems during postnatal development and in the differentiated nervous system in vertebrates are as follows:

• Several reports have centered on the changes in the number of synaptic connections during postnatal development in experimental animals subjected to various environmental conditions. Accordingly, it was shown in various species (rodents, cats, fish) that rearing of animals under conditions of *light deprivation* generally results in a decrease in dendritic nerve terminals associated with altered formations in synaptic contact areas. For example, a reduction in the layer thickness of the optic nerve terminals in the midbrain was evident in fish deprived of light as compared to a control group that was maintained in conditions of light (Fig. 9.14a). This is accompanied by a decrease in synapse thickness (Fig. 9.14b). Parallel to this, the region of actual synaptic contact is less well defined (Fig. 9.14c) and the vesicle thickness is increased considerably (Fig. 9.14d and e). The last observation might

represent a buildup in the vesicles resulting from inactive synapses. These ultrastructural disturbances in the formation of synapses in the optic tectum (ophthalmencephalon) in fish can be correlated with a serious impairment of visual power in light-deprived animals. It could be shown in determinations of *visual acuity* that gauged *optomotor reactions* in the visual distinction of striped grids of varying width that values of visual acuity improve with age postnatally (as is the case, of course, in the control group), but that the acuity values subsequently decline drastically to the point of total, irreversible loss of sight (Fig. 9.14f).

• Structural-analytical studies of the formation of synapses during later stages of development have shown considerable differences in the differentiation of the forebrain cortex in rats that were raised in a *complex environment* (a cage equipped with facilities for play) as opposed to those kept in pairs or in isolation (Fig. 9.15). Not only was the forebrain thicker, the size of the neuronal bodies larger, and the thickness of the glial cells greater in rats that were raised in the variable-option environment, but the number of thorn, or spine, synapses overall, and, thus, the synapse density per neuron, was substantially greater (by about 20%) (Fig. 9.15). Even in learning experiments (maze learning in rats, imprinting in newborn chicks), pronounced morphological changes were observed in the respective brain regions being utilized (forebrain cortex, hippocampus). These manifested themselves in increased ramification of distal dendritic regions, an increase in the number of synapses, and an enlargement of the synaptic contact zones. The examples presented above also delineate the effects of subjecting young vertebrates to environments of total or extensive stimulus deprivation. They report the consequences of such deprivation relative to synaptic differentiation in specific brain regions and underscore the immense importance that must be ascribed the earliest possible onset of stimulation-dependent activity in neuronal structures for later corresponding behavioral development. These findings certainly can be applied to humans, as well.

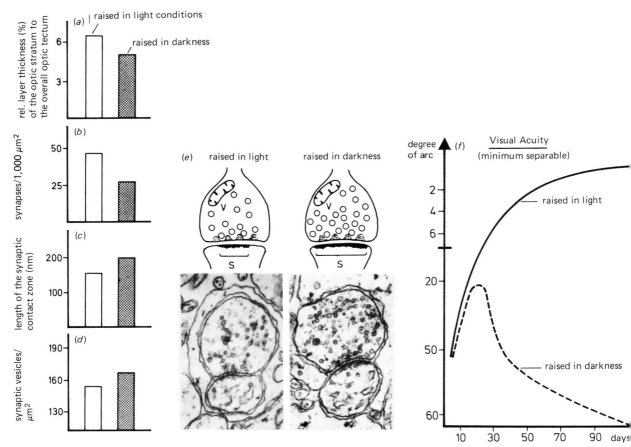

Fig. 9.14. Effect of synapse formation in the layers of the optic nerve terminals in the optic tectum resulting from raising the cichlid fish Sarotherodon for 100 days under conditions of total light deprivation: (*a*) relative thickness of the layer of optic nerve endings in the tectum; (*b*) number of synapses per 1,000 μm^2; (*c*) length of the synaptic contact zone; (*d*) density of the synaptic vesicles; (*e*) electron microscopic overview; (*f*) formation of visual acuity determined by optomotor reactions to a rotated black/white striped pattern.

9.2.3 Synaptic Plasticity in the Electrical Response Behavior of Neurons

Synaptic plasticity, in the sense of an experience-based change in neuronal function, can be established utilizing methods borrowed from the area of electrophysiology insofar as neurons are able to change their *bioelectrical response behavior* relative to stimulation. A presynaptic action potential effects changes in postsynaptic potential through the event of stimulation. Two categories of presynaptic mechanisms can account for this. In the first instance, it is the activity of the presynaptic terminal itself that

causes a more or less short-term, stimulation-dependent change in the postsynaptic cell. In the second case, long-term changes in synaptic function are determined by the effect of modulator substances in the area of the synaptic contact zones (see Chap. 8.2.4).

9.2.3.1 Short-Term Bioelectrical Response Reactions in Synapses

- Short-term *synaptic facilitation* can be shown at the motor end plate by the fact that the triggering there of a second synaptic potential—after the first has subsided completely—causes

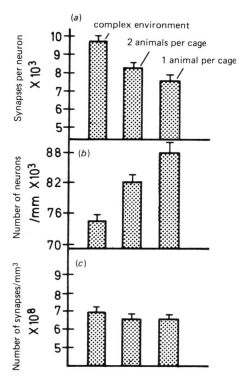

Fig. 9.15. Effects of raising young rats in various environments (complex environment, in pairs in unstimulating cages, individually) on synapse and neuron thickness in the prosencephalic cortex between the 23rd and 55th day. (Reprinted with permission from Greenough, 1984).

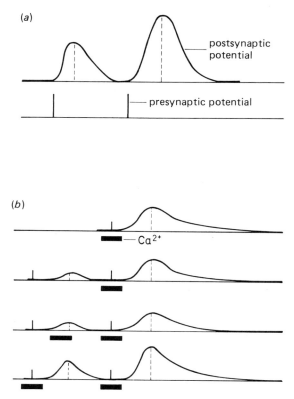

Fig. 9.16. Synaptic facilitation: (a) upon the fading of a first synaptic potential, a second, heightened postsynaptic response reaction (amplitude) is triggered [postsynaptic potentiation (PSP)]; (b) Facilitation of a second synaptic potential is dependent upon the presence of extracellular calcium during the first presynaptic action potential. (Reprinted with permission from Eckert and Randall, 1986).

a greater response reaction (amplitude) in the postsynaptic cell than the preceding potential did [*postsynaptic potential (PSP)*; Fig. 9.16a]. Katz and Miledi were able to show as early as 1968 that this facilitation was based on an increase in the concentration of intracellular, free Ca^{2+} ions in the presynapse. Ca^{2+} must be present extracellularly in the region of the synaptic terminal in order to gain access to the presynapse during the event of an action potential. The additional Ca^{2+} ions that penetrate into the presynapse through repeated stimulation cause a greater postsynaptic response reaction (Fig. 9.16b).

• If a presynaptic neuron is stimulated tetanically, i.e., with great frequency, then, given a normal extracellular concentration of Ca^{2+} (1 to 2 mM) upon cessation of tetanic stimu-

lation, there occurs a *posttetanic depression (PPD)* of the end-plate potential followed by a very rapid *posttetanic potentiation (PTP)* (Fig. 9.17a), i.e., responses to subsequent impulses are potentiated up to several minutes after the cessation of the stimulus. However, in the case of a lower concentration of extracellular Ca^{2+}, the initial depression phase comes about faster and the PTP drops off again more rapidly (Fig. 9.11b). When the concentration of Ca^{2+} is high, the transmitter units contained in the vesicles can be released more rapidly than they can be replaced ($\hat{=}$ depression). If the renewal of extracellular

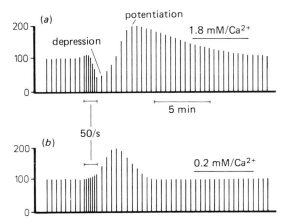

Fig. 9.17. Posttetanic depression (PTD) and potentiation (PTP) of end-plate potentials. With normal concentrations of Ca^{2+}, a long-term potentiation follows a high-frequency stimulation of a depression in the end-plate potential (*a*). Only a brief potentiation occurs in the presence of lowered Ca^{2+} concentration (*b*). (Reprinted with permission from Eckert and Randall, 1986).

Ca^{2+} is sufficient, the intracellular Ca^{2+}-binding centers become filled very quickly and remain there until they are pumped out at that point in time when Ca^{2+}-ATPases are activated (see Chap. 7.2). When the concentration of Ca^{2+} is depressed, the limited quantity of Ca^{2+} leads to diminished synaptic vesicle release that still is accompanied by the availability of transmitter units. Depression ensues; posttetanic potentiation is, however, of briefer duration.

• Rapidly repeated electrical stimulation of brain regions, e.g., of sections of the hippocampus studied in vitro, leads to amplification (potentiations) of several hours in the neuronal response behavior of those cells that were excited by the stimulated fiber system [*long-term potentiation (LTP)*]. Facilitated by the lead of electrically triggered sum potentials from the layer of granular cells in the hippocampus in the guinea pig (see Fig. 3.18), potentiations of this kind can be achieved in the postsynaptic amplitude response that are several times their initial responses (Fig. 9.18). D.O. Hebb had presumed as early as 1949 that potentiation of a synapse that was triggered by the simultaneous activation of a

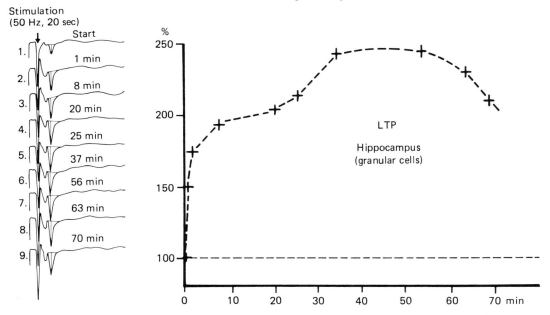

Fig. 9.18. Long-term potentiation (LTP) triggered by repeated electrical stimulation (50 Hz for 20 sec) of granular cells in the hippocampus of in vitro sections of the forebrain of a guinea pig: (*a*) amplification of the response characteristic of sum potentials; (*b*) relative change in the LTP over a period of 70 minutes.

pre- and postsynaptic neuron (*coactivation*) was the basis for memory storage. Since the LTP effect is based on the long-term potentiation of a synaptic process in which transmitters are released on the presynaptic side and depolarization is triggered on the postsynaptic side simultaneously and in greater than normal degree, the electrophysiological correlate of an incipient memory formation is rendered observable, as Hebb predicted.

9.2.3.2 Long-Term Synaptic Responses

• In addition to the LTP phenomenon, synaptic plasticity in neuronal networks has been documented in the form of substantially prolonged *adaptations in bioelectrical responses* in the nervous system as well as in long-term *adaptive changes* in higher associative performance in the *processes of learning and memory* that parallel those prolonged adaptations. It is especially noteworthy that the following occurs in poikilothermic vertebrates (fish) that, subsequent to long-term acclimatization at a constant environmental temperature, are transferred abruptly (i.e., without transition) to another water temperature (Fig. 9.19): after a brief period (ca. 5 min) of hyper-excitability and violent, convulsive swimming movements, the fish sink to the floor of the aquarium and

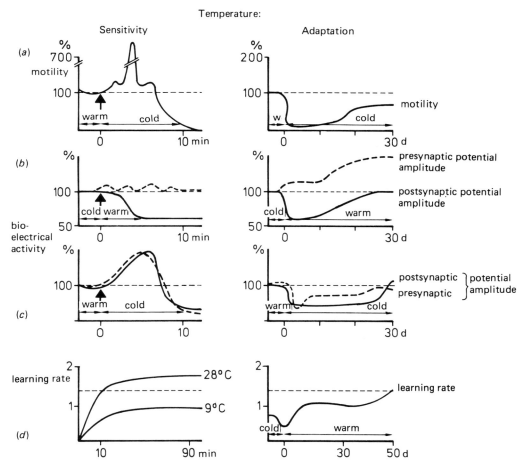

Fig. 9.19. Effect of drastic temperature change upon the sensitivity and adaptability of responses in fish: (*a*) motility (swimming activity); (*b*) and (*c*) bioelectrical activity (by strobe induced pre- and postsynaptic potential amplitudes of sum potentials in the optic tectum; (*d*) learning rate (training of electrical avoidance conditioning).

remain lying there, absolutely motionless, for about two weeks, i.e., apparently dead. Only very gradually do they regain their ability to swim, albeit on a different level (Fig. 9.19a). Parallel to the brief initial phase of spasmodic swimming behavior, it can be established by utilizing the lead of *sum potentials* in the *optic tectum* triggered by *light stimulation* that, in particular, the *postsynaptic* (and less so the *presynaptic*) *potential amplitudes* undergo very strong temperature-sensitive changes (Fig. 9.19b, c). The elapsed time of these changes is correlated directly with that of motility. The changes lasted over a period of 1 to 2 weeks. Thereafter, the event of potentiality reverted to the original level in a compensatory manner over a period of 5 to 6 weeks (i.e., to the pretemperature change level). It is especially noteworthy that these long-term adaptive changes in *bioelectrical activity* can be correlated with the gradual return of *learning capability* in the fish which was interrupted by the temperature change. Although goldfish that are maintained under warm conditions can be trained readily within 10 minutes in an electrical avoidance exercise, this is not possible in those maintained under cold conditions. If, however, goldfish that have adapted to the cold are placed in warm water, then their learning capability returns only very gradually over a period of 5 to 6 weeks.

9.2.4 Structural and Biochemical Aspects of Synaptic Plasticity

The previous section dealt with various forms of electrical response behavior (postsynaptic potentiation, posttetanic potentiation, long-term potentiation) in stimulated neuron systems that documented synaptic adaptation of a more or less short-term nature. Moreover, quite long-term adaptive changes in elicited potential amplitudes were described that show the bioelectrical event apparently is accompanied by gradual processes of adaptation that might be based in structural or, ultimately, biochemical modulations.

Such findings lead one to question precisely how functional synapses are formed. Is it that

during a life-long process of formation, only those synapses that correspond in their momentary state of maturation to a stimulus are selected to develop out of a surplus of still unstable developing synaptic structures?

9.2.4.1 Structural Aspects

Recent findings suggest that stimulation-dependent synapses form shortly after successful stimulation. It can be shown in *hippocampus* sections in vitro that a significant increase occurs in the number of stabilized synapses subsequent to short-term *LTP* stimulation (Fig. 9.20). In certain cases, electrostimulation of only 3 seconds causes lasting morphological changes in the corresponding nerve terminals. Stimulation that does not trigger LTP responses, however, does not contribute to the stabilizing of synapses.

Beyond the mere increase in the number of synapses, numerous changes in synapses can be shown in long-term adaptations of neuronal processes. In the case of fish becoming acclimatized to altered water temperatures over a period of several weeks, as reported in the previous section, distinct changes occur in the synapses of the optic tectum that affect the thickness of the presynaptic vesicles, the mitochondria of

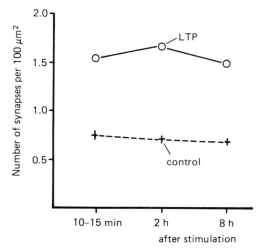

Fig. 9.20. Increase in synapse density in brain sections in vitro of the hippocampus in rats as soon as 10 minutes after an LTP trigger (electrical stimulation, 100 Hz/sec). (Reprinted with permission from Greenough, 1984).

glycogen grains, products of the Ca^{2+}-ATPase reaction, and the localization of calcium in the synaptic contact area (Fig. 9.21a; see also Fig. 7.20). These changes occur parallel to the previously cited changes in swimming behavior, changes in the formation of elicited potentials, and the reacquisition of learning ability (Fig. 9.19).

Overall, such findings relative to structural changes in the fine composition of synapses that result from an altered functional state, clearly address the extraordinary flexibility of nerve fiber terminals.

9.2.4.2 Biochemical Aspects

The question arises as to whether the previously cited structural changes in the region of the

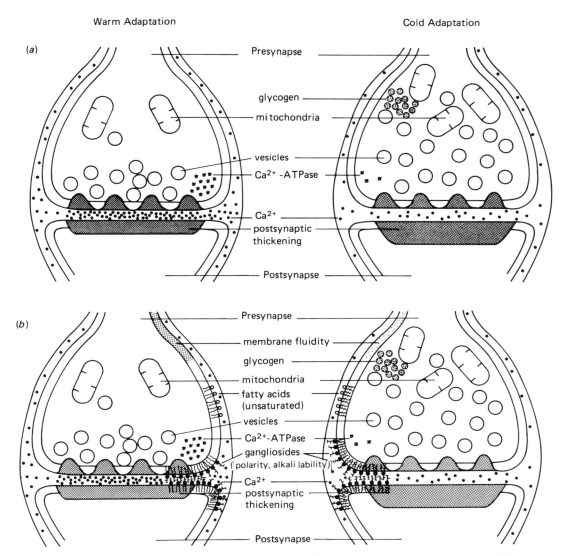

Fig. 9.21. *a:* Diagram of structural changes rendered by electron microscope (vesicle density, glycogen storage, Ca^{2+}-ATPase activity, CA^{2+} localization, membrane formation) *b:* Biochemical and physico-chemical changes (degree of saturation of fatty acids, ganglioside polarity, membrane fluidity) in brain synapses in fish maintained under warm and cold conditions.

synapse can be correlated with biochemical and/or physicochemical adaptations by which the efficiency of synaptic impulse transmission would be adapted optimally to the altered conditions. In this regard, several adaptive changes already have been identified, parallel to the above-mentioned structural findings, on the model of temperature adaptation in fish. Information yielded by electron microscopic analyses with respect to morphological adaptations in the ultrastructural area (Fig. 9.21a) can be correlated effectively with adaptive changes in membrane fluidity. These, in turn, can be traced back to molecular adaptations that pertain to the degree of unsaturation in the fatty acids in membrane-bound phospholipids and, especially, to polarity changes in the synapse membranes of concentrated gangliosides (Fig. 9.21b). Before this backdrop, two questions arise: what roles are ascribed the presynaptic and postsynaptic sides in the emergence of activity-dependent, functional adaptations (as, for example, in the LTP reaction), and which modulatory mechanisms might be involved?

Pre- or postsynaptic induction of long-term potentiation (LTP)? The question as to whether the pre- or the postsynaptic side is more critical in LTP formation cannot be answered conclusively based on the present state of research. Indeed, there is much to support the notion that the LTP phenomenon is induced postsynaptically (see Chap. 11). However, other studies indicate that the presynapse plays an equally decisive role inasmuch as a stimulus-dependent increase in the release of transmitter substances can be observed.

Broadly viewed, a synapse is, indeed, a functional entity. Not only does an exchange of substances take place in it, such as the release of transmitter substances from the pre- to the postsynapse, but specific substances (polypeptides: *retrophins*) such as the *nerve growth factor* (NGF, see Chap. 2.2.3.1) are secreted from the postsynaptic cell, as well. These are absorbed by the presynaptic membrane and delivered to the presynaptic cell body by means of retrograde transport (see Sect. 9.1.3) to facilitate appropriate trophic reactions there. The cessation of such trophic feedback from the post- to the presynaptic

cell would mean the demise of the presynaptic cell.

With respect to a gradual building of communication between two neurons in accord with an LTP phenomenon, fine-tuning or adjustment of each of the various steps is necessary. The previously described mechanisms of neuromodulation (see Chaps. 7.1.3.2 and 8) assist in just such a process of adjustment. This ensures the process of signal conversion in which the voltage-dependent activity of ion channels, ion pumps, and receptor molecules play a decisive role, even though the environmental conditions (physical and chemical parameters such as temperature, pressure, pH, ion milieu) possibly could change. Against this background, membrane lipids (especially *gangliosides* in vertebrates), in addition to certain *neuropeptides* (see Chap. 7.1.3), are critically important as *modulator substances*. As membrane matrix molecules, they surround the protein function molecules mentioned above and are able to ensure an efficient, stimulation-dependent synaptic transmission due to their modulatory variability. They are capable of doing this for the long-term, as well, in the sense of synaptic facilitation, or memory formation (see Chap. 11.2.4).

9.3 Degeneration in the Nervous System

Cell formation and decomposition (turnover) are essential to embryonic development and to the differentiation of structures. Accordingly, these antithetical processes constitute the basis of development during the entire individual cycle of living things until their natural death. Cells may die in a number of ways during early development:

1. So-called *phylogenetic cell death* refers to the demise of cells of entire structures during embryonic development that had been formed previously for functional efficiency and utility (the tadpole's tail, for example) or were present during an intrauterine phase of development (the human gill structure, for example).
2. *Morphogenetic cell death* occurs when certain

tissues assume their ultimate form (for example, the necrotic areas between fingers and toes in the plate-like structures).

3. *Histogenetic cell death* refers to a certain number of cells that die in wholly differentiated tissues. To this category belongs neuronal cell death during the early development of the nervous system. In the nervous system, the number of neurons, with very few exceptions, is not increased beyond that of its original structure. Of course, a *surplus of cells* is produced at first that is reduced in very specific ways in the course of development. This *embryonic cell death* is limited to specific areas. It is limited to specific brain centers and to several ganglion centers in the peripheral NS. Cells are decomposed in these centers within rigidly circumscribed periods of time. Generally, this occurs when the nerve fibers establish contact among themselves and with the peri-

phery; thus, when they begin to function. Surplus, unconnected cells are eliminated. The *rate of attrition* varies from region to region and can amount to between 30 and 70% of the total neuronal matter. In chicks, for example, about 40% of the original retina ganglion cell pool dies off during early development of the optic nerve.

In addition to embryonic cell death in the differentiated nervous system, there are also many causes of unnatural neuronal degeneration. Among these are traumatic injury to the nerve fiber or neuron body, damage due to toxins, drugs, medication, alcohol, genetic diseases, infection, and the absence of trophic factors (nerve growth factor, for example). Degeneration often follows *traumatic injury* to a neuron. It happens in several stages (Fig. 9.22) and quite consistently. Degeneration occurs first in the

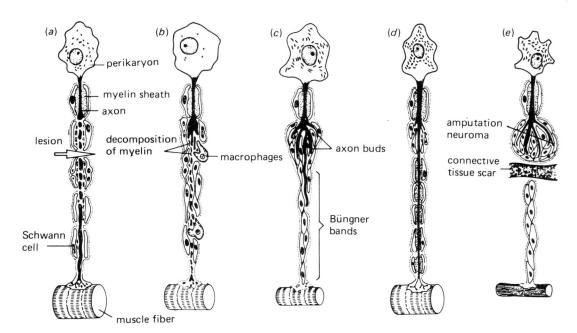

Fig. 9.22. Phases of de- and regeneration in a peripheral, myelinated, mammalian nerve fiber subsequent to severance: (*a*) wallerian degeneration (retrograde, in the stub located proximally to the severance; incipient chromatolysis in the cell body); (*b*) total degeneration of the distal branch; (*c*) progressive regeneration with axon buds growing at a rate of 1 to 3 mm/d, pronounced denervation atrophy of the

muscle fiber; (*d*) reinnervation of the target organ, neoformation of the myelin sheath, neoformation of the Nissl substance in the cell body; (*e*) neuroma formation resulting from cicatrization (scar tissue formation) of glial cells and fibroblasts, the Hanken-Büngner bands are not reinnervated, the muscle fiber undergoes fibrous metaplasia.

nerve fiber and then in the cell body (*wallerian degeneration*):

- During *primary degeneration*, macrophages migrate to the point of injury and break down injured tissue sections in the myelin sheath, proteins, and lipids (Fig. 9.22a).
- In the course of subsequent *secondary degeneration*, the distal section of the nerve fiber that was severed from the cell body region decomposes from the point of injury. This is followed by enzymatic degradation of the cytoplasm and the membranes. Ultimately, only the sheaths of the nerve fibers remain, which are referred to as *Hanken–Büngner bands*. These possibly serve as guide structures for subsequent processes of regeneration and consist of proliferating Schwann cells in longitudinally oriented cell columns (Fig. 9.22b, c).
- The proximal section of the fiber also decays during *retrograde degeneration* and forms Hanken–Büngger bands if the degenerating neuron was fully mature. In the best possible case, retrograde degeneration will not progress beyond but a few segments of the fiber, node by node, sparing the cell body. The fibers will then regenerate (Fig. 9.22d).
- Regeneration is impossible following serious damage. A fibrous connective tissue sheath forms between the distal and proximal sections that ultimately separates the two (Fig. 9.22e). A so-called *amputation neuroma* can form at the end of the proximal stub at which myelinated axon collaterals collect.
- The cell body will die as a result of very serious injury. Connecting neurons subsequently can die in the wake of a cell body death (*transneuronal degeneration*). The functional demise of the receiving cells results from the absence of trophic data coming from the degenerated cell (*inactivity atrophy*; Fig. 9.22e).

The following morphological changes occur on the cellular level during each of the above-cited degeneration events. These cytoplasmic changes in the distal section of the degenerating nerve fiber are the first to become visible: the neurofilaments become hypertonic, i.e., they swell from excess pressure; the number of mitochondria increases; the smooth endoplasmic reticulum and the mitochondria also swell; the neuro-

filaments aggregate; and, finally, the axon cylinder swells. The transport phenomenon effectively ceases. Swelling also occurs in the proximal section of the nerve fibers and in the cell body as a result of the increase in axoplasm and neurofibrils. However, the transport system remains intact on the proximal side. Since, however, the substances can no longer reach their goal, transmitter enzymes accumulate in the cell body. Note that they are returned via rapid retrograde transport (see Sect. 9.1.3). The question remains as to whether the signal for activating the regeneration program in a damaged axon, i.e., the triggering of regenerative and restorative functions, might be a factor of the altered rate of retrograde transport.

In the case of progressive degeneration, chromatin in the Nissl substance, i.e., the rough ER, breaks down (*tigrolysis*, Fig. 9.22b). The synthesis of RNA is reduced and, as a result, the synthesis of proteins is halted. Finally, the cell nucleus and the plasma membranes break down.

Beyond the causes of neuronal degeneration suggested above, hormones can play an important role, also. For example, a thyroxine deficiency during early embryonic development can impair normal development of the brain so drastically that severe mental disorders (*cretinism*) can result. On the other hand, even minimal doses of thyroxine can trigger neuronal cytolysis.

Beyond the unnatural causes of degeneration in the nervous system which were mentioned earlier, brain research over the last 30 years has assumed almost dogmatically that, in the aging human, one million cells, of the approximately 14 billion neurons of the cortex present at birth, die off daily. Recent, more precise measurements of a great number of brains of varying size have led to the understanding that *brain cells in the healthy human do not die*. Indeed, the individual cells become smaller; however, their count remains unchanged. During the first 60 years of life, brain wieght and volume remain relatively constant. Thereafter, both factors will have decreased, on average, 7 to 8% by age 90. This *degeneration of brain substance* is solely a factor of *neuronal shrinkage*, presumably brought on by a general loss of water. Since the extent of this shrinkage is far greater in young brains than in older ones, there arose the false impression that the cell count in younger brains must be greater than in older ones. Although the cell count in the various brain regions also remains constant, the shrinkage rate in the individual brain sections can vary greatly. For example, between the ages of 20 and 110, cell size in the *visual cortex* (area 17) remains effectively constant. Likewise, there is no drastic reduction in volume in area 7, the sensory processing region for visual impressions. However, a reduction in cell volume of 30 to 35% can be observed in the motor region after age 45. It is not yet clear whether or not these differences in shrinkage are genetically programmed or triggered with age by the abatement of certain functional requirements. In support of the latter notion, the loss of one's nimbleness of mind, induced either by old age or extended stays in the hospital, for example, can be regained by appropriate psychophysiological exercises. That would not be the case if the cells had died.

Finally, the understanding that the cell count in the brain does not decrease with age has extraordinary consequences when *rehabilitating deficits in brain efficiency*. It can be assumed that specialized exercises, "*brain-jogging*," if you will, can increase the efficiency of the brain vis-à-vis failing recall of that which was learned previously, or for new learning. Even in old age, the human brain retains a considerable degree of efficiency and, thus, remains capable of learning.

9.4 Regeneration in the Nervous System

Cell formation and decomposition (turnover) is a normal phenomenon in the cycle of a living organism. Individual organ systems may have different rates of turnover. For example, the turnover rate of human blood cells or epithelial cells is especially high. The nervous system is of particular interest in this regard as neuronal division in the higher vertebrates is limited almost exclusively to the phase of embryonic development. Subsequent to age 2, cell division in the human CNS can be observed only in glial cells. Only in the lower vertebrates (fish and amphibians, for example) do a few so-called *matrix zones* remain intact postembryonically.

These zones pertain to brain areas that can replace epithelial cells of the olfactory system and, in particular, to periventricular structures that exhibit an especially high rate of mitosis during embryogenesis. In adult amphibians and fish, these matrix zones can be reactivated subsequent to injury. They account for the relatively great capacity of nerve tissue to regenerate (the regrowth of lost extremities, for example). Neuronal regeneration does not occur in higher vertebrates. In general, a neuron dies if the cell body has been damaged, since this injury is attended by an irreversible cessation of function in the neuron.

If, however, only the axon is affected, it can regenerate by way of the intact cell body (*axonal regeneration, synaptic regeneration*). This holds true for peripheral axons and for the CNS both in lower and higher vertebrates, albeit with distinct differences between the two vertebrate groups:

Following regrowth, for example, regenerating motoneurons in a salamander would have their former end plates exactly in the original fiber. This is not at all the case in mammals; rather, the muscle fiber would become innervated at the first, best location.

Sensory axons behave similarly in both cases; however, it can occur (as observed in axons of the optic tectum in the goldfish) that they make "detours" en route to their intended area of innervation. These pathway digressions are corrected over time.

Fiber regeneration occurs from a fine axon tip that emerges at the distal stub of the axon (Fig. 9.22b, c). It has the appearance of a neuronal growth cone during embryogenesis. A new axis cylinder grows in anterograde fashion from the perikaryon at a rate of about 1 to 3 mm/day. It is likely that pioneer fibers also emerge that serve as guidance structures, or pilots for axon outgrowth. The *Hanken-Büngner bands*, which result from a proliferation of Schwann cells, also serve as guidance, or pilot structures. In regenerating neurons, they grow in the shape of a mushroom from the degenerating stub of the nerve fiber and then function as a pilot structure for the axon buds growing into them during continued regeneration. The possibility for an outgrowing nerve fiber to reconnect is limited temporally by the rate at which the degenerating, separated fiber decomposes. Thus, the Hanker-Büngner bands probably function as pilot structures only in cases of mildly damaged axons (Figure 9.22c, d). In cells that have suffered total destruction, degeneration proceeds far too rapidly for regenerative processes to take place.

Although the new axon cylinder emerges at the normal growth rate of 1 mm/day, a significant increase in the synthesis rate of certain subatances can be observed in the cell body (perikaryon). This can be recognized morphologically in an enlarged cell body, enlargement of the nucleoli (i.e., an increase in RNA synthesis), and in an increased synthesis of protein. Naturally, this is attended by an increase in the amount of substances moving by anterograde transport that ensures regeneration of the axon.

Since the outgrowing fiber plays no functional role in neuronal transmission, there is, at first, a reduction in the synthesis of transmitters accompanied by an increase in the rate at which other enzymes are synthesized. These latter enzymes control the synthesis of membrane proteins and accelerate the formation of other soluble proteins such as actin and calmodulin.

It has been shown in rhesus monkeys that even the CNS in higher vertebrates is partially able to regenerate fibers and to form new synapses. The projections of the fibers can be reconstructed, as well. In such experiments, a specific region in the frontal lobe of the cortex is removed. It has been determined, by labeling transported proteins with radioactive amino acids, that the fibers gradually regenerate over time and new synaptic contacts are estabished. This experiment demonstrates the impressive *plasticity of the nervous system.*

Although the outgrowing neuron is controlled by the body of the neuron, special signals must emanate from the cell that is to be innervated. These signals must be eminently clear and unambiguous so that the outgrowing nerve fiber recognizes the cell with which it must establish contact. For example, each muscle cell must be marked differently so that it can be recognized. The research of Weiss (1931) showed that these signals have to do with the given developmental state of the cell and with its function. Students of Weiss found in their studies of the optic nerve

that the principle of this chemical emission of signals applied not only to the peripheral region (nerve-muscle), but to the CNS, as well. They could show that severed optic nerves in amphibians were restored and that blinded animals regained complete visual efficiency within weeks or months. Sperry (1963) studied *chemoaffinity* in the development of neuronal connections with regard to possible regeneration in lower vertebrates. Repair of the severed optic nerve in higher vertebrates can be achieved only to a very limited extent. The same holds true for functions in the peripheral neuromuscular area. Successful restoration of functions is limited to certain developmental stages in the development of the individual. Thus, in humans, restoration of the optic nerve following damage during a very early embryonic phase leads to normal visual capability. If, however, a period of 26 days is exceeded, then the untimely repair will result in reversed vision in the adult.

Biochemical analyses have shown that a sudden reduction in repair will result in reversed vision in the adult.

Biochemical analyses also have shown that a sudden reduction in the rate of DNA synthesis occurred in the retina during this 26 day period. In this regard, Sperry had established as early as 1944 that the chemical specificity of neurons develops much along the lines of a road map. Morphogenetic substances from the epidermal area affect the developing nervous system in various ways during embryonic development. Their effect is that of chemical diffusion gradients that proceed from the epidermis and influence the formation of characteristic features. Glycomacromolecules that are present at the outside of the membrane, most especially gangliosides, have more than a passing involvement in cell-to-cell recognition in the regeneration of nerve fibers.

The answer to the question of why the restorative capacity of severed nerves in higher vertebrates (especially in mammals) is so limited might be found in the diverse configuration of the *astroglial cells* in the higher versus the lower vertebrates.

Numerous studies of the CNS in goldfish indicate clearly that only the *oligodendroglial cells* are involved in regeneration, but that astrocytes are essential to the growth and regeneration of fibers. Moreover, the astrocytes are capable of producing *laminin*, a substance that is thought to promote neurite growth, especially in the presence of lesions. Presumably, astrocytes in higher vertebrates do *not* synthesize laminin. This has been established for *astrocytes* in the mammalian brain. Another difference between lower and higher vertebrates lies in the membrane properties of astrocytes (see Fig. 1.19). "*Orthogonal arrays*," or "*assemblies of particles*" (AOP) are found in chicks, pigeons, and rats. These rod-like particles are absent in the membranes of astrocytes in lower vertebrates.

Presumably, both differences (laminin and AOP) are significant factors contributing to the capacity to regenerate. One hypothesis in this regard would suggest that, in general, all astrocytes are capable of producing laminin; in the phylogenetic series, however, some astrocytes "unlearn" the process of laminin synthesis (probably postnatally) in conjunction with the introduction of the AOP; thus, the capacity of degenerated neurons to grow out is reduced and an impenetrable fibrous connective tissue barrier can form between the neuron stub and the degenerated neuron section, effectively preventing regeneration.

10

Behavioral-Physiological Basis of Memory

It is essential to be entirely familiar with the behavioral spectrum of an organism in order to understand the phenomenon of memory. The determination as to whether an organism has or has not formed memory can be established only on the basis of lasting changes in behavior in response to specific environmental stimuli.

In general terms, *behavior* is understood to mean the totality of observable (measurable, perceptible) responses, or changes in condition, in organisms relative to stimuli. In the narrower sense, behavior is the transformation of neuronal activity into outwardly detectable responses. Among these are motor activity, sound production, odor exudation, etc. These neuronal activities can be effectuated either by an inherent condition or by changes that occur within the organism resulting from altered environmental factors.

Behavioral responses can serve the process of communication between an organism and its environment. Normally, such responses are intended to improve the organism's chances of survival. Moreover, they serve inter- and intra-species communication, since they, in turn, elicit responses in others.

Information responsible for the development of behavioral expression can be stored in the genome (species-specific memory) and passed on to the next generation. It is likely that the resulting *innate behavior* predominates both in lower vertebrates and invertebrates because their nervous systems are less plastic than those of higher vertebrates. Information that leads to the formation of behavior also can be acquired from the environment in the course of individual development (*individual memory*). *Acquired behavior* is particularly characteristic of higher vertebrates whose nervous systems allow for a great degree of behavioral adaptability vis-à-vis changes in the environment.

Within the context of behavioral development, both innate and acquired, the following section will address the fundamentals of modern behavioral physiology. Accordingly, Fig 10.1 illustrates how the spectrum of innate versus acquired behavior patterns shifts with increasing phylogenic development in animals. It reveals that, in addition to the fundamental program of inherited behavior, behavior acquired individually is characteristic of groups of higher phylogenic development. Acquired behavior is based on the capacity of the organism to learn and to store the individual experiences for the long-term as memory. Thus, in the course of ontogenic development in animals and humans (both on the phylogenic and individual level), higher organisms have improved their capacity to adapt to changing environmental conditions. Consequently, an individual's responses to the environment can be rooted in three formats: (a) innate programs (program structures) that function reliably and stereotypically upon maturation without the benefit of corrections acquired individually through learning, (b) specifically adapted responses predicated upon individual experience, and (c) a combination of these two formats.

In each instance, an organism receives information about changes in the environment select-

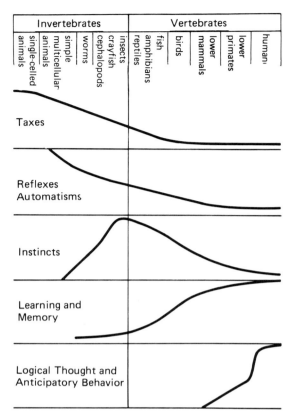

Invertebrates							Vertebrates						
single-celled animals	multicellular animals	simple animals	worms	cephalopods	crayfish	insects	reptiles	amphibians	fish	birds	lower mammals	lower primates	human

Taxes

Reflexes
Automatisms

Instincts

Learning and
Memory

Logical Thought and
Anticipatory Behavior

Fig. 10.1. Relative shifts in various modes of behavior in the course of phylogeny.

ively and actively through sensory organs. These data are processed and transformed in the functional structures of the brain in such a way that the resulting response is intended to increase the individual's chances of survival against the backdrop of changing conditions. Storing the characteristics of the new situation supplementally to those characteristics that were acquired previously ultimately enables the organism to adapt its model constantly to new environmental characteristics. More highly developed species even may be able to anticipate future events, i.e., to behave according to a plan and, thus, to achieve an ever-greater degree of behavioral independence from the environment.

10.1 Phenomenology of Memory

Earlier chapters promoted a step-by-step approach to understanding the structural basis of the nervous system and its underlying neurobiological mechanisms. Upon that footing, the following section will present briefly the most essential modalities of innate and acquired behavior to be encountered in the animal world. It will become apparent how important the role is that memory plays in the spectrum of behavioral expression. The phenomenology of memory deserves comment, therefore, here at the outset. Chapter 11 will focus in detail on the present state of knowledge relative to the functional modality and functional mechanisms of memory.

Memory is the capacity to store and recall (ekphoriate) individually acquired information. It is comprised of numerous processes in which the following sequential components are involved:

- Reception of information from the environment through the sensory organs,
- Selection (filtering) of the information to be stored,
- Permanent storage of information (*formation of engrams*),
- Networking with other stored information,
- Reactivation of stored information (*ekphoriation*) in the sense of remembering.

All multicellular animals having neurons have memory. Evidence of memory in single-celled animals has yet to be produced. Human memory functions on three levels (Fig. 10.2a, b):

- Regarding *short-term memory*, all information arriving from the sensory organs is stored for ca. 6 to 25 seconds. It includes our active *consciousness* which is comprised of information that the nervous system obtains directly through the sensory organs or that is reactivated from long-term storage (*ekphoriation*) in the sense of *remembering*. Approximately 10^9 to 10^{11} bits of information reach the nervous system every second from all of the sensory organs (1 *bit* represents the amount of information that can be obtained by a yes–no decision). However, only about 16 bits per second can flow into one's consciousness (short-term

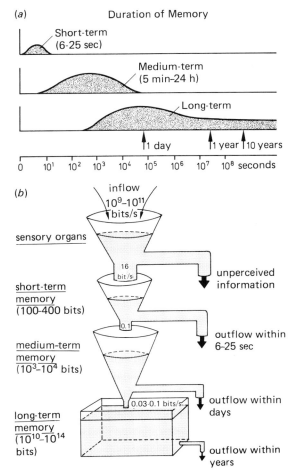

Fig. 10.2. Diagram of the three stages in human memory: short-, medium-, and long-term. (*a*) Duration of memory in seconds. (*b*) Conceptualization of information inflow, outflow, and capacity.

flood of incoming information only the most essential to the individual.

- *Long-time memory*, receives only very few bits of information (0.03 to 0.1 bits/sec) per unit time. These are passed on from medium-term memory subsequent to selective control processes and are to be stored for the long-term, i.e., for days, months, or even a lifetime. The capacity of long-term memory is thought to be 10^{10} to 10^{14} bits.

Several factors mentioned above that pertain to the flow of information responsible for regulating human consciousness are presented in the overview illustrated by Fig.10.3. Even though many details are known today that pertain to the areas of information reception and output, our knowledge of information processing and storage in the brain is still quite spotty. In regard to the emergence of the three levels of memory, various mechanisms are under consideration that are discussed in greater detail in Chapter 11. A plausible explanation for the phenomenon of *remembering* has yet to be advanced.

The problem of *forgetting* remains unexplained, as well. This process runs counter to memory formation, i.e., retaining and recalling. At issue is the fact that an organism becomes unable to recall (reproduce), either totally or in degree, something that was learned or perceived.

Even though we assume that the ability to recall a stored memory event potentially lasts a lifetime, or as long as the participatory neuronal structures are intact, we are confronted continuously with the phenomenon of forgetting. The following points apply to "forgetting":

- With regard to stored information (i.e., engrams), more will be forgotten the longer the information is stored and the less the information is called upon during storage.
- Unimportant, less meaningful information will be forgotten before basic experiences are forgotten.
- Impressions following a learning process greatly affect the extent of the forgetting.
- The motivational milieu at the time of storage as well as during the attempt to remember greatly affects retention.

Presently, the precise manner in which acquired information is stored for life, i.e., through which

storage); the remainder of the information goes unperceived. Consequently, short-term storage has a capacity of about 100 to 400 bits; these values, however, may vary according to the age and condition of the individual.

- *Medium-term memory* takes only a fraction (0.3 to 1 bit/sec) of the information from the short-term reservoir. The extent of its storage time ranges from several minutes to several hours. Medium-term storage capacity is thought to be ca. 10^3 to 10^4 bits (Fig. 10.2b). It represents the unstable precursor stage to long-term memory. It filters from the continual

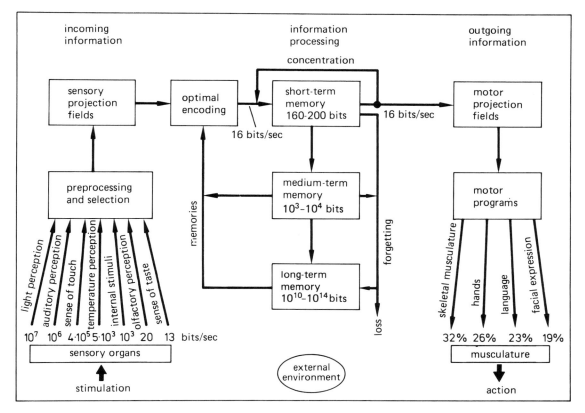

Fig. 10.3. Diagram of the information flow underlying consciousness in humans. (Reprinted with permission from Frank, 1969).

structural-molecular mechanisms stimuli are fixed in the form of *memory tracks* (*engrams, mnemes,* or *residues*), is understood only in broad terms (see Chap. 11). More intensive study in the area is required, for memory research has been challenged in recent decades in ever-increasing degree by a startling surge in *memory disorders* (dysmnesias) encountered worldwide within general populations. These disorders concern temporary and permanent changes in memory capacity that manifest themselves as illnesses of an organic and/or emotional nature. For example, total or partial *amnesia* occurs subsequent to severe cerebral concussion. Memory deficits relative to events that are experienced under heightened emotions and limited consciousness (*hypomnesia*), on the other hand, are usually psychogenic in nature. Serious memory deficit can result from arterial calcification or other diseases of the brain that are accompanied by a

progressive partial breakdown of neuronal matter and *senile dementia.*

Even these few cursory comments serve to illustrate the urgent need to advance basic memory research with the greatest possible commitment. One essential requirement in this regard is to understand fully and with utmost clarity the totality of innate and acquired behavior in human beings and in animals.

10.2 Innate Behavior

The following sections will address the various expressions of innate and acquired behavior. The two categories will be delineated separately even though it is apparent today that a clear distinction between the two is not always possible. This is especially true on the level of the higher vertebrates, since it is not always feasible to determine with absolute certainty whether the manifestation of a specific behavioral characteristic is innate in origin, or can be traced back to environmental factors.

Innate behavior can be identified by using a number of research methods:

1. *Behavioral-genetic studies*: certain behavior ceases when individual genes or gene groups are eliminated. The primary subject of research in this regard is the fruit fly (*Drosophila*).
2. *Rearing progeny under conditions of experience deprivation* (*Kaspar Hauser experiments*, see Sect. 10.3.1.1). The so-called Kaspar Hauser experiments offer information about innate versus acquired behavior in animals. (Kaspar Hauser, supposedly born 30 April, 1812 and murdered in Ansbach 17 December, 1833, was a foundling of unknown origin who appeared one day in 1828 in Nuremberg. Apparently, he had grown up in nearly total isolation). If certain behavior appears during or subsequent to *Kaspar Hauser experiments*, i.e., conditions of stimulus deprivation, then they may be considered to be innate. However, if characteristic behavior is not demonstrated under such conditions, then one may not assume that these behavioral components must be acquired because isolation experiments often lead to degenerative symptoms.
3. *Cross-breeding experiments*: the species specificity of behavior and its inherited nature are represented especially well by means of cross-breeding experiments that are combined with Kaspar Hauser experiments. For example, it has been found in cricket hybrids that their song is a blending of the two parent species. In studies of this sort, it is necessary first to study the individual species and their song in soundproof surroundings.
4. *Determining form consistency in behavior*: when studying inventories of behavior and the sequence of individual responses in members of a species, many forms of behavior can be filtered out that occur consistently in the same fashion without being influenced by external stimuli.
5. *Uniformity in the emergence of first-time behavior* can be observed in young animals in

many species. Various methods are used in this type of study and great cross sections of species with numerous individual representatives are involved.

10.2.1 Taxes

Taxes are genetically predetermined movements of orientation that are directed toward external sources of stimuli. They are evident in all animals, including monocellular organisms. Movements of orientation can occur vis-à-vis light (*phototaxis*), shade (*scototaxis*), temperature (*thermotaxis*), gravity (*geotaxis*), or objects that stimulate through contact (*thigmotaxis*). Such movements

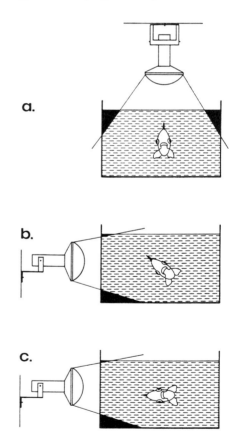

Fig. 10.4. Combined orientation of a fish to light and gravity (photo- and geotaxis): (*a*) normal vertical position with light from above; (*b*) slight lateral position toward side-lighting; (*c*) full lateral position in response to side-lighting with total separation from the otoliths of the vestibular system.

of orientation can be both positive and negative in nature (positive and negative phototaxis, for example; Fig. 10.4). In multicellular organisms, taxes are associated with special stimulus reception systems, i.e., sensory organs, and stimulus-processing neuronal structures that control motor response.

10.2.2 Reflexes

Unconditioned reflexes are genetically determined motor (or secretory) response reactions that elapse in a stereotypical fashion to simple sensory stimuli. The anatomical basis for a reflex response is the reflex arc. It consists of an afferent fiber that receives impulses from an effector (sensory) cell, conducting them via synaptic connection to an efferent fiber that, in turn, affects the effector cell (*monosynaptic reflex arc;* see Chap. 5.1). If one or more interneurons are linked into such a reflex arc, then impulses from other neuron chains can affect the course of the reflex process, or they can conduct impulses further to other neuronal assemblies (*polysynaptic reflex arc*). A discussion of the reflex wiring can be found in Chap. 5.

Depending upon the type of wiring, one can distinguish in humans between *animal (somatic)* and *vegetative reflexes. Animal reflexes* are subject to control by the consciousness, whereas vegetative reflexes occur without influence of the will. Animal reflexes can be subcategorized as either proprioceptive reflexes or reflexes in response to external stimuli.

In the case of *proprioceptive reflexes,* the beginning and end of the reflex pathway are found in one and the same organ (skeletal musculature, for example; see Fig. 5.3). They are connected only by a monosynaptic crossway and conduct impulses of very brief reflex and refractory periods. Proprioceptive reflexes are exemplified by the patellar reflex, the Achilles tendon reflex, and eyelid reflex (in response to physical contact with the lid).

The beginning and end of the reflex pathway in those *reflexes* that *are in response to external stimuli* are located in separate organs, a condition requiring polysynaptic linkages with substantially longer reflex and refractory times (see Fig. 5.5). These external response reflexes

primarily serve functions of protection (e.g., coughing, eyelid closure), defense (wiping, scratching), approach (sucking), and regulation (body posture) in the organism.

Vegetative reflexes regulate those body processes that normally transpire without conscious intervention. The circuitry for these processes runs through specific vegetative centers in the spinal cord, the myelencephalon (medulla oblongata), the diencephalon, and remote nerve centers (ganglial nodes) in the periphery, for example in the region of the intestines. Some products of vegetative reflexes are sweat secretion, "goose bumps," a reflexive increase in the rate of breathing due to chemoreceptors signaling O_2 deficiency or CO_2 excess in the blood, swallowing, vomiting, and salivating.

Imagination reflexes have special significance in higher associative brain functions. These represent a phenomenon whereby an external stimulus, one that was required initially to stimulate a vegetative reflex, is replaced by the thought of the imagined stimulus. For example, salivation can be brought on not only by seeing or smelling savory sustenance; the mere thought of a "mouth-watering" morsel can, indeed, make the mouth water! Conversely, aversive reflexes, such as vomiting, can be triggered in an analogous way by summoning associations of revulsion.

Generally, then, emotional conditions such as fear, horror, joy and the like can be triggered by imagining them and in doing so, organ functions such as heart rate, breathing, perspiration, vascular dilation, etc. can be altered considerably. Such phenomena find practical application in the area of applied psychology (lie detection, for example).

Psychosomatic causal relationships frequently are rooted in the fact that emotional conditions can affect organ functions via the vegetative centers of the diencephalon. The connection becomes clear by example of *autogenous training,* a process by which the normal process is reversed through intensive practice in order to affect vegetative functions deliberately.

10.2.3. Instincts

TINBERGEN regards *instincts* as "hierarchically organized nervous mechanisms which respond to cautionary, arousal and orientating impulses, both internal and external, with well coordinated movements intended to protect life and preserve the species."

IMMELMANN defines instincts as "innate mechanisms of behavior which manifest themselves in organized sequences of movement (hereditary coordination), and which are activated by specific stimuli via a neuronal release mechanism."

These definitions would suggest that an instinct is an immensely complex neuronal mechanism which is comprised of various component parts. The factors of appetency behavior, innate release mechanism and culminating behavior all must be synchronized carefully in order that an instinctive behavior can take place in a properly coordinated manner.

10.2.3.1 Appetency Behavior

Instincts are based on internal factors which interact in such a manner as to promote an energy build-up in the CNS which places the entire organism in a state of readiness commensurate with a behavior (appetency, mood, drive). *Appetency* may be defined as a specific and adaptable craving for a situation that offers a triggering stimulus for purposes of introducing culminating behavior.

As a rule, various factors are involved in originating an appetency behavior, such as responses to sensory stimuli (internal or external), spontaneous behavior induced by activity in central nervous automatism centers, and hormonal changes.

Motivation (see Sect. 10.3.3) can be viewed as an internal factor that influences behavior. In stark contrast to culminating behavior (see Sect. 10.2.3.3), appetency behavior is characteristically highly plastic and variable in nature. It may consist of a simple taxis (see Sect. 10.2.1), or a complex chain of behavior (bird migration, for example). Appetency behavior must not result necessarily in culminating behavior; rather, it can be replaced, or superseded by yet another special appetency behavior. It can even be interrupted and discontinued. Appetency behavior need not always be a positive craving; it is just

as likely to be an avoidance response manifesting itself in aversion behavior.

10.2.3.2 Innate Release Mechanism (IRM, Releaser)

The term *release mechanism* is a collective one that applies to all peripheral and neuronal filtering mechanisms that ensure that not all stimuli which act upon an individual trigger a behavioral response; rather, only those stimuli, the so-called key stimuli (those of biological importance to the individual) will result in behavioral response. A *key stimulus* is understood to mean an external stimulus or combination of stimuli that sets in motion or maintains a specific behavior.

Within this framework, a key stimulus is characterized as a *releaser* if it has a triggering function within the context of species-specific communication.

Its origin can be either experimental or innate in nature (IRM, *releaser*). The properties of the releasing stimulus often must be learned in detail. Thus, so-called following behavior in chicks is innate, whereby they chase after the first moving objects they see upon hatching; detailed differentiation of the object is a capability that is acquired in time.

The IRM can be analyzed by using mock-up experiments in the course of which the effective key stimuli must be distinguished by studying neutral objects (Fig. 10.5). Results have shown that, relative to its effect, a releaser can be disproportionately simple and less specialized. If one abides by the metaphor of a lock and a key (the origin of the concept of "key stimuli"), then the key would not be one that is specialized to fit only one lock, but rather would be a sort of master key. Moreover, as the releaser for several extremely complicated sequences of behavior, the stimulus can behave quite abstractly and generally. The capability of more highly developed animals (in particular, the human being) to generalize and to abstract as part of the learning process (see Sect. 10.3.1.10) (i.e., not merely imprint details) may be rooted in this phenomenon. It is interesting to note that natural objects do not always elicit the optimal effects that can be prompted by artificial "*supranormal*

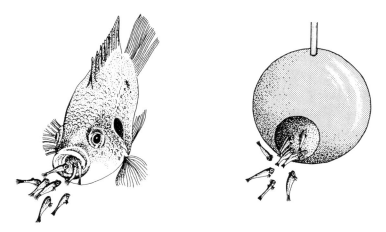

Fig. 10.5. The young of the orally gestating cichlid (*Tilapia sp.*), when threatened with danger, will seek an opening in a spherical simulation of the mother's mouth.

releasers." A natural orange-red spot on the beak of feeding sea gulls, for example, brings forth in younger birds only minimal begging behavior when compared to an artificial black spot placed on the beak of a mock-up (Fig. 10.6).

Releaser mechanisms (it cannot always be determined whether they are innate or acquired gradually) are evident in human beings, as well. With regard to the recognition of facial expressions, a baby seems to possess an innate releaser

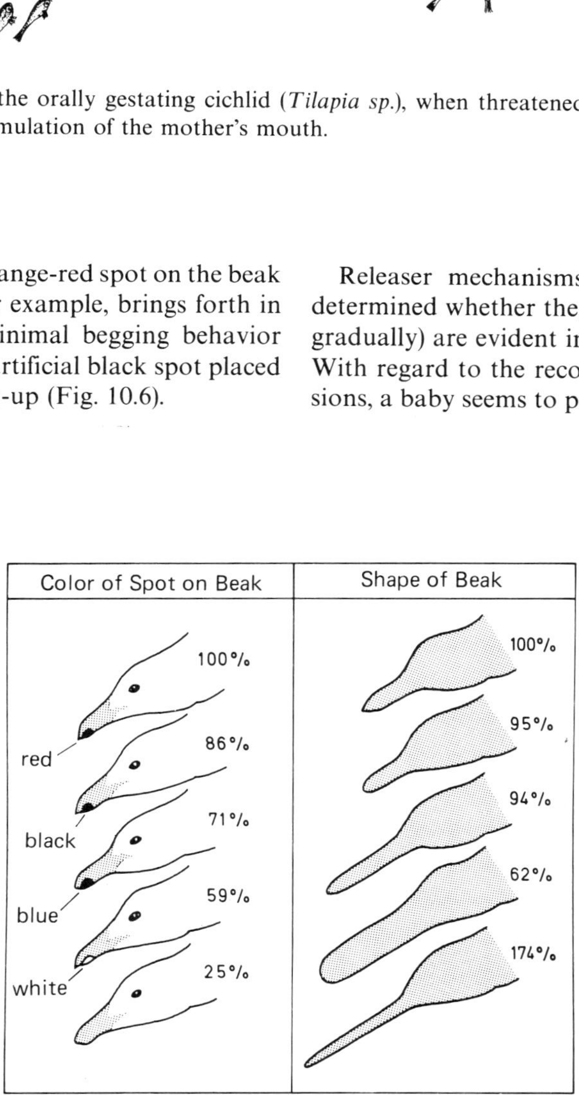

Color of Spot on Beak	Shape of Beak
red — 100%	100%
86%	95%
black — 71%	94%
blue — 59%	62%
white — 25%	174%

Fig. 10.6. Shape of the parent's beak and color of the spot on the parent's beak as stimuli in eliciting pecking responses in silver gull chicks. The figures indicate relative effectiveness of the mock-up in the selection test. (Reprinted with permission from Tinbergen, 1966).

mechanism for recognizing a smiling expression as opposed to one of anger on the parent's face.

The youngest of children respond negatively to a vertically furrowed, "angry" brow, whereas a high, raised (i.e., "laughing") oral fissure elicits a positive response. Conversely, parental stirrings are evoked by the so-called innocent babe model, i.e., by short faces, domed foreheads, round eyes, and chubby cheeks (Fig. 10.7). Even in the choosing of one's partner, releaser mechanisms may be responsible for the fact that partners in personal relationships frequently resemble one another (often more so with increasing age); or they are often diametrically opposed in appearance.

10.2.3.3 Culminating Behavior

An instinct event concludes with an instinct-specific *culminating behavior*, a rigid, genetically fixed manifestation of behavior. The organism is satisfied in its drive only upon manifestation of that specific behavior. Swallowing, for example, represents the culminating behavior of the instinct to self-nourish, whereas copulation is that of the instinct to reproduce.

The rigidity of culminating behavior has also been documented in terms of ineffectual, or neutral, behavior under certain conditions. A dog, for example, "digs" itself an imaginary hollow in which to sleep, even on smooth household flooring. On the other hand, the progression of an instinct can be suppressed when culminating behavior is repeated frequently and in rapid succession. A chaffinch ultimately refrains from displaying warning behavior when a mock-up of an owl is presented to it repeatedly, presumably because it has learned that the inanimate impostor presents no real danger.

Conflicting situations can lead to *inappropriate behavior* when, for example, the appetencies of fight and flight are equally urgent. In such instances, the impulse cannot be directed toward an unambiguous response and the result is often unreasonable or absurd behavior, such as yawning, scratching, stroking one's hair, or biting one's lower lip.

10.2.3.4 Stimulus-Response Chains (Instinct Chains)

Based on his behavioral studies of ducks, Lorenz indicated that instincts emerged gradually in the course of phylogeny, apparently through the linking of sequences of behavior and movement that exist independently of one another. Over time, mechanisms of coordination developed for modes of behavior that were controlled by the CNS in either a facilitative or an inhibitory manner. Thus, many instinctive activities prove to be complex systems of chain reactions (*instinct chains, stimulus-response chains*).

It is often the case that the last phase of culminating behavior can act as the releaser of new

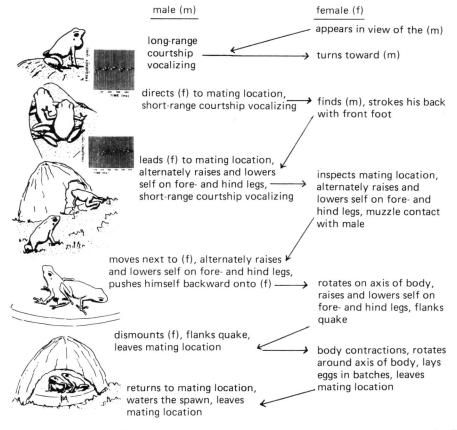

Fig. 10.8. Stimulus-response chain during courtship of the South American curare toad (*Phyllobates terribilis*). (Reprinted with permission from Zimmermann and Zimmermann, 1985).

appetency behavior. The result of this sort of instinct chain-reaction can be observed in stimulus-response behavior between two sexual partners of a species who are engaged in communication during courtship. The classic example of instinctive stimulus-response behavior during courtship is offered by sticklebacks, although many other species exhibit well-defined response chain reactions within the context of courtship. Figure 10.8 illustrates the stimulus-response interplay of a pair of South American curare toads (*Phyllobates terribilis*) during courtship and egg-laying. A specific behavioral characteristic in one partner of the pair is the key stimulus that triggers a behavioral response in the partner. The mating call of the male is particularly instrumental insofar as the female will not be distracted in her response to it, even when it is broadcast in recorded form over loudspeakers. Courtship and mating behavior, when properly coordinated with parenting behavior, assure survival of the species.

10.3 Acquired Behavior

With increasing developmental complexity in animals, acquired behavioral elements obtained through individual experience become integrated with innate behavior. Acquired behavior becomes increasingly important with regard to the overall behavior of the individual. Indeed, it affords higher animals quite plastic (i.e., adaptive) behavior that provides them ever-greater independence from their environment. To varying degrees, individual experience enables an animal to anticipate future events and, thus, allows for a systematic structuring of its near or distant future.

Generally, noninnate forms of behavior result from learning and, as such, are manifestations of memory events.

In the following section, the various modalities of acquired behavior are addressed successively. It should be noted that the classification put forth is more or less arbitrary. The diverse forms of learning, for example, are often intertwined or transpire simultaneously. The categorization that is suggested here was selected solely for purposes of clarity in overview.

10.3.1 Learning Processes

"*Learning*" is a collective term that applies to behavioral changes or events of rather long duration that result from individual experience. Learning can be understood as a process that enables many organisms (and technical robots, as well) to react with greater efficacy than was possible prior to having interpolated new experience.

In general, *learning* can be categorized as the product of *habituation* (*adaptation*), *imprinting*, *conditioning*, *trial-and-error activity*, *training*, *imitation*, and, lastly, *insight*.

For humans, learning primarily represents the judicious, active acquisition of skills and knowledge, modes of behavior and beliefs resulting from social interaction. The learning process can be divided into four *learning phases*:

- The *preparation phase*, which engenders attention to perception and differentiation of stimuli through specific motivational situations;
- The *acquisition phase*, during the course of which the new experiential data are recognized and processed by means of associative linkages;
- The *storage phase*, during which the individual experiences are stored;
- The *recollection phase*, during which the stored information reenters the consciousness as the basis for response.

Disorders can occur in each of the phases (learning, memory, recall) such as insufficient intelligence, partial deficits (e.g., congenital alexia), developmental disorders, apathy, environmental trauma (stimulus overload, dearth of social contact, for example), etc.

In general, a learning process refers to a specific stimulus of considerable duration that triggers identical responses that are not established by heredity. Responses formed associatively in this way usually are triggered only after frequent, successive repetitions of the same stimulus; they are less likely to be originated through a single, intense stimulus.

The following *types of learning* can be differentiated depending upon the modality of stimulus and response:

- *Perceptual learning*, whereby primarily visual, tactile, and auditory perception is altered through training (ear training, for example);
- *Motor learning*, whereby sequences of movement are acquired through repetition as, for example, in learning how to play the piano or drive a car;
- *Verbal learning*, whereby language acquisition is made possible by the repetition of words and texts;
- *Cognitive learning*, whereby loftier knowledge is acquired (concepts, rules, systematic categories, problem solving);
- *Social learning*, whereby social structures (social rules and laws) and social behavior is conceptualized.

Learning preparedness, i.e., a positive orientation vis-à-vis the targeted learning event, is critical to the desired result of learning in humans; it is a concept that is central to the psychology of learning, one task of which is to encourage the *will to learn* and the *motivation to learn* by identifying appropriate *learning goals* (for example, a reasonable curriculum).

10.3.1.1 Learning Through Imprinting

Imprinting refers to species-specific information acquisition during sensitive periods of development that conditions irreversible responses. Although, as a rule, these responses do not become manifest at the time of imprinting due to the immature state of the individual, imprinting enables an individual to recognize the identical configuration of characteristics over extended periods of time and such configuration recognition always elicits the same behavioral responses.

From the standpoint of neurophysiology, imprinting must be understood as that event resulting from the first connection of nerve fibers in neurons at precisely the moment when those neurons reach maturity. It is most likely, also, that they undergo a concomitant structural morphosis in certain sections (see Chap. 11.3).

Stimuli that trigger an imprint are often disproportionately simple and nonspecific; more often than not, they are quite general. Moreover, imprinting is not rigidly defined genetically; rather, it is an event transmutable in nature that

must not necessarily correspond to normal species-specific behavior.

The imprinting phenomenon in ducklings is especially interesting (Lorenz). Immediately upon hatching, they are instinctively predisposed to leave the nest and follow the adult birds. However, at that point, i.e., immediately subsequent to hatching, they are capable of imprinting anything, even mechanical mock-ups, and they will choose to follow them at a later point in time even if they encounter adult birds of their own species.

Numerous other imprinting phenomena have been ascribed to birds that leave the nest prematurely: migratory birds imprint their place of birth; cuckoos and African widow birds (*Viduinae sp.*) imprint their host; chaffinches imprint song patterns (Fig. 10.9), etc.

It can be assumed that there are sensitive developmental periods in humans, too, during which, and only during which, specific patterns of perception and behavior can be formed. Certain neuronal networks are especially flexible relative to the perception, processing, and storing of information during very early perinatal periods of development. Irreversible formations take place in them within the CNS only during these specific periods. These formations are evoked by the first inner images of the environment. Doubtlessly, these images vary from one person to the next. Accordingly, the *fundamental imprinting patterns* that they condition are greatly diverse. However, essential elements of the fundamental pattern of perception refer to the environment. To varying degrees, this pattern can be shaped naturally or artificially, but it is found to be relatively common within spcecific cultural regions. Thus, from the time of infancy, typical elements of the environment in which the individual is ensconced are imprinted. One result of this is the fact that the basic imprinting patterns in an "average" European baby from a highly technological society differ radically from those formed under natural (i.e., less technologically altered), or otherwise different climatic conditions.

The first impressions of imprinting often form the basis for response patterns in individuals, and possibly even the basis for the many misunderstandings that arise when members of

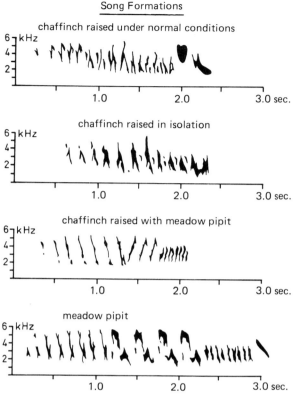

Fig. 10.9. Song imprinting in the chaffinch; chaffinch male, raised in acoustic isolation, develops only partial song patterns. It acquires the essential song elements of a different species (the meadow piper) upon exposure to it. (Reprinted with permission from Thorpe, 1964).

different cultures gather without first familiarizing themselves with the fundamental structural pattern of the other groups. Indeed, it is known that in later life, people, whether consciously or not, attempt to fit all new perceptions and information into their original pattern of perception. Thus, among the various races of humankind, manifestations of group-specific, cultural behavior can appear quite strange. Europeans, for example, are surprised by so many facets of other cultures, facets that are, obviously, self-evident to the people of those cultures who, in turn, do not fully understand the European mind-set in its myriad cultural orientations. It begins with habits of daily life—eating customs, for example—and culminates in attitudes toward law and morality. The importance of the individual's initial and otherwise early social experience in regard to the formation of a human social

order and, indeed, its ethical underpinnings has been illustrated repeatedly.

Imprinting occurs without inner resistance since no other potentially counteractive impressions have been defined that might act as impediments. The *social learning experience* is critically important for normal development in all more highly organized social vertebrates, including human beings. The field of *pedagogy*, then, would have to be concerned with the diverse array of sensory perceptions that are imprinted during early childhood. Abundant patterns of experience promote adaptability in later adulthood. A wide range of environmental imprints, especially with regard to social and cultural considerations, will lead later to fewer circumstances in which inner conflicts arise, or for which reformulation and relearning are necessary. This results from the fact that the earliest input relative to hearing, touching, smelling, seeing, tasting, etc. apparently is stored in as permanent a fashion as are innately determined behavioral responses (reflex behavior, etc.) and certainly more firmly than most input that is acquired at a later time. Severe developmental disorders (*hospitalism*, for example) can occur as a result of stimulus deprivation within the arena of social interaction, such as is experienced by children who grow up without mothers. Serious physical and emotional problems become manifest in children who reside in nurseries without the benefit of consistent parental, or caregiver, models. Among these grim afflictions are retarded growth, greater susceptibility to infectious diseases, stereotyped movement, diminished play and curiosity behavior, speech disorders, deficits in the ability to establish social contact, etc. In humans, *deprivation syndromes* develop mostly during the unstable phase of imprinting, the onset of which is about three months after birth and which radically decreases after the age of three. It seems to be during this period that the foundation is laid for social behavior (*socialization*).

An infant who is deprived of human nurturing during this period, whether by chance or intention, can develop symptoms of the so-called *Kaspar Hauser complex*. In the field of social psychology, this condition is characterized by such deficit disorders as a lack of emotional feeling and the inability to establish contact with others. It stems from isolation and, in the case of animals, experiential deprivation.

10.3.1.2 Learning Through Habituation and Sensitization

Habituation represents the simplest form of learning. It refers to the slow abatement of a response triggered by a stimulus (Tembrock, 1974). *Habituation* is also the phenomenon whereby the initial neurophysiological counter-response to a stimulus gradually disappears when repeated stimulation is shown to be inconsequential to the individual. A snail (*Helix sp.*) reacts, for example, with extraordinary sensitivity to an initial touch by retracting its feelers and contracting its body. After numerous repetitions of the same stimulus, however, the response soon diminishes, only to recur with full vigor after a brief interruption of stimulation. The decrease in the response is stimulus-specific. If the quality of the repeated stimulus is altered, even minimally, then the initial response recurs with full intensity.

Habituation (also referred to as fatigue, negative learning, and afferent fatigue) is not a factor of musculature fatigue; rather, its origin lies in the neuronal structures involved. Thus, it has been shown electrophysiologically that a new, regular potential pattern consisting of theta waves is induced in the mammalian hippocampus during the reception of new information. This new wave pattern gradually disappears, doing so in proportion to the extent to which the stimulus loses its characteristic newness.

Learning through habituation might contribute substantially to the phenomenon whereby the neurophysiological component, which is responsible for processing and selecting new information, is cleared relatively quickly as it makes way for new information. This enables the individual to concentrate on the new, unknown data.

In contrast to habituation, *sensitization* is a rather more complex form of learning and, indeed, the strength of a response formed in this way is reinforced if an additional, possibly painful stimulus follows. An individual, for example, who is awakened by a loud noise will react to a subsequent, less blatant sound with a greater degree of sensitivity. Sensitization is of biological importance inasmuch as alertness to sources of danger is heightened. Sensitization and habituation can occur alternately. Thus, the male fighting fish (*Betta splendens*) defends its territory with exceptional vigor against newcomers while it gradually becomes accustomed to recognizable "neighbors" (Fig. 10.10).

10.3.1.3 Learning Through Conditioning (Conditioned Response; Conditioned Reflex; Classical Conditioning)

In contrast to the relatively simple forms of learning discussed above, the one that follows addresses more complex *associative learning* in which a neuronal connection is in place between a neutral stimulus and a second stimulus, the latter being either a reward or a punishment. The formation of memory events that result from

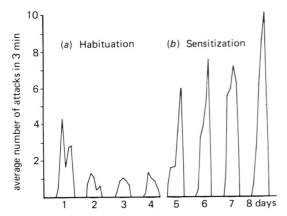

Fig. 10.10. Habituation and sensitization of the attack behavior of a male fighting fish upon daily exposure to another of his species at 3-minute intervals for a periods of 15 min. (Reprinted with permission from Peeke and Merz, 1973).

this can be traced in laboratory experiments under conditions that can be standardized and reproduced. Consequently, it is extremely useful in the field of experimental learning psychology. The most important distinction to be made here is between learning that is based on classical conditioning via the formation of conditioned reflexes, and so-called operant conditioning, or training.

The model of learning by way of classical conditioning is the *conditioned reflex* that was first studied by Pavlov (1889), the founder of experimental learning psychology. The theory suggests that if a neutral (*conditioned*) *stimulus* (CS) were presented together with a natural

(*unconditioned*) *stimulus* (UCS), then, at a later time, the conditioned stimulus alone will trigger the reflex upon frequent repetition of that CS.

It was shown in the original Pavlovian experiments using dogs that, after a period of time, dogs would show signs of increased salivation at the mere signal of a bell that had been rung previously at the moment when a piece of meat was presented to them (Fig. 10.11). Thus, in place of a natural stimulus, the conditioned stimulus triggered a response, i.e., a conditioned reflex. An associative connection took place that was established as neuronal *facilitation* in the sense of a *memory track* (*engram, residue*) in the nervous system (see Chap. 11).

10.3.1.4 Learning by Trial and Error

According to *stimulus-response (S–R) theories*, learning processes occur through a connection of stimulus constellations and response modalities. Depending upon whether specific additional mechanisms are a part of the learning process, a distinction is drawn between S–R theories of reinforcement and S–R theories of contiguity. Among the *S–R theories of reinforcement*, Thorndike (1898) promoted the theory of connection in which he suggested that learning is acquired by the process of *trial and error*. Thus, at first, young adult ravens use everything that they possibly can accumulate for nest building. They soon learn by trial and error to take only those branches that can be interwoven to suit their construction needs. Other young animals, too (whether birds, frogs, or mammals) at first hunt

Fig. 10.11 Unconditioned (*left*) and conditioned stimulation of salivary secretions in a dog (Pavlov's experiment).

all forms of insects until they learn to avoid un-palatable species that, in many instances, warn of their inappropriateness for consumption by way of their coloration.

The theory of instrumental learning, conceived by E.G. Guthrie, is among the theories of S–R contiguity. Guthrie suggests that the essential factor between stimulus and elicited response is temporal contiguity, whereby the reinforcement of an S–R interplay is ascribed only a secondary role.

Theories of the types mentioned above lead inevitably to learning theories by way of operant conditioning.

10.3.1.5 Learning Through Operant (Operative or Instrumental) Conditioning

Although the test animal in experiments of classical conditioning is merely a passive partici-pant that reacts to stimuli, it must solve problems or make decisions during tests of *operant condi-tioning*. B.F. Skinner developed the theory of operant conditioning on the basis of numerous experimental findings. Today, the theory has found lasting significance in the area of *pro-grammed instruction*. Within this context, learn-ing achievement frequently is documented in terms of *learning by succeeding*, which can be quantified by using various pieces of experi-mental equipment (Skinner box, labyrinth, two or multiple choice devices for distinguishing visual characteristics; Figs. 10.12 and 10.13). In each of these testing devices, the animal under observation is required to act; it must become active (operative) in order to receive a reward or to avoid a penalty stimulus (electroshock, for example). Thus, trial and error is the learning modality that is in effect.

Learning through operant conditioning (learn-ing based on an individual's own activity) can be enhanced by motor movement sequences, for example, rhythmic behavioral sequences that are made possible by the basic kinesthetic ability in higher vertebrates (Rahmann–Esser, 1964). Over time, and subsequent to sequences of operant conditioning behavior or decision-making beha-vior, a given situation, more or less, will come under control. The overall process can be repre-sented effectively in the form of so-called learning and memory curves.

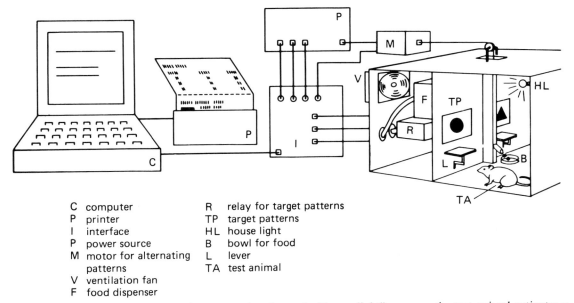

C	computer	R	relay for target patterns
P	printer	TP	target patterns
I	interface	HL	house light
P	power source	B	bowl for food
M	motor for alternating	L	lever
	patterns	TA	test animal
V	ventilation fan		
F	food dispenser		

Fig. 10.12. Skinner box electronic accessories for computer-controlled, automated training of mice and rats utilizing a food reward in the differentiation of visual patterns (here, a circle and a triangle). When the "house light" goes on, the test animal activates a lever under the positive pattern (target pattern) and receives a food reward (F).

 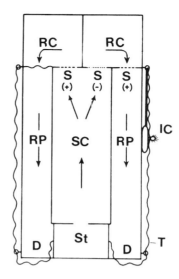

Fig. 10.13. Selection apparatus for training small mammals to distinguish between pairs of visual patterns presented on scuttles (S), the negative one which does not open. St: starting chamber; SC: selection chamber; RC: return chambers for food reward; RP: return paths; D: dividing doors; IL: indicator light; T: thread.

10.3.1.5.1 Learning Curves

The decrease in the number of errors made, for instance, in successive runs through a maze can be represented effectively by way of *learning curves*. One further criterion of learning, the length of time required to completely run the task in a maze, can be indicated, as well. Associative learning of optical signal patterns (for example, associating a triangle with a reward of food as opposed to a square with no such reward) can be quantified by using Skinner boxes, or choice-directed movement experiments (see Figs. 10.12 and 10.13). In such experiments, the test animal learns to activate the lever of a food dispenser when the positive signal pattern (illuminated triangular pattern) is presented. In training of this type, the test animal must make 30 to 50 decisions of choice daily with 75 to 80% accuracy in order to meet the so-called *learning criterion*. If the learning performance lies above this statistical threshold of significance on three consecutive days, then a *learning test* is administered in which both signal patterns are rewarded with food. *Duration of memory* can be ascertained by administering a further memory test at rather great intervals (about every 10 days) that rewards both responses. The animal has forgotten the

task when memory performance falls below the significant threshold.

These double-choice experiments reveal much about the higher associative capability of animals. Thus, it can be shown that not all animals within a test group respond uniformly; rather, as in the case of humans, the individuals can be divided into groups of "good" and "poor" learners (Fig. 10.14). Under these test conditions, young animals can learn to distinguish simple pairs of patterns more quickly than older animals who are more facile when confronted with more complex structures. The effect of psychotropic drugs or other test substances on the rate of learning and the duration of memory can be tested using these methods.

Figure 10.15 illustrates that hamsters, under the influence of methamphetamine (pervitin) administered within the peritoneal cavity in a concentration of 0.5 mg/kg of body weight, retained the memory of a learned visual stimulus differentiation exercise (distinguishing a horizontal pattern from a vertical one) about four times longer (up to 200 days after concluding the test) than control animals. A fourfold increase in the dose of pervitin, however, had a completely opposite effect.

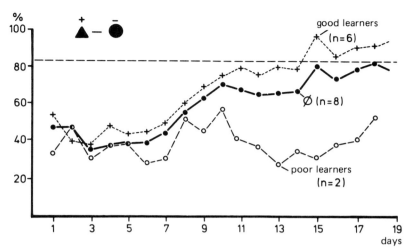

Fig. 10.14. Average duration of learning of eight white laboratory mice in simultaneous training to differentiate a triangle (rewarded by food) from a circle (no reward). Y axis: percentage of correct choices. *Broken horizontal line*: threshould of statistical significance (82% correct selections of 20 per day per test animal).

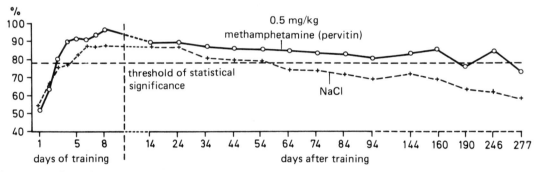

Fig. 10.15. Effect of the psychotropic drug pervitin on hamsters while distinguishing between a horizontally striped pattern and one vertically striped. X axis: days after the beginning of training (with pervitin) and termination of training (without pervitin); ordinate: average percentage of correct selections from each of 12 test animals. (Reprinted with permission from Rensch and Rahmann, 1960).

10.3.1.5.2 Learning and Memory Capacity, Duration of Memory

Learning and memory capacity, i.e., the limits of the ability to distinguish different patterns of stimuli, can be ascertained on the basis of double stimulus differentiation tests. Capacity appears not to be so much a factor of the animals' phylogenic placement as of its absolute brain size and complexity, i.e., the total number of its neurons and their circuitry (Fig. 10.16).

Duration of memory can be understood in similar terms. Depending on the organizational complexity of the nervous system, it is limited in invertebrates to a few days or weeks. It is equally dependent upon the organizational level of the nervous system in vertebrates and can last for many months or years (Fig. 10.17). Chapter 11 will address in detail the physiological and biochemical bases of memory formation and storage.

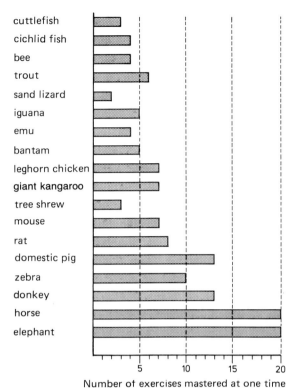

Fig. 10.16. Differing learning and memory capacities in animals based on the number of optical pairs of patterns mastered at one time.

10.3.1.6 Learning Through Imitation, Observation

This represents a form of learning that all too frequently is given short shrift. A commonly cited example is the manner in which birds learn their song. Only a basic song pattern is innate to the chaffinch, for example. Young chaffinches learn to vary the song by imitating the warbling of the father (see Fig. 10.9). Learning through imitation is especially marked in primates. Young chimpanzees in the wild, for example, learn to use tools, etc. from their kinship group through imitation. Humans rely heavily on imitation in order to acquire an *education in tradition*, i.e., those nongenetic bits of information that are collected on an individual basis and passed on from generation to generation. An extremely great number of conventions (tool making and usage; many modes of behavior) are learned

through imitation. There is much evidence to support the fact that learning through imitation occurs during early development even before the intended behavior can be performed by the still immature functional structures. This is referred to as *learning through observation.*

10.3.1.7 Learning Through Repetitive Activity

Learning through repetitive activity may well be the basis for most forms of learning, especially operant conditioning (see Sect. 10.3.1.5). The issue of repetitive activity centers on the refinement of raw, undeveloped abilities. These can be innate in their nature, such as types of movement (flying, swimming, etc.) that only have to be refined, or they can be behavior that first must be acquired and then refined, such as the special techniques of playing the piano or the new experience of walking upright. They can be associative and cognitive experiences, as well (language learning, arithmetic processes, etc.).

10.3.1.8 Learning Through Play and Inquisitiveness

Behavior of play and inquisitiveness, which is so often observed in higher animals such as fish, birds, and mammals, usually is encountered among the young. It represents an activity that is performed without conscious purpose, but rather for sheer enjoyment. It seems to be associated with an *emphasis on positive feelings*. Playful activity may have its basis in the natural drive for activity and movement, or in an energy surplus. It also may serve as practice for species-specific modes of behavior. The neurophysiological basis of play and inquisitive behavior may lie in an elevated state of synaptic instability during early developmental periods.

Playful behavior that mimics *fighting and hunting* is very common among animals. It can be distinguished from earnest aggression by virtue of specific inhibitory mechanisms that are evident (e.g., relative to biting). Through such playful behavior, animals begin to familiarize themselves with their own physical capabilities. A bird learns to land on a branch, for example, through this type of activity.

In addition to developing cognitive abilities, the play activity of a child allows it to develop

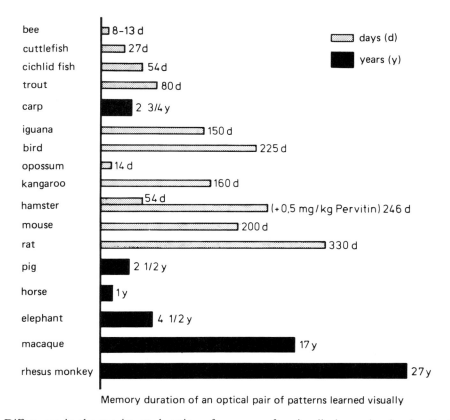

Memory duration of an optical pair of patterns learned visually

Fig. 10.17. Differences in the maximum duration of memory of a visually learned pair of optical patterns. Note the effect of the psychoactive agent on the duration of memory in the hamster (see Fig. 10.15.)

a social identity. During its first two years, the child practices physical functions through the frequent repetition of movements (functional play). Thereafter, the phase of *fictional play* begins (role playing, imaginative play, interpretive play) during which the child attempts to imitate the activity and behavior of others by mimicking facial expressions and gestures. In general, the behavior of human play is rooted in many diverse factors. Clearly, it is crucial for the healthy interaction between the individual and society.

General indicators of play might be: behavior without serious point of reference; the appearance of completely new behavior and/or freely combined modes of behavior from diverse functional areas (for example, catching prey, attack or flight behavior); exaggerated, perhaps nonsensical behavior; behavior that requires a considerable expenditure of time or energy; self-

oriented behavior (as an ape playing with its tail).

Inquisitive behavior is related to play behavior and may be the basis for reconnaissance or *exploration behavior* with which a maturing higher animal claims its territory. Exploration behavior can manifest itself in response-readiness or alertness that an animal exhibits when confronted with unknown test objects. In this regard, considerable differences can be observed that are commensurate with the organizational level of the brain: primates and predatory animals are very much more receptive to new test objects than rodents or marsupials, whereas reptiles behave with nearly unrestrained indifference in comparable test situations (Fig. 10.18).

In contrast to the rigid nature of instinctive behavior, the normal sequences of appetency behaviors, key stimuli (innate releaser mechanism), and culmination behavior are not always evident in behaviors of play and inquisitiveness.

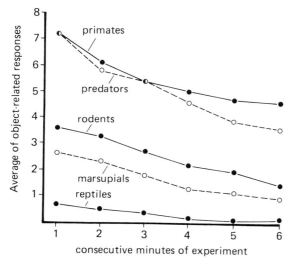

Fig. 10.18. Inquisitiveness and exploration behavior in vertebrates of varying developmental complexity. The average response readiness toward new test objects was tested within a period of 6 minutes. The number of times a test animal turned toward the object was observed in 5-second intervals. (Reprinted with permission from Franck, 1985).

Often, individual components of a total behavioral complex are repeated. As a rule, play among adults is relatively rare and occurs only under utterly relaxed circumstances, i.e., under conditions in which the adult is relieved of pressures from external forces, such as threats from an adversary or the drive to search for food.

10.3.1.9 Learning Through Insight (Cognitive Learning), Methodical Behavior

Aside from the individual acquisition of skills and behaviors, human learning, to a great extent, consists of acquiring information and adopting convictions in a cognitive manner, actively and socially. Based on cognitive theories of learning (W. Köhler, E.C. Tolman) problem situations can be overcome through the gaining of insight that allows for the formulation of concepts and mastery of organizational patterns, rules, and systems. To a certain degree, *learning through insight* represents a logical extension of behavioral alternatives that are founded in perceived environmental qualities and past experiences that have been stored on an individual basis. It results in behavioral adaptation to new condi-

tions. Learning through insight also can be characterized as learning by trial and error based on one's own inner model. Correcting a developing memory track which is being formed during this process occurs internally by means of "internalized activity" (*thinking*) that is based on the comparison of current data with past experience.

Learning through insight can be verified conclusively not only in humans, but in higher apes, as well. Wolfgang Köhler performed the first experiments in this area. He proved that chimpanzees possess a certain anticipatory capability that goes beyond a primitive understanding of correlations. A chimpanzee is capable, after a period of goal-directed thinking, of using tools and other helpful devices to attain a reward (food) that hitherto was beyond attainment.

In considerably complex learning experiments with a young chimpanzee female, Bernhard Rensch and his colleagues proved her to be capable of anticipating varied actions during an extended period of planning relative to an anticipated goal. The chimpanzee learned, in a series of experiments, to open a box, with a key, in which she found a tool for opening yet another box of differing design, etc., until she finally obtained a food reward from the last box. After mastering the basic task, these and several additional boxes were placed randomly in the chimpanzee's cage. Eight of the boxes contained tools to open boxes; one box was empty and one contained a food reward. At the same time, the chimpanzee was presented with a "selection" box that contained two opening tools, only one of which would open the box containing the food. Before the ape set out to select one of the two tools from the box with the double offering, it sat itself upon an elevated observation seat in its cage and planned its course of action for a while. It had to think through, or anticipate methodically, the most reasonable way in which to attain the food reward (Fig. 10.19). After deliberating briefly prior to each box-opening event, the chimpanzee, almost without fail, discovered the most efficient way to proceed and rapidly worked its way through opening the entire series of boxes.

In a second series of experiments, the female chimpanzee learned to draw an iron ring, which

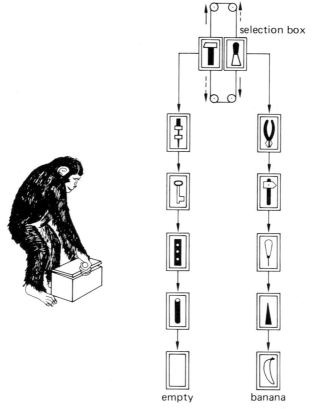

empty   banana

Fig. 10.19. Planned, or anticipatory behavior: a female chimpanzee learned to open in succession boxes in which the tool to open the next box in succession was located. The boxes shown in the diagram were arranged randomly. In taking possession of variously "useful" openers, the female chimp, after a period of consideration, selected the key that would best expedite reward acquisition (a banana). (Reprinted with permission from Rensch, 1973).

was located beneath a plexiglass barrier, through a maze using a magnet, then to exchange the ring for a food reward from a food dispensing machine. The device was designed in such a way that the ape had to survey the desirable path through the newly arranged labyrinth in order to draw out the iron ring successfully, since one false move at the start would interrupt the process (Fig. 10.20). In this case, too, the female ape scrutinized the individual maze channels that she had to negotiate, planned her actions carefully in advance, and executed them in one movement with extreme accuracy. Students who were tested in identical experiments did not fare as well.

Such methodical behavior can be verified not only in the laboratory. It can be observed under natural conditions, as well. Chimpanzees will begin to accumulate sticks and the like at distances of up to several hundred meters from termite nests. The sticks are used to poke into the nests in order to draw out the termites.

These experiments and observations show that, in addition to humans, at least anthropoids (apes) are capable of understanding simple causal relationships and of incorporating long sequences of activity methodically into their behavior.

The ability of animals to comprehend causal and, ultimately, logical correlations might be based, in part, on a *nonverbal analytical capability*, i.e., on the ability to draw relationships between two complexes of perceptions or concepts, as is the case in comprehending similarities and discrepancies. Moreover, it might be rooted

Fig. 10.20. Planned, or anticipatory, behavior: after a long period of training, a female chimpanzee succeeded in guiding an iron ring through a complicated maze without making a mistake even when the path-ways were changed from test to test. Prior to each test, the chimpanzee planned her actions. She then executed them smoothly and, usually, without error. (Reprinted with permission from Rensch, 1973).

in the ability to draw *nonverbal conclusions*, as often occurs spontaneously when learning new material. Such processes of association are expressions of *nonverbal thought*.

The manner in which past experience is correlated with current information, such that insightful, possibly future-oriented, and methodical activity results, has eluded experimental science. It is possible that, in each case, a sufficiently large backlog of experience is required and, as such, a differentiated model of the environment. The latter might well be associated with memory structures that were acquired through simple forms of learning.

10.3.1.10 Abstraction, Generalization, and Extrapolation in Learning

It can be concluded from the results of learning experiments with higher vertebrates that test animals (and humans) do not remember every detail of the characteristic features being imprinted (total recall), but rather focus on essential qualities. Having learned particular characteristics, animals can recognize them at a later time even if these features are only separate, albeit prominent components of the whole. Thus, ani-

mals, too, have the ability to abstract, generalize, and extrapolate.

Within this context, *abstraction* means the ability to differentiate and separate essential features from nonessential or coincidental features. In principle, this ability may underlie all learning processes in animals. In vertebrates, however, it may lead to nonverbal complexes of images (i.e., complexes to which word symbols do not apply) and, thus, to nonverbal *concepts*. In the case of visual stimulus differentiation training, a test animal first must abstract from the environment in the training device and concentrate on the characteristics of the test pattern. A strong stimulus effect is produced by the patterns, since a reward value can be placed on them after only a few selections are made. This, in turn, focuses attentiveness. Pattern imprinting teaches the animal the task, i.e., to receive food upon making a correct decision. At the same time, it perceives the difference between a positive and a negative pattern. In learning the actual pattern, the animal normally does not imprint all of its characteristic features, rather only those that especially draw its attention. Thus, a kind of abstraction or generalization takes place in this regard. When fish are trained to differentiate "horizontally positioned shapes, either squares or triangles," they continue to react according to their training even if the upper and lower parameters of the square are presented at an angle.

The special intellectual qualities of humans is attributable, in large part, to our extraordinary ability to generalize. In the field of learning psychology, the term *generalization* refers to the phenomenon whereby learned responses (aside from those that result from conditioned stimuli) can be elicited by stimuli that are similar to conditioned stimuli. In other words, to generalize means to reduce specific experience to generalities, or general concepts.

If the original pattern is altered gradually at the end of a training period involving a pair of features, then a test animal still will differentiate the pair correctly, even if it perceives only a few components of the pattern that it originally

#	Initial Pattern		
1.	• ● ●•		
	Test Patterns	Number of Tests	Percentage of "Correct Choices"
2.	▴▲ ▲▴	100	70
3.	● ' ⸒ '	95	87
4.	˙ ɢ ❙❙	70	99
5.	❵ ⟨⟨	50	90
6.	𝟜⟨ +⁺	60	80
7.	❙❙ 99	60	85
8.	⌐ ◢ ⋏ ◢	110	75
9.	◣▪ ▪ ◪	50	78
10.	▪◼ ▪ ◼	60	73
11.	★ •● ⌃ •	50	90
12.	⬥˅ ▪▪	50	80

Fig. 10.21. Percentage of spontaneous selections made by a civet (*Viverricula sp.*) during initial training in response to equal and unequal circular areas in a series of new, unfamiliar pairs of patterns. (Reprinted with permission from Rensch, 1973).

learned to recognize. The animal behaves as if it had formed a generalized concept. Thus, figures are recognized if they are enlarged, reduced, rotated, presented as components, or in other colors. In each instance, some characteristic features must have remained unaltered and, thus, recognizable. Rhesus monkeys learn to distinguish the nature of "similar" and "dissimilar" by learning only "dissimilarity" in their training. The testing involves three test objects that are presented simultaneously, of which two are always the same and one is different. One further example is offered by a Tibetan civet (*Viverricula sp.*) that was trained in circumstances utilizing two similar and two dissimilar circles. In subsequent memory tests, the basic pattern was altered continually until nothing of the original pattern remained except for the basic principle of "similarity" versus "dissimilarity." The cat was able to demonstrate with amazing accuracy that it had learned a likeness of relation between the two patterns contained in each new test (Fig. 10.21). The ability to generalize implies, to a certain degree, the ability to *extrapolate*, i.e., to be able to determine functional values by means of approximation, those values actually lying outside the realm of an individual's experience. This ability ultimately expresses itself in methodical, future-oriented behavior.

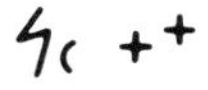

10.3.1.11 Formation of Prelinguistic Value Concepts in Learning

The ability of higher vertebrates to form abstractions and generalizations might well be the basis for all learning processes. The formation of conceptual complexes (i.e., *abstract, nonverbal concepts*) is one result of this ability. These are not characterized by symbols such as words.

The examples illustrated previously, in which apes and the Tibetan civet learned through training to distinguish similarities in a series of visual test objects, demonstrate in these animals the ability to form a *concept of similarity* that has been corroborated in tests that stimulated other senses. A trained gray parrot, for example, responded correctly when optical signals were changed to acoustic signals, and it did so spontaneously. In other experiments, chimpanzees were able to identify by feel, i.e., without visual control, the one object among several three-dimensional objects that resembled an object previously introduced to them by sight only. This represents an even higher degree of generalization relative to the concept of similarity. Furthermore, when presented with two feeding bowls covered with cardboard lids on which were painted arrangements of either three or four dots, jackdaws were able to learn to throw off only the lid with three dots in order to find a food reward. Neither the arrangement of the dots nor the intensity of their presentation caused the birds to falter in attaining their reward. Even when the ink spots were replaced with mealworms, the jackdaws spontaneously selected only the pattern of three, effectively ignoring the pattern of four. At a later point in time, the birds chose the numerical dot patterns even subsequent to the patterns having been presented by way of a cue card. Thus, they had acquired the ability to assess objects that were presented simultaneously solely by their number (recognition of identical numbers; comprehension of the *concept of number*).

Moreover, "dealing with numbers" has been addressed in random sequence tests of birds whereby it has been shown that 8 is the highest number of entities that the gray parrot can distinguish. The number for ravens, magpies, and jays is 7; for budgerigars and jackdaws 6; for pigeons 5. It is fascinating to note that humans cannot get beyond the number 8 without conscious counting. In general, we have the same prelinguistic ability to comprehend numbers as the higher vertebrates. The human ability to associate numbers with language probably developed from the nonverbal concept of numbers that already has been established for higher vertebrates. (It is interesting to note that the Japanese language originally possessed individual words only for the numbers 1 to 9; beyond that, the word for "many" was used.)

Prelinguistic value concepts are evident even in primates. A rhesus monkey was trained to choose three metal rings from a variously colored grouping of 12. These three rings could be exchanged for differing quantities of food from a food dispenser (Fig. 10.22). A yellow ring returned 15 peanuts; a white one, 6; a green one, 3; a blue one, 1; a red one was of no value in

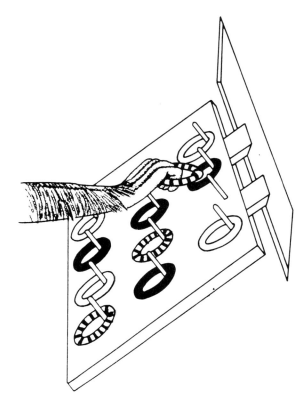

Fig. 10.22. Concepts of value in apes: of 12 differently colored rings, a female rhesus monkey consistently chose the three rings whose value was the greatest when they subsequently were exchanged for peanuts from a food dispenser. (Reprinted with permission from Rensch, 1973).

the peanut economy. The frequency and arrangement of the rings on the selection board were changed constantly subsequent to the necessary training period. The rhesus monkey almost always chose the three rings with the highest peanut values and retained the knowledge of the stratified value system (yellow–white–green–blue) for nearly three years. Only the red ring, for which it had shown a spontaneous preference initially, and which did not equate to a food reward, prevailed over the blue ring after this extended period of time.

In general, experiments of these kinds indicate that higher apes are capable of developing prelinguistic value concepts with respect to material value.

It is particularly important within this context to consider whether animals, in a manner similar to humans, are able to form a nonverbal concept relative to their individual identity, i.e., a *concept of self*. The presence of such a concept should be assumed, at least for the higher apes, on the basis of the structure and function of the sensory organs, the nervous system, physiology, and behavior. Cleaning and caring for one's own body, as well as the double sensation of feeling one's limbs as they are moved with intention and seeing the effects of that movement in relation to one's body might serve to define "self" vis-à-vis the environment. One's place within one's group, one's position of rank, one's display patterns, at times one's claims of ownership and response to one's own name all indicate that, at least in the case of apes, the formation of a self-concept can be assumed (Rensch).

10.3.2 Creativity

In general, *creativity* refers to the human ability to think productively and to fashion the results of this thinking, i.e., the newly formulated, original cognitions, into concrete terms. The creative process is one that consists of several phases:

- Tracking down and recognizing problems (or shortcomings, deficiencies, or inconsistencies) in extant theoretical or practical systems,
- Defining problems and formulating pertinent questions,
- Formulating new hypotheses,
- Seeking solutions in support of these hypotheses,
- Communicating the newly obtained information,
- Applying the new information in light of existing information.

The neurobiological basis of creativity is, above all else, the ability of an individual to learn and form *memory*, i.e., to store and recall individual experience.

Necessary for the eventual emergence of creative behavior is the *undisturbed morphological differentiation* of the neuronal systems of facilitation during early ontogeny. This ensures that genetically determined modes of behavior (taxes, reflexes, automatisms, instincts) can occur as they are intended to occur.

It is also necessary for the emergence of creative behavior that the first, individual patterns of experience occur during the prenatal stage of development. They can manifest themselves in distinct responses of the embryo to tactile or acoustic stimuli that may have an imprinting effect upon later behavioral norms. In the further course of development, indelible imprints of the individual's early experiences are established during unstable postnatal periods in the course of which basic patterns are emplaced for the recognition of faces, spacial relationships, and, especially, language as the basis for verbal thought.

The *ability to learn* allows for the progressive acquisition of memory content (be it through habituation or sensitization of neuronal pathways, or through practicing behavioral sequences, through imprinting or conditioning through trial and error). This occurs parallel to the strengthening of genetically determined behavioral structures. Individual memory forms from and emerges out of all of these learning events (see Sect. 10.1 and Chapter 11).

In all of these learning processes, particular significance is placed on the *predisposition* of the individual (i.e., on the genetic foundation, on early perinatal experience) and on motivational conditions (positive or negative feeling relative to learning). An intensification of training in those areas in which a natural talent is recognized should be encouraged by fostering *play and inquisitive behavior* as well as developing *learning* through insight (cognitive learning). An effective learning environment can be tailored to the individual by blending the *pressure to perform* well with adequate periods of leisure.

It may be acknowledged commonly that perfecting steps in the training process alone (*virtuosity*) does not lead necessarily to creative behavior. However, one's powers of abstraction, generalization, and extrapolation of individual experience may be enhanced by intensive training in various areas.

In conclusion, we raise the questions once again: from which processes does the human being derive creative abilities? How do these processes emerge? The answer: creativity is a function of the ability to learn in various modalities, i.e., to collect individual experiences, to abstract, generalize, extrapolate, and store the experiences over long periods of time. Ultimately, this enables the human to generate new information from existing information (theoretically available to all) in the form of productive thought. These abilities, moreover, must be developed through proper and sufficient training.

10.3.3 Motivation and Emotion

The bioelectrical and biochemical mechanisms that underlie the learning processes, memory formation, associative and creative thinking and, of course, recollection do not function independently of an individual's corresponding motivational state at the onset of these functional processes. Within this context, *motivation* is the sum of those motives that precede specific modes of behavior or actions and affect their course (either pointing the way, in a stimulatory fashion, or acting in an inhibitory manner in the absence of motivation). Thus, motivation provides the *impulse* or the *drive* for the course of behavior. This drive either can be stimulated externally or it can result from internal physiological conditions (initial hormonal conditions, metabolic state). Pertinent key stimuli can feed back from the environment and affect the internal activity patterns of the nervous system that correspond to such conditions.

Effective learning requires positive motivation. Appreciation of and interest in the material to be learned, or merely the acquisition of knowledge, individual experience, or reward, establish *positive emotional associations*, whereas fear (of punishment or pain, for example) result in negative emotional associations.

It is recognized widely that the external motivational factors both of reward and punishment can result in learning. Indeed, learning can occur more quickly under the threat of punishment, although the *neuroses* that can manifest themselves in the wake of such negative stimulation often negate any learning that has been achieved in this way. Thus, *emotions* play a significant role in processes of learning and association.

The *motivational complex*, which can consist in humans of diverse drives (e.g., play, manipulation, exploration, learning, etc.), can be categorized as follows (each component, in and of itself, may influence the way in which a memory storage event transpires):

- Basic physiological needs, such as hunger, thirst, sleep, relaxation, sex;
- Avoidance of threat and danger;
- Feelings of fear, love, jealously, sympathy, and apathy;
- Emotions, affects, stress, illness;
- High self-esteem, prestige, perfectionism;
- Self-affirmation, self-realization;
- Intellectual stimulation;
- Influence of upbringing, religion, experiences in youth;
- Special interests;
- Aesthetic needs;
- The need to be free;
- The need to understand;
- Prejudices;
- Social influence of the family, society, imprinting of traditions and customs, friendships;
- Frustration.

Within this context, drive is understood to mean a functional system that is based on neuronal structures and can be activated spontaneously. Furthermore, drives can be set off by external triggers, learned experience, and internal feedback mechanisms, the results of which can be vegetative, motor, and, especially, cognitive modes of behavior.

Neither the causal mechanisms that control the origination and emergence of the various forms of motivation as cited above nor their neurobiological principles are understood very well. This may be due in large part to the complex interactions that exist among the constituent parts. Presently, detailed information is available only in regard to the origin and development of relatively simple, albeit enormously important, motivational factors, such as hunger, thirst, stress, pain, sex, parenting drives, and *emotions*. Control centers for these lie primarily in the *hypothalamus*. Detailed information about the interaction of the individual neuronal regions and about the neuroendocrine system are just beginning to emerge. The neuronal basis of the more complex forms of motivation, such as fear, sympathy, antipathy, and jealously is extremely elusive; one's basic genetic disposition and individual experience clearly are significant factors in these regards. The structures of the *hippocampus* are ascribed increasing importance with respect to these areas inasmuch as they coordinate the integration of vegetative and somatic responses and seem to be involved extensively in controlling motivational and emotional conditions as well as one's attention span.

What then is the significance of the motivation-drive complex for learning and memory formation?

It is recognized that a learned event is never stored in isolation, but rather always in connection with numerous other bits of accompanying information. Thus, in regard to establishing a memory event, one ought to consider that not only is the engram of a specific event retained, but that the accompanying circumstances and, especially, the motivational milieu, with its *positive or negative valence*, is subject to the process of becoming an engram. As such, the intensity and valence of the motivation can be a decisive factor in the longevity of the engram. Thus, aside from the predisposition of the individual (including the genetic basis, early perinatal experience, and one's social milieu), the learning process is subject to the intensity and valence of *motivational factors*.

During the act of recollection (or even free association), the real-time emotional state might well correspond with the one under which the memory event was established. It has been shown in psychological studies in humans that during the event of ekphoriation (recollection), memories first are evoked of recent events and only thereafter of earlier ones. Indeed, the latter do not necessarily assume the order of their original sequence; rather, they occur in an order that reflects the strongest emotional associations to the real-time condition. Thus, one concludes that a balance must be struck between the motivational milieu and its intensity of the moment when the engram was formed and the moment of recollection. It should be emphasized that selective memories cannot be evoked at will, regardless of their physical-biochemical foundations, if a balance between the earlier and the real-time emotional state is not established.

Memory disorders, even if they manifest themselves in only one manner, can have various causes, i.e., they can result from the irreversible loss of certain neuronal structures (from injury to the brain, for example), from reversible disorders of a pathological nature (*stroke*, for example), and even from reversible shifts in motivational milieu.

A negative shift in the real-time motivational milieu can be triggered in humans, for example, by *excessive emotional encumbrance* or by its opposite, a *deficit relative to emotional value*. This can be evidenced in well-educated/productive individuals who find themselves underutilized as a result of job loss, retirement, or underemployment. They can become so negatively motivated that their efficiency wanes.

Thus, a negative motivational milieu can cause defects in higher neuronal activity, e.g., physiological activity. Physical changes in the nervous system (hypothalamus, hippocampus?) might occur from changes in the motivational state as might more easily identifiable physiological changes. Accordingly, therapeutic measures undertaken for patients who suffer from deficits in brain function the causes of which are unknown should reflect a thorough examination of motivational history.

10.3.4 Social Behavior

Social behavior is a collective term that applies to modes of behavior in animals who live in groups and to humans in reference to our social nature. In contrast to the modes of behavior discussed previously that were individual in nature, social behavior in animals is not related to any one, special functional circle; rather, it is comprised heterogeneously of diverse behavioral components. Social behavior brings about the formation of *societies*. These can differ greatly in size and structure and are differentiated from mere gatherings of animals in which no behavioral associations are evident.

Certain aspects of social behavior serve social communication. Highly ritualized modes of behavior have developed in this regard (visual, acoustic, and chemical communication, specific body movements such as those reflecting humility, aggression, display, etc.).

Social behavior in animals is controlled largely by instinct and induced by specific signals (key stimuli, releasers); however, with increasing levels of phylogeny, social learning processes play a significant role. In contrast, *social behavior in humans* is largely a product of traditional standards (cultural symbols and norms).

10.3.4.1 Social Organization

Social animals develop organizational structures that facilitate living together. These structures are maintained by means of appropriate systems of communications among the participating members of a species. Living together may not necessarily last a lifetime; it is often limited to the mating season. This pattern is exemplified by species whose members do not establish social connections of long duration, but rather live a solitary existence most of the time.

Social organization in the animal world can be predicated upon environmental factors (*aggregations* of animals that occur at watering areas or wintering locations) or factors of *social attraction* (socialization). The latter refers to a social drive that is genetically predetermined, i.e., a striving for community that generally is present among all individuals of a species. For example, young animals react to isolation with cries for social contact, cries of fear and following behavior. Normal behavior resumes only when group contact is established. As one further example, birds that have been driven from their flock constantly attempt to regain their group standing.

Social attractions are categorized as either anonymous or individualized groupings. There is no individualized familiarity within an *anonymous group* (flocking of birds, spawning of eels or salmon, migrating of lemmings). The group members are essentially interchangeable within the group, which is held together by virtue of group characteristics (insects colonies, rat families). The limits of *sociability* among individual animals within the anonymous grouping are quite broad ranging and, excluding interchangeability among their members, even allows for the intermingling of species (social flocking of jackdaws, crows, and starlings; herd form-

ations that consist of a mixture of plains ungulates such as zebras, gnus, gazelles, and giraffes).

In contrast to the anonymous grouping, the *individualized group* is characterized by individualized familiarity among its members who are involved with one another on numerous levels of social interaction. Accordingly, the absence of one of its members triggers search responses (ape family, wolf pack). Individual animals can draw together very closely (moving together to the point of physical contact in sleeping groups of wrens, golden-crested wrens, and tree-creepers).

Within such flock or group formations, the actions of the group are determined either by one initiator (dominant member) or by the behavior of the majority (a flock of birds taking off en masse, for example). *Social imitation* refers to the phenomenon whereby the behavior of an initiator is imitated by other individuals one by one. In each instance, social imitation brings about a coordination and a strengthening of behavioral activity within the society.

Sociability within *individualized groups* is most pronounced within *families*. Families are particularly prevalent among birds and mammals and differ from the larger group by virtue of a well-ordered coexistence among parents and offspring. This exclusive, *temporary society* serves the rearing of the young and begins with the caring for the eggs, as is practiced, for example, by Suriname toads (*Pipa pipa*) and South American curare toads (*Phyllobates terribilis*). Genuine family units even exist among fish that gestate orally (cichlids) in whom a period of guidance and protection follows the phase of oral gestation. The *merging of families* occurs when family isolation and exclusivity is diminished or eliminated completely. Penguins, for example, establish integrated groups when the self-centered behavior associated with breeding abates. The young, then, are cared for collectively in "nurseries." *Extended families* evolve as the offspring mature and remain with the family group, contributing their offspring to the clan. This is a rare occurrence, however; it is typical of the brown rat who will defend its territory, reproduce within its family, and kill members of other families.

The *biological function of the merging of social families* might have its basis in:

- Protection against predators through mutual warning or common defense (musk oxen),
- Acquiring food through hunting in groups and the reciprocal exchange of information,
- Avoidance of predators during reproduction (breeding in colonies, communal parenting),
- Taking advantage of what communal living has to offer (insects that build colonies).

10.3.4.2 Social Modes of Behavior

Characteristic *social modes of behavior* emerge among social animals. These occur only among the animals that live in a group and do not occur among those existing in isolation. In biology, the term *social* refers to any interactive behavior between members of a species. Thus, the term refers to both agonistic (aggressive) and defensive modes of behavior and, as such, is applied in a nonjudgmental manner. It cannot be equated with the decidedly positive value that is associated with the human social sciences. Essentially, social behavior is comprised of four areas of interest: agonistic behavior, sexual behavior, parenting behavior, and group behavior.

10.3.4.2.1 Agonistic Behavior (*Fight and Threat Behavior*)

Intraspecific aggression, or agonistic behavior, has developed frequently in social animals. The biological significance of such behavior can be understood in terms of the even distribution of individuals throughout a given area, thus ensuring optimal use of the food supply. The factor of sexual selection is similarly one of balance and ensures that only the healthiest and strongest individuals multiply. Since animals that engage in fight can sustain injury or, at least, become weakened in their quest for territorial and mating privileges, certain modes of behavior have developed supplementally to aggressive behavior and effectively circumvent the disadvantages of aggression.

Thus, threat and *display behavior*, which can be explained in terms of a surplus of fight and flight tendencies, serve to intimidate an opponent in order to avoid an actual fight. The outward manifestations of this behavior are, first, an enlarged presentation of the body's contour (the raising of hackles or ruffling of feathers, the dis-

tending of body parts such as the gill coverings and laryngeal sack), and second, the demonstrative display of weaponry (teeth, beak, antlers).

Gestures of supplication and appeasement are antithetical to threatening behavior: the body becomes as small as possible, weaponry and aggressive triggers (color signals in fish) are presented in a nonbellicose manner and disengaged, respectively. *Gestures of supplication* are especially pronounced in animals capable of fighting and occur in situations when flight is not an option. *Gestures of appeasement* (between sexual partners or between parent and child) are rooted in the activation of behavioral tendencies that are incompatible with aggression. They often occur as gestures of greeting in couples or group members and enable closer physical proximity. They also allow for members of a species to live together.

Territorial behavior, as an agonistic mode of behavior, is exhibited in order to avoid severe injury due to aggression. It allows for distinct territories to be established that provide the occupants with an adequate food supply, promote effective mating activity and offer familiar locations of refuge when danger threatens. The territorial imperative is common to all vertebrates and is known to exist among certain invertebrates, as well (spiders, crabs, insects).

Marking behavior serves to identify parcels of territory. It establishes boundaries and helps to reduce violent encounters. Territories can be marked in various ways: visually (presenting conspicuous body signals, occupying exposed areas), acoustically (birds, seals, gibbons, nocturnal presimians), through olfaction (using excrement: fox, rhinoceros; urine: dogs, cats, prosimians; by glandular secretions: marmots, antelopes, racoons, deer); and electrically (mildly electrical fish).

Hierarchical rank-order behavior within higher societies presupposes individual recognition of its members. Normally, hierarchical organizations are linear in nature with an animal of highest ranking, the so-called α-animal and an animal of lowest ranking, the so-called ω-animal. There also exist multistaged hierarchies within which the members of both sexes are organized in a segregated manner. Rank sometimes changes through "marriage." In jackdaws, for example, a low-ranking female can assume the position of its higher ranking male mate. The rights and activities of the high-ranking individuals, who also lay claim to a greater proportion of space than the others, extend to numerous social activities. High-ranking males, for example, have a greater number of females with whom to mate than do low-ranking males. This often results in the forming of a harem. In birds, there exists a negative correlation between social rank and mating frequency: α-females chase away lower ranking females from the vicinity of the α-male; they also lay more eggs, a function of their superior physiological constitution that, in turn, is a result of a more plentiful diet. The high-ranking members of a social group have certain duties: they explore the territory, provide the group with food, protect it from its enemies, and settle disputes within the group. As a rule, high-ranking members behave aggressively toward lower-ranking members. However, they defend the lowest-ranking, or ω-members, preventing their untimely demise. Clear hierarchical instincts exist in humans, also. They manifest themselves in behavior that reflects certain levels of power. This is expressed in titles, awards, jewelry, etc., and is "worn," either figuratively or literally as the case may be, as a status symbol for all to see.

Struggles for sexual dominance among members of a species (rutting among even-toed ungulates) are also expressions of aggressive behavior. Normally, these behaviors are extremely ritualized (*ritual combat, codified combat*) since inhibitory mechanisms are built into the behavior that protect the opponents from serious injury. Such *ritualized behavior* is characterized by threatening behavior with the tendency to fight, or by display behavior that is intended to intimidate the opponent. Gestures of supplication and appeasement, mentioned above, belong to this category of behavior, as well.

Injurious combat, sometimes leading to the death of a member of the species, does occur occasionally in spite of the numerous mechanisms that are intended to reduce aggression. For example, adults who are members of another group are often victims of unrestrained aggression and are killed in altercations with rival group members (to prevent overutilization of the food supply, as is the case in chimpanzees). In addition, the young sometimes are killed (infanticide) when they are developmentally impaired, or they can succumb to stress in captivity as occurs among tree shrews and chimpanzees upon the death of the α-male and in the ensuing struggle for his females).

10.3.4.2.2 Sexual Behavior

Certain modes of behavior have developed that promote sexual activity. Their purpose is to ensure procreation. Two courses have emerged in this regard: first, to produce as many offspring as possible in order to hedge against losses; second, in species with relatively few progeny, to protect the young against enemies and environmental hazards.

In addition to necessary morphological and physiological preconditions having been met, behavioral–physiological prerequisites must be satisfied in order to ensure success in reproduction. The sexual partners must recognize one another by species and gender, they must find one another, they both must be ready to mate at the same time, and they must overcome the intraspecies aggressions they commonly harbor.

Attracting a partner is usually the job of the male, and he carries this out in much the same way as he marks his territory. Environmental factors (the season, changing length of the day) are largely responsible for the synchronicity between the partners whereby one partner reacts more strongly to the altered environmental stimuli and, in turn, stimulates sexual willingness in the other (courtship behavior of male songbirds stimulates maturation of the ovaries in the female). Overcoming intraspecies aggression in animals that live as solitary members of a species is particularly interesting. The mechanisms are the same, as evident in the gestures of humility and appeasement that were cited earlier. Movements are borrowed from the prematuration behavioral repertoire, as are vocalizations and posturing.

Mating behavior can be forced by the environment (cohabitation of still immature male and female crabs in their shelters whereby increased body size due to maturation prevents them from exiting, thus sealing their fate as mates), or it can result from individual rapport between the partners (instinctive crossing in courtship and parenting behavior; duet and antiphonal vocalizing in the African shrike and in gibbons; reciprocal grooming of the fur or feathers).

10.3.4.2.3 Parenting Behavior

To ensure the survival of their relatively small number of offspring, species that produce only a few progeny have developed highly complex parenting behavior. Such behavior consists of protecting and feeding the young as well as passing on special information such as a bird's song, knowledge of food sources, warning of rival attackers, etc. Communication of this sort is the result of long-lasting relationships between parent and offspring that, in turn, often are rooted in a family structure. The modes of organization within such families can vary greatly. For example, in a family with two parents, both can attend to the young simultaneously, or they can alternate their attention giving. The latter case applies to most bird families. A division of labor between the sexes usually is encountered in families in which both the mother and the father are present (hornbills). There are also mother-dominant families in which the female parent assumes responsibility for raising the young (most mammals) and there are those in which the father exclusively takes on the parenting role (stickleback, emu). Finally, there are extended families, or kinship groups in which other members of the species (siblings, relatives) become involved in caring for the young (barn owls, dwarf mongoose).

10.3.4.2.4 Communication Behavior

Communication behavior is essential to harmonious social interaction. It refers to the exchange of information and messages among members of a society and is rooted in several mechanisms. *Hormonal communication* coordinates physiological conditions among group members by the release of species-specific exo- and sociohormones (*pheromones, telergones*). One such condition is the drive to engage in sexual activity during significant physiological cycles. Other examples are the release of pheromone inhibitors by the queen bee which prevents the ovaries in female workers from maturing, and the release of sexual attractants by rutting mammals. The feedback principle probably controls hormonal interactions within a group much as it controls organ functions within the individual.

Neuronal communication is based on the perception of specific stimuli by the sensory organs, the conduction of the sensory impulses to the CNS, their transformation there into specific patterns of impulses in various coordination centers, and their conversion into specific behavioral changes via the motor centers. Three different types of neuronal communication have developed in the course of phylogeny:

- Insects exhibit a rigid pattern of behavior upon which learning processes have virtually no effect. Among these forms of communication are possibly the most complex in the animal world (the communication "*dance*" used by bees, for example).
- Vertebrates are far more likely to incorporate learned experience into their behavioral patterns (*instinct-training interplay*). Learning processes are often necessary for communication behavior and are connected to specific, sensitive phases (imprinting, see Sect. 10.3.1.1). The extraordinary plasticity of vertebrates is achieved only when the genetically predetermined patterns of behavior are augmented with patterns of acquired behavior.
- Humans have an extensive repertoire of *nonverbal means of communication* (mimicry, gestures, pheromones, etc.). In combination with the development of a *verbal language*, the exchange of information among members of a group has attained a level of perfection that some with an overview of animal communication abilities might consider sublime. This verbal ability, which is associated with the formation of specific brain structures in the cortex (the Broca and Wernicke areas), allows humans to express experiences, feelings, and thoughts by means of vocal utterances and to communicate among one another. The experiences of others can be used to one's own advantage. Experiences can be passed on from one generation to the next and, thus, the groundwork for *establishing tradition* is laid,

a phenomenon that can be observed in animals only in its incipient stages (primates, birds). The development of *writing* and of other means by which information is stored (film, tape recording) have enabled humans to preserve diverse bodies of information (language oriented, etc.) such that other members of our species can gain access to them at will in the form of "*extracerebral chains of association*" (Rensch). The ability to store information leads ultimately to the exponential growth of the collective human experience and, thus, to establishing traditions as the actual basis for cultural evolution.

11
Neurobiological Models of Memory

Discussions concerning the essentials of information processing were presented in preceding
chapters. Cellular, morphogenetic, and functional principles were addressed concerning the
ways in which organisms receive, conduct,
transmit, and store information. The remarkable
plasticity (i.e., adaptability) of neuronal processes,
especially relative to information transmission
and synaptic events, bears reemphasizing. In
light of these discussions, it is appropriate now
to reflect upon the neuronal basis of long-term
memory and examine the existing memory
hypotheses in terms of soundness, efficiency, and
effectiveness.

To gain historical perspective, this chapter will
touch upon several early molecular models of
memory, most of which have been superseded
by later research. This will afford us a reference
point for the discussion of hypotheses commonly
held today. A discussion then follows that
addresses the concept of memory formation as
a result of molecular facilitation in synapses.
Gangliosides play a crucial role in this model of
memory formation. References to it have been
made throughout the book, particularly in
Chap. 8 in regard to experimental findings that
pertain to the modulation of neuronal information transmission.

The question of engram localization, or
memory content can be addressed most effectively in terms of specific memory models.

11.1 Historical Overview: Early Models of Memory Formation

The concept of molecular encoding of acquired
information, i.e., of nongenetically determined
memory, can be traced back to Hydén et al. in
the early 1960s. Considering that the human gene
pool comprises only a few million (10^6) genes
that function as carriers of genetic information,
and that the circuiting capacity of the human brain
(10^{14} to 10^{15} synapses) is incomparably greater,
it was felt that the carrier of individual memory
was the *neuronal RNA*, and not the DNA of the
nucleus. Indeed, Hydén and numerous other
authors found stimulus-dependent changes in
the basic composition of neuronal RNA that
were equated with the formation of specific
memory molecules. However, the identified
changes were ascribed to other factors, such
as altered physiological activity and/or stress
phenomena during the learning process. The
hypothesis that suggests that neuronal RNA is
the carrier of specific individual memory content
is no longer considered valid.

With renewed vigor, research into the
molecular basis of memory then turned to the
general protein metabolism of the nervous system
(ca. 1970–1975). Here, too, many findings indicated that, in general, protein synthesis in the
brain is altered in a region-specific manner as a
result of adequate stimulation or by special
learning events. As anticipated, interruption of
protein (or RNA) synthesis by appropriate
synthesis blockers (puromycin or cycloheximide,

for example) prevented the formation of long-term memory, although the learning process itself was not compromised. Indeed, strictly localized elevations in general protein synthesis could be seen in specific regions of the brain subsequent to their stimulation by means of radioactive tracer techniques that utilized labeled amino acids or sugars. These elevations persisted for up to two days after stimulation. However, this did not point conclusively to the formation of memory-specific molecular changes, rather only to altered intensity in the rates of metabolism.

Following this period, research was directed for a period of time toward seeking *specific proteins* that might function as *memory molecules*. However, the spectacular "molecular transfer of memory" by means of specific proteins (scotophobin, for example, the darkness-avoidance substance) from the brains of trained donor animals to untrained recipients, which caused such a sensation in the early 1970s, was found to be scientifically untenable.

Nonetheless, since then, the possibility that highly specific proteins might be involved in the formation of memory has inspired further research. Proteins such as the *S-100 protein* (a soluble protein of glial cells), the *14.3.2.-protein*, *vasopressin* (see Chap. 7.1.3.2) and the *synapse membrane protein* (SMP) were attributed specific properties associated with memory storage because, for one, their antisera prevented the formation of memory when administered to test animals that were involved in training exercises. Future studies would have to prove the extent to which those substances produce specific reactions, or whether the antisera brought on some kind of acute or delayed toxicity.

11.2 Memory Formation Through Molecular Facilitation in Synapses

The notion of memory-specific protein compounds was received with skepticism for a number of reasons. In the nerve cell, the encoding-dependent metabolic events of *transcription* (RNA synthesis) and *translation* (protein synthesis) are known to be limited to the cell body (perikaryon). Since there is no rough endoplas-

mic reticulum present in the nerve fibers and, thus, no ribosomal RNA (see Chap. 1.1.3), neuronal transport (see Chap. 9.1) supplies the nerve fiber terminals with the products of synthesis, in particular, with proteins and membrane lipids. Because the nerve fibers involved in learning processes are often of considerable length, the direct participation of the encoding-dependent metabolism probably is not required, at first, for the necessarily short-term, stimulation dependent changes in the region of the synapses. This means that the biochemical, stimulation-dependent changes described earlier (on the level of RNA or proteins) have to result from *indirect* metabolic reactions that, upon completion of the primary processes in the region of the synapse, ensure that the *immediate* molecular and, ultimately, structural changes endure, these having been triggered by a stimulation of the pre- and postsynaptic membranes.

At this point, it is appropriate to cite various alternative models of *memory formation through molecular facilitation in synapses*. In general, the ideas of Ramon y Cajal (1911), Sherrington (1940), Hebb (1949), Lashley (1950), Eccles (1953), Kandel and Spencer (1968), Szentagotaj (1971), Changeux (1976), and Rahmann (1975, 1976) suggest that the phenomenon of *long-term memory* is encoded in the form of *synaptic connectivity*.

On the basis of these considerations, it would have to be possible to relate the commonly described, function-dependent morphological changes in synapses (synaptic plasticity, see Chap. 9.2.1 and 9.2.2) and the adaptive electrophysiological responses of neurons (see Chap. 9.2.3) with stimulation-dependent, biochemical changes in the region of the synapses. Various models have been developed for biochemical and functional mechanisms that might be responsible for the long-term storage of information in neuronal networks.

The following section presents some of the more important models under discussion today. They have been selected from the numerous hypotheses of memory formation that are contending for predominance or that compliment one another. Under consideration here are

- The *Aplysia* model of memory,
- The protein kinase C model,

- The hypothesis of the significance of extra-cellular proteins for learning and memory formation,
- The hippocampus model of memory,
- The ganglioside model of memory.

11.2.1 The *Aplysia* Model of Memory

Invertebrates and vertebrates have the ability to learn and to form memory. This ability has been proved not only in highly developed mollusks such as the snail (*Helix pomatia*) or marine snails of the genus *Aplysia* and *Hermissenda*, but also in nonsocial insects such as grasshoppers, common house flies, and the fruit fly (*Drosophila*). Invertebrates, too, store individually acquired experience for the long-term by means of associative learning. The marine snails *Hermissenda sp.* and *Aplysia sp.* are known by neurobiologists to be ideal subjects for electrophysiological and neurochemical studies. This is due largely to their relatively few, clearly identifiable neurons and the fact that the snails are so well suited for the study of short-term habituation and sensitization. They are eminently trainable, as well, and, thus, lend themselves to analyzing the formation of long-term memory.

In *Aplysia sp.*, sensory neurons can be identified from among the relatively low total count of 20,000 neurons that comprise the nervous system. The sensory neurons surround the siphon and are connected to six motoneurons that innervate the branchial retraction musculature. Upon repeated, mild physical stimulation of the siphon, *Aplysia sp.* responds with less and less vigor in retracting its branchia (*habituation*, see Chap. 10). If the *Aplysia sp.* is stimulated in the caudal region with strong electrical impulses, the entire nervous system of the animal is alerted. In this condition, even mild physical stimulation of the siphon causes a disproportionately strong retraction of the branchia (*sensitization*, see Chap. 10).

Aplysia sp. can be trained according to the principles of classical *conditioning* (i.e., *conditioned reflexes*; see Chap. 10.3.1.3) to associate mild physical stimulation of the siphon with strong electrical shock stimuli in the caudal region. The snail learns, after only a few combinations of the stimuli, that an electrical shock follows the mild stimulation. Thus, it reacts with a strong branchial retraction immediately upon mild stimulation of the siphon: it has learned.

The question now is which mechanisms are responsible for this memory formation. Extensive biochemical and electrophysiological measurements of the participatory neuronal systems of *Aplysia sp.* by Kandel and Schwartz (1985) have led to the formulation of the following model (Fig. 11.1).

During the initial stimulation, the sensory neurons of the siphon react to the mild stimuli of physical contact (1) in the corresponding pre-synaptic terminals by establishing a reflex in response to external stimuli (see Chap. 10.2.2)

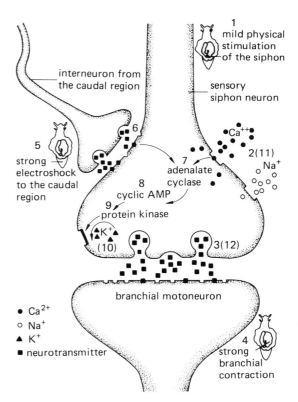

Fig. 11.1. Diagram of the manner in which conditioned reflex responses are formed in the marine snail *Aplysia sp.* The biochemical changes that occur in the nervous system are represented, as well. Together, they offer a model of molecular facilitation in the area of the presynapses. See text for details. (Reprinted with permission from Kandel, 1984).

by opening K^+ and Ca^{2+} *channels*. This allows both types of cations to flow into the presynapses at an increased rate (2). Ca^{2+}, acting as a *secondary messenger*, then causes the release of the *neurotransmitter* (*serotonin*) into the synaptic cleft (3) such that the receiving postsynaptic motoneuron, which controls the branchial response, is induced to react (4).

If, during the formation of a conditioned reflex, the caudal region of *Aplysia sp.* is stimulated by means of strong electrical shock immediately upon the mild stimulation of the siphon (5), then the release of transmitters (6) from the caudal neurons (and interneurons) triggers a torrent of secondary biochemical reactions in the neuron of the siphon (7–12). Since Ca^{2+} still is present in the sensory neuron of the siphon from the preceding stimulation of the siphon (2), the new stimulus amplifies the effect of its secondary messengers: the activation of a membrane-bound *adenylate cyclase* (7) leads to an increase in the level of *cyclic AMP* (adenosine monophosphate) (8). Cyclic AMP activates an enzyme (*protein kinase*) that carries phosphate groups to specific K^+ channel proteins in the membrane (9). This prevents an outflow of K^+ from the cell (10). Since the Ca^{2+} channels are still open at this point in the process (2, 11), transmitters are released in copious amounts to the motoneuron (3, 12) resulting in a strong contraction of the branchial musculature.

Upon repeated stimulus-response chains of this kind, a mild stimulus induced by physical contact triggers a branchial response from the siphon equal in intensity to the response induced by electrical shock to the snail's caudal region: a *conditioned reflex* arc has been established.

The example of the *Aplysia sp.* demonstrates how the activation of adenylate cyclase is triggered in a stimulus-dependent manner and how the torrent of secondary messengers is released in the presence of Ca^{2+} ions into the presynapse. This results in an increase in the release of transmitters and, thus, to the formation of a conditioned reflex, i.e., to a learning event. This *Aplysia* model is based on the supposition that a single training process only can lead to the formation of a *short-term memory*. Repeated response behavior is necessary to establish a *long-term*

memory. The processes that serve the storage of information are time-dependent and cause gradual functional changes in the membrane-bound ion channels and in the mechanism of transmitter release. Ultimately, they cause structural changes in the synaptic contact zones and changes in the properties of the receptors of the postsynaptic membranes.

- *Short-term memory* emerges through covalent changes in preexisting proteins. General constituent elements involved in cytoplasmic signal processing (membrane-bound receptors, transfer proteins, intensifying enzymes such as adenylate cyclase and phospholipase C, cytoplasmic signals such as cAMP and Ca^{2+}, and, finally, protein kinases) are used to bring about a change in a functional protein (K^+ channel protein, Ca^{2+} channel protein, transmitter release mechanisms) and, thereby, transfer information to a receiving neuron.
- *Medium-term memory* can be explained in terms of the previously described processes, which occur in the periphery of the neuron (synapse), becoming reinforced through repeated stimulation over a brief period of time.
- *Long-term memory* might result from repeated, or sufficiently strong, activation of participatory neurons whereby secondary messenger information activates *effector* and *regulator* *genes* in the nucleus of the neurons. These, in turn, ensure that the above-mentioned changes in the region of the synapses are maintained for the long-term.

By virtue of its clarity, the *Aplysia* model, which has been substantiated in detail through extensive biochemical experimentation, offers the opportunity to conceptualize similar molecular mechanisms in the formation of memory content in the considerably more complex neuronal assemblies in vertebrates. It begins with the triggering of secondary phenomena on the inside of the synaptic membranes. Questions still remain to be answered as to (a) the primary processes involved in the internalization of electrically encoded impulse patterns on the outside of the presynaptic membrane and (b) the processes that may be involved in the ekphoriation (reactivation) of previously stored information.

11.2.2 The Protein Kinase C Model

Another molecular model of memory that deserves comment is the hypothesis set forth by Routtenberg (1987). It suggests that regulation of synaptic plasticity and memory formation in vertebrates must be closely associated with the activation of *protein kinase C* and with the phosphorylation of one of its substrates, the so-called F_1 *protein*, both of which are present in the synaptic region. This model is based on experimental findings suggesting that the phosphorylation of the F_1 protein is increased in corresponding brain preparations following long-term potentiation (LTP) stimulation (see Chap. 9.2.3). Furthermore, this occurs simultaneously to the activation of membrane-bound protein kinase C and at the same time that protein kinase C, found in the plasma, is reduced. It has been shown in the optic tract in apes that phosphorylation of the F_1 protein occurs with varying intensity and, moreover, that it corresponds to the topography of the individual cortical fields of the prosencephalon that are involved with processing the visual event. The conclusion that can be drawn from this is that these centers vary in the degree to which they are capable of storing information by means of synaptic changes.

11.2.3 The Hypothesis of the Significance of Extracellular Proteins for Learning and Memory Formation

This model was developed for vertebrates and borrows heavily from studies of the CNS in fish. It suggests, as was mentioned in the introduction to this chapter, that short-term molecular mechanisms must become engaged at the onset of activity in neuronal assemblies (i.e., subsequent to adequate stimulation) in order to stabilize the hitherto unstable synapse formations. One of the foremost proponents in this area, Shashoua, points to the presence of various extracellular factors in the brain that apparently act as mediators in the learning process.

Thus, the effect of the hypothalamic-hypophyseal *neuropeptides, vasopressin* and *ACTH* (see Chap. 7.1.3.2), come into consideration. They are released into the extracellular space and are thought to have a modulatory effect, during the learning process, on specific target cells in the brain that lie far from the origination point of the neuropeptides. Furthermore, the *s-100 protein*, which is comprised of glial cells, also is released into the extracellular fluid where it, too, is thought to be involved in learning events.

The functional hypothesis of Shashoua centers on the possible effect of *ependymines* (i.e., proteins that, in goldfish, are formed from the *ependyma*, or ventricle wall cells, and normally are present in soluble form in the extracellular space of the brain). During a learning event, the ependymines are thought to be transformed from a soluble state to an insoluble, fibrillary state at those locations where *extracellular Ca^{2+}* is released momentarily following a stimulation event, i.e., in the region of the synaptic contact zones. Subsequently, the resulting extracellular matrix serves to stabilize a hitherto unstable synapse as a point of attachment.

Thus, the role of the ependymines is to reinforce that region in which, by virtue of stimulation dependency, a new synaptic contact should become established. An indication of the correctness of such a functional correlation between ependymines and the formation of a memory can be drawn from two facts: first, the turnover rate of the ependymines is increased during learning training, and second, administering ependymine antisera prevents the reactivation of stored memory content.

The *ependymine hypothesis* is helpful in analyzing the possibilities that should be considered in a synaptic connection. Future studies must determine the degree to which findings, obtained through work with goldfish (and, most recently, mice) are applicable to vertebrates in general. The hypothesis still does not explain the stimulation-dependent specificity in the linkage of neuronal assemblies. Nor does it attempt to clarify the possibility of reactivation (ekphoriation) of the previously stored memory content.

11.2.4 The Hippocampus Memory Model

As a model for the formation of memory, the *hippocampus* is considered to be an ideal neuronal structure for analyzing the manner in which information is stored. There are three structures in the hippocampus that lend themselves to this application by virtue of their relatively simple monosynaptic modality of wiring (see Chap. 5.3.3 and Fig. 5.9). First, fiber bundles supply the dendrites of the *granular cells* within the *fascia dentata* with *afferences* from the cortex of the prosencephalon. Second, in the form of *mossy fibers,* the axons of these granular cells establish contact with the dendritic ramifications of the *pyramidal cells* in the so-called C_3 region. Third, the axon ramifications of these pyramidal cells, known as *Schaffer collaterals,* are connected synaptically to the dendrites of pyramidal cells that are located in the C_3 region.

These three interconnected circuitry systems can be studied effectively by utilizing appropriate in vitro sectional preparations through the hippocampus as well as implanted electrodes in vivo. A discussion in Chapter 9.2.3 referred to the phenomena whereby the various forms of adaptive (i.e., stimulation-dependent) bioelectrical response reactions can be triggered in these three monosynaptic wiring systems in the hippocampus. To review, these phenomena are (a) *postsynaptic potentiation (PSP),* (b) *posttetanic depression (PTD),* or *potentiation (PTP),* and (c) *long-term potentiation (LTP)* (see Fig. 9.18), the last of which is especially crucial for the interpretation of memory. The duration of an LTP formation can last up to 10,000 times longer than that of a PTP. All of the response systems are dependent upon the presence of Ca^{2+} (see Chap. 9.2.3). The postsynaptic *spine synapses* of the neurons of the hippocampus are characterized by two different types of *glutamate receptors.* The first type is the *NMDA (N-methyl-D-aspartate) receptor.* Upon strong stimulation, these bring about increased flow of Ca^{2+} into the post-synapse. The second type is the *non-NMDA receptor* (*kainite and quisqualate* types), which control the postsynaptic K^+ and Na^+ ion channels (see Chap. 7.1.3.1.2).

The formation of an LTP also appears to be associated with the activation of the NMDA receptor synapses. The NMDA receptors are not activated during normal excitation of neurons, but rather only during LTP stimulation. However, if these NMDA receptors specifically

are blocked by selective antagonists (*aminophosphonovaleric acid, APA,* for example), then neither an LTP event nor a memory formation will occur, even though the process of synaptic transmission is not hindered.

Through intensive research in recent years, scientists have been able to refine their understanding of events that trigger an LTP in the hippocampus as a result of high frequency electrical stimulation. The following chain of events may be involved:

1. Suppression of inhibitory postsynaptic potential (IPSP) activity;
2. Concomitant intensification of excitatory postsynaptic potentials (EPSP);
3. Activation of NMDA receptors at the post-synapse;
4. Flow of Ca^{2+} into the postsynaptic cell;
5. Ca^{2+} activation of secondary messenger systems in the postsynapse by Ca^{2+}-calmodulin or other systems (calpain-protease effect on the cytoskeletal protein spectrin);
6. Reinforcement of the postsynaptic membrane thickenings;
7. Elevation of glutamate receptors to more efficient binders of transmitters.

The hippocampus model of memory attempts to explain the LTP events, which can be measured electrophysiologically, in terms of adaptive biochemical and structural changes that occur primarily in the postsynaptic cell. To date, science has not correlated adequately all the triggering events that occur at the presynapse, although there is evidence to the effect that the release of transmitters from the presynapse is elevated during an LTP event. Insight still remains to be gained into the primary processes that occur at the presynapse when an impulse pattern is internalized. The same holds true for the questions surrounding the reactivation (ekphoriation) of previously stored memory content. It should be noted that the dependence of the LTP event (as a basis for memory formation) upon the presence of NMDA receptors may be viewed only in reference to the hippocampus. Formation of memory is possible in other regions of the brain without NMDA receptor synapses, at least in lower vertebrates that do not possess a hippocampus (fish, amphibians).

11.2.5 A Brief Evaluation of Earlier Molecular Models of Memory

The common thread running through the models of memory presented above (in contrast to earlier theories) is that they assume a necessary factor in the emergence of memory to be that of marked adaptability, especially in the synaptic terminals of neurons. This *modulation capability of the synapses* ensures, for one, that the transmission of electrically encoded information from one neuron to the next always will occur in the same manner (subsequent to the onset of activity) even though important parameters, such as temperature or the ion milieu, will vary. Secondly, this ability of the synapses to modulate, or adapt, also allows information, which possibly was stored for years (long-term memory), to trigger the same feelings and behavioral responses when it is remembered (i.e., through processes of ekphoriation) as were attendant to the formation of the engram (i.e., the memory formation) originally. All of the models presume that hitherto unstable synapses become stabilized as a result of a sufficiently intense onset of function in neuronal wiring assemblies.

According to the *ependyma hypothesis*, the stabilization of new synaptic connections emerges as a result of the stimulation-dependent transfer of specific proteins (ependymines). These proteins are thought to be present in soluble form in the extracellular fluid up to the moment of stimulation, whereupon they are transformed into an insoluble, fibrillary state (see Fig. 11.2). The main emphasis of the *hippocampus model* centers on synaptic plasticity, i.e., the great adaptability of the postsynaptic membrane where a stimulation-dependent increase in receptors ultimately causes transmitters to bind more efficiently and, thus, transmit information more efficiently. The *Aplysia* model suggests a covalent change in the proteins that are present in the presynapse, in general terms, just as it occurs analogously during the processing of signals in the cytoplasm of nonneuronal cells. Then, increased functional demand placed upon the participating neurons causes an incremental shift in protein modifications from the synapse to the cell body, such that altered proteins can be formed for the long-term. These ensure the stability of the newly established synaptic contact points. In this regard, the *protein kinase C model* offers an explanation for the fact that the phosphate group, which is taken up by the F_1 protein and is assisted by membrane-bound protein kinase C, brings about adaptive, stimulation-dependent changes in the synapses that, upon repeated stimulation, produce a molecular facilitation.

All of the above-mentioned models assume that *extracellular calcium* is necessary to trigger a series of partial reactions (for example, the activation of adenylate cyclases, phospholipases,

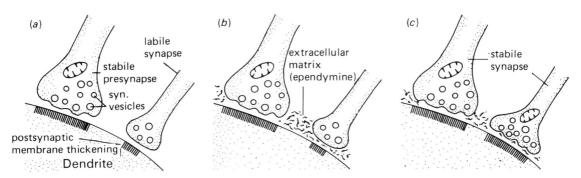

Fig. 11.2. Representation of the possible role of extracellular proteins (ependymines) in learning and memory formation. See text for details. (Reprinted with permission from Shashoua, 1985).

protein kinases, etc.). Ca^{2+} (which flows into the synaptoplasm via electrical stimulation), the initiator of adaptive molecular modulations, is believed to originate in various locations in the synapse. Its origin can be postsynaptic, presynaptic, or even extracellular. Of extraordinary significance in this regard is, for one, evidence of the Ca^{2+}-dependent release of transmitters at the presynaptic terminal and, secondly, Ca^{2+}-dependent transmitter binding at the receptors on the postsynaptic side.

It is important to emphasize, for the development of an *integrated model of memory*, that the voltage-dependent flow of extracellular calcium into the presynapse, which precedes release of the transmitter substances, is not only dependent upon the nature and functional modality of channel proteins, but is also dependent upon the nature of the lipids that surround these channels. Both elements work together in the process of synaptic transmission and, thus, ensure an activity-dependent, measured release of the neurotransmitter.

11.2.6 Memory Formation Through Molecular Facilitation in Synapses with Gangliosides

An integrated functional model of the mechanisms involved in the short-term process of synaptic information transmission as well as the long-term event of information storage must fulfill many diverse requirements. Among such consideration are:

- Extracellular concentrations of Ca^{2+} of up to 5 mM occur in the nervous system of lower vertebrates whereas concentrations in higher vertebrates can vary between only 1 and 2 mM.
- Ca^{2+} functions not only as a secondary messenger in intracellular synaptic metabolism, but might perform a crucial role as a primary messenger (a) for internalizing electrically encoded information on the outside of the presynaptic membrane, and (b) in the interaction between transmitters and receptors on the outside of the postsynaptic membrane.
- Membrane-bound lipids modulate the function of ion channels and receptor proteins in the synapse.

- In synaptic information processing, the processing of the primary signals at the presynaptic membrane must trigger adaptive molecular changes within milliseconds, whereby secondary manifestations are expressed in the neuronal cell body in terms of adaptive metabolic responses.
- Molecular feedback must occur from the synaptic membranes to the cell body upon sufficiently intense and/or frequently repeated stimulation. In the cell body, the feedback must trigger the activation of effector and regulatory genes which ensure that the initial adaptive changes in the region of the synapse become stabilized for the long-term.

It was before this backdrop that the *hypothesis of memory formation through molecular facilitation in synapses with gangliosides* was developed (Rahmann, 1976, 1983, 1987). Gangliosides have been discussed in this book in varying degrees of detail with regard to their chemical composition, occurrence in the nervous system, physiological adaptability, and their physicochemical properties (see Chap. 2.2.3.4 and 8.2). They appear to fulfill all of the requirements mentioned earlier. For one, they have a modulatory effect in the short-term process of synaptic transmission (see Chap. 8.2 and Figs. 8.15 and 8.16). Second, on the basis of their biochemical and physicochemical properties (long biological half-life and a great affinity for forming complexes with Ca^{2+}), they appear especially well suited to stabilize synaptic connections over long periods of time. Thus, they are apt to perform well in the facilitation of neuronal wiring, i.e., memory formation.

The possible functional significance of gangliosides in short- and long-term neuronal processes is discussed below. The emphasized perspective is that of their essential involvement in molecular facilitation in synapses as is required for the formation of memory tracks (Table 11.1 and Fig. 11.3):

- The content of neuronal gangliosides increases dramatically even during *early development of the nervous system* as the individual brain regions gradually are taking shape according to genetic control programs (see Fig. 2.12) and especially prior to critical developmental

TABLE 11.1. Overview of the most important steps involved in memory formation through molecular facilitation in synapses by gangliosides (see text and Fig. 11.3).

1. Adaptive changes in biosynthesis of individual gangliosides during characteristic stages of **neuronal differentiation**

2. Function involvement of gangliosides in **cell-to-cell recognition** and cell contact

3. Formation of local, as yet unstable **synapses** in conjunction with Ca^{2+}-ganglioside complexes

4. Upon electrical stimulating impulses (change in electrical fields), initial **transmission** of an unstable synaptic assembly via changes in configuration of gangliosides as a result of the disassociation of Ca^{2+}-ganglioside binding points. Changes include that of viscosity and the activation of channel and receptor molecules in the pre- and postsynaptic membrane

5. Formation of **short-term memory** through reciprocal reinforcement of activity-dependent changes in ganglioside configuration and the correlated change in functional synaptic proteins

6. Upon repetition of an identical stimulation pattern, the formation of **long-term memory** in participating neuronal facilitation chains through molecular feedback of synaptic membranes to the cell body via retrograde transport; activation of effector and regulatory genes in the cell body enabling synthesis of substances that are transported via anterograde transport to the synapse where they ensure that the synapse becomes stabilized functionally and increase the efficiency of the functional proteins (ion channels, receptor proteins)

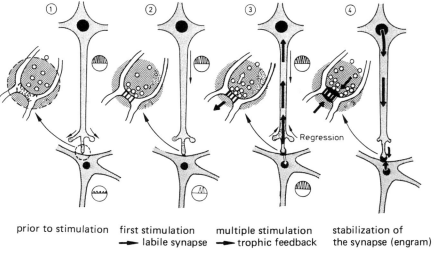

prior to stimulation   first stimulation   multiple stimulation   stabilization of
➤ labile synapse   ➤ trophic feedback   the synapse (engram)

Fig. 11.3. Model of memory formation through molecular facilitation in synapses by gangliosides. See text for details. (Reprinted with permission from Rahmann, 1985).

phases such as hatching, birth, opening of the eyes, etc. This increase correlates to a simultaneous increase in enzyme activity that regulates ganglioside metabolism (sialyl transferases, neuraminidases).

• The increase in ganglioside concentration does not occur uniformly for all gangliosides. Rather, the synthesis of individual fractions correlates to the degree of *neuronal differentiation* such that each of the various stages of

neurogenesis (cell division and migration, fiber growth and ramification, synapse formation and myelination) is characterized by the synthesis of highly specific gangliosides (see Chap. 2.2.3.4 and Figs. 2.13 and 2.14). This should be considered a sign that each individual ganglioside fraction might perform specific functions in the course of neurogenesis. Gangliosides, which contain neuraminic acid, could warrant recognition for the part they play in *cell-to-cell recognition* and cell contact (especially during the *phase of neuronal migration* and in the *formation of synaptic contacts*) on a par with the glycoproteins (which also contain neuraminic acid). The latter also are referred to as *cell adhesion molecules* (*CAM*; see Chap. 2.2.3.3).

- The negative charges of both membrane-bound macromolecules might well be of great functional significance in the area of the synaptic membrane surface where they occur for purposes of *cell recognition* and *cell adhesion*. These negative charges might affect the *formation of electrical fields in a special way*, whereby a reversible, unstable binding of cations, especially of Ca^{2+}, significantly affects the surface-oriented gangliosides insofar as the gangliosides–in accordance with the current model–occur in molecular clusters in the outer membrane.

- The functional model, which focuses on Ca^{2+} ganglioside complexes in *synaptic transmission* (see Chap. 8.2.4) suggests that, at first, gangliosides are present in a largely disorganized state (Fig. 11.3.1) as integral components of the synaptic membrane (see Fig. 7.5) at the membrane surface of a still unstable synapse formation (see Fig. 9.11). *Lipophilic Ca^{2+}-ganglioside complexes* contribute here to increased rigidity and, thus, to the sealing of the synaptic membrane. According to this model, at the moment of the first functioning of the unstable synapse, i.e., upon the first conduction of an electrical stimulating impulse, there occurs a sudden *change in the configuration* of the ganglioside constellation. A simultaneous change in the action potential at the neuronal terminal causes Ca^{2+} to be released by virtue of the changes in the electrical fields of the ganglioside binding points. The changes in configuration in the ganglioside clusters cause the corresponding local membrane region of the presynapse to become more fluid. Secondly, the change in configuration of the Ca^{2+}-ganglioside clusters (which are known to be in the environs of the ion channels) also might bring about a sudden *release of Ca^{2+}* which, in turn, triggers a vacuum-driven flow of calcium. This excess Ca^{2+}, which is present in the extracellular

space, flows into the presynapse. Thus, up to this point, the calcium that has been mobilized by the gangliosides has functioned as a *primary messenger* involved in internalizing the electrically encoded signal.

- Upon Ca^{2+} having permeated the presynapse, it assumes myriad secondary *messenger functions* that generally are recognized in terms of cytoplasmic signal processing, i.e., as steps in the activation of membrane-bound *phospholipases* that are necessary for the release of transmitters and, in turn, supply *inositol triphosphate* (ITP) *diacylglycerine* (DAG). This substance then activates *protein kinase C* which is capable of phosphorylation and, thus, is able to activate various proteins. Secondly, it affects the *fusing of the synaptic vesicles* to the presynaptic membrane and, thus, triggers the release of transmitter substance (for details, see Chap. 7.2 and Fig. 7.19).

- By virtue of the Ca^{2+}-induced *release of the neurotransmitter* into the synaptic cleft and its subsequent interaction with *postsynaptic receptors,* ion channels of the postsynaptic membrane are opened, either directly or indirectly, by means of *adenylate cyclases* (see Fig.7.14). It is likely, beyond this, (as has been proved in physicochemical experiments with acetylcholine, in particular) that Ca^{2+} is released by the transmitter that is discharged from its postsynaptic ganglioside complexes and, in a fashion comparable to the event that took place at the presynaptic membrane, now flows into the postsynapse to trigger a plethora of Ca^{2+}-dependent, secondary metabolic changes there, as well. The resultant opening of the *postsynaptic Na^+ channels* causes *local depolarization* in the postsynaptic membrane which leads to the further conveyance of the transmitted, electrically encoded information in this cell—that is, if the intensity of the impulse is sufficient (see Fig. 8.15).

- The original condition of the membrane ultimately is restored upon the enzymatic deactivation of the transmitter in the synaptic cleft and by the activation of membrane-bound *ion pumps* (*Na^+/K^+-ATPase Ca^{2+} ATPase*) by means of which Ca^{2+} is pumped back into the extracellular space where, once again, it can accumulate loosely at the negative neuraminic acid residues of the gangliosides. This causes the change in configuration from the previously fluid state to a regained state of rigidity in the corresponding membrane section which harbors accumulated ganglioside clusters. In turn, this membrane section is restored to the state of being impervious to the extracellular Ca^{2+} that has reestablished itself in very high concentrations (for details, see Chap. 8.2.4 and Figs. 8.15 and 8.16).

In light of these concepts, the Ca^{2+} ganglioside complexes well might ensure an optimal constellation of physicochemical parameters of the synaptic membranes (viscosity, associated molecular field for channel and receptor proteins, electrical field properties, etc.) for the process of short-term transmission. This would allow for the modulatory release of transmitters and, thus, an adaptive transmission of an electrically encoded signal from neuron to neuron. (Numerous experimental findings in reference to the extraordinary adaptability of gangliosides in interaction with Ca^{2+} were reported in Chap. 8.2.3; the results obtained through physicochemical simulation experiments lend particular credence to the hypothetical consideration discussed above.)

Thus, if viewed within the larger framework of events, Ca^{2+}-*ganglioside complexes* might be regarded as *neuromodulator substances* inasmuch as they, perhaps more so than any other class of substance, are capable of affecting the function of synaptic ion channel and receptor molecules by virtue of the fact that they are involved in controlling the basal functional properties of the synaptic membrane by means of interaction within the outer lipid layer. They manage this by means of activity-dependent changes in configuration. By virtue of the extremely flexible organization of the ganglioside mixtures that are found in the form of molecular clusters and are concentrated particularly in neuronal membranes, this class of substance is extremely well suited to adapt the lipid environment of the functional synaptic proteins to changing ecophysiological parameters (temperature, pressure, ion milieu, etc.).

Just such a functional model of neuronal gangliosides gains ever greater plausibility when considered in light of the following facts that have been gained through experimental research. Gangliosides:

- Generally contribute to the stabilization of plasma membranes,
- Act as modulator substances in the process of cell adhesion in that they affect specific protein interactions (GM3 and GD3 affect fibronectin and laminin),
- Regulate cell growth,
- Control cell-to-cell contact during cell differentiation,
- Are involved in the transfer of information between the cell surface and substances of the extracellular fluid. This is exemplified by the modulatory functions that gangliosides perform with regard to receptors of growth factors (EGF) and for receptors of bacteriotoxins and myxoviruses.

In light of these concepts, the phenomenon of *short-term memory* could be explained in terms of the *activity-dependent* (*voltage-dependent*) changes in configuration that occur at the outer membrane of hitherto *unstable synapses,* particularly in the region of the developing synaptic contact points. A second explanatory clue lies in the evident ordering of the molecular arrangements of the Ca^{2+}–ganglioside complexes. It, too, is activity-dependent. This makes possible the first flow of extracellular Ca^{2+} into the synaptoplasm which, in turn, activates the functional synaptic proteins for the first time (Fig. 11.3.2).

Within this context, the following applies to the formation of *long-term memory*:

- The phase of molecular stabilization of synapses in the participative neuronal facilitation chain is initiated if the electrical signal impulses that arrive at the presynapse are sufficiently strong and/or are repeated often enough and in a consistent fashion (*training*) (Fig. 11.3.3).
- By means of *retrograde transport* (see Chap. 9.1.3) information in the form of chemical substances reaches the neuronal cell body via the molecular changes in the region of its synaptic terminals, these having been triggered by the torrent of secondary messenger

inductions. *Effector* and *regulator genes* then are activated in the neuronal cell body subsequent to which substances are synthesized in the cell body that are necessary for maintaining activity-dependent changes in the region of the synapse. These substances then are transported continuously via *anterograde transport* (see Chap. 9.1.2) from the cell body to the synapses that have been prepared in a molecular sense and where these substances ensure long-term stabilization of the synapses (Fig. 11.3.4). Maintaining functional synaptic stabilization of this sort could be brought about by the fact that (a) gangliosides are replicated and synthesized in specific arrangements in the neuronal cell body (subsequent to their axonal transport, they are incorporated into the synaptic membrane), and/or (b) the modulatory function of gangliosides enables the functional proteins of the synaptic membrane (ion channels, receptor proteins) to function more efficiently. At some future time, the resulting synapse that has been "facilitated" in this manner will be able to transfer an incoming electrical signal impulse more efficiently from cell to cell.

According to these views, the phenomenon of *memory formation on the basis of molecular facilitation* is rooted in the fact that the affected synaptic membrane regions in a neuronal network, a network that is stimulated in a sufficiently strong and/or in a sufficiently consistent and repeated manner, are modulated by physicochemical changes in the configuration of the varyingly polar Ca^{2+}–ganglioside complexes (possibly stimulation-specific in the sense of an "electrical resonance of the ganglioside structure"). Moreover, only those signal impulses are transmitted efficiently that correspond to the original impulses, i.e., to those that occurred during the first memory storage event.

11.3 Aspects of the Formation of a Neuronal Information Processing System

The question arises, in regard to the model of memory formation through molecular facilita-

tion in the region of the synaptic contact zones in neuronal terminals, as to the ways in which such complex neuronal networks take shape, i.e., how they form the nervous systems in higher animals and in the human.

Insofar as the development of a system for processing and storing information is concerned, it can be assumed that such a multileveled system emerges as a result of epigenetically, i.e., individually, acquired experience. It can be assumed, against the background of a *neuronal information system that has been programmed through phylogeny*, that a certain degree of knowledge about the environment that affects an individual is innate. During mating periods, for example, animals recognize members of their species, even if they have never before encountered them. They have at their disposal a kind of identification mechanism to enable them to recognize their own species. The mechanism is formatted in the brain as a kind of *innate triggering mechanism (ITM)*, and although it is limited to the recognition of only a few marked features (*key stimuli*), the ITM, nonetheless, is adequate for the task of recognition (see Chap. 10.2.3.2). In similar fashion, most animals have an innate repertoire of individual behaviors that enables them to signal effectively to a member of their species upon a first encounter. The signals are appropriate for a given situation and, in turn, serve as key stimuli for others of the species. Ultimately, this means that information from the environment is being infused into the genetic code of species in a constant interplay of *mutation* and *selection,* that it gradually manifests itself genetically and, thus, is passed from one generation to the next.

The transmission of information that manifests itself genetically can take unusual turns. Through *mimicry,* for example, one species copies certain characteristics of another species for the purpose of deterring adversaries (for example, the special manifestation of owl's eye markings in butterflies).

The appropriate question to be asked is, How is the *original, genetically determined information pattern translated from the genome into behavior* such that the innate programs and abilities within an animal species unfold in a manner true to that species? This question cannot be answered definitively given our present state of understanding. Discussions are presented in Chap. 2, however, that address important aspects of this *genetic-embryological information pathway*. Particular emphasis is placed on the discussion of *molecular differentiation in neuronal assemblies*. It underscores the significance of nerve growth factors, neuritogenic substances, and gangliosides as marker substances of functional neuronal differentiation for the code-dependent emergence of the nervous system. Chap. 5 presents models of possible wiring principles in neuronal information processing. On the one hand, it is demonstrated that one must assume rigid, genetically determined reflex circuits that serve to control life maintenance systems. On the other hand, more complex neuronal circuitry is presented, such as that of the cerebellum, hippocampus, or neocortex of the forebrain, which exemplifies the opposite extreme, i.e., great plasticity and complexity in neuronal circuitry that enables greater variation in behavioral response. In light of these understandings, it becomes clear that in the course of phylogeny, and especially in regard to vertebrates, impressive functional-morphological differentiations occurred in individual regions of the brain, each of which assumed control of specialized tasks (see Chap. 3).

With the intention of supplementing concepts that were presented in Sect. 11.2.5 regarding the notion of memory formation through molecular facilitation, the following comments are in reference to the third kind of neuronal information flow, i.e., the *experience-dependent processing and storing of information* from signals from the environment that are perceived by the sensory organs, communicated to the brain, and returned to the environment in the form of motor responses.

In a manner similar to the way in which *electronic computers* receive signals via input lines, the brain constantly receives vast amounts of information via a profusion of neuronal fibers each of which is capable of sending signals independently of the others. Although the signals in the technical world of computers are merely *electrical pulses,* neuronal signals are complex, physicochemical events (see Chaps. 6 and 7). Ultimately, however, both the brain and the computer do share a comman trait relative to the transmission of information: discrete, quantifiable events are in play that can be distinguished from one another very clearly. Therein lies one advantage of pulse conduction when compared to the continuously changing flow of information encountered in an electrical potential: the latter can become distorted during its transmission, thus, altering its content. This is not true for the *transmission of impulses,* however. The pulse, or a specific sequence of pulses (*pulse volleys*) represents, as it were, the actual symbol for what is to be conveyed. Each symbol is encoded into individual pulse volleys that are distinguishable by their frequency. In this sense, the brain represents a signal processing device that selectively receives diverse input stimuli from the environment much as information processing organizes individual letters of the alphabet into words and these symbols into sentences. Thus, from the flood of the most diverse sensory input, individual events are given specific symbols of brain activity in the form of pulse patterns in the participating neuronal systems, pulse patterns that are organized according to their various frequencies.

It is possible that a mechanism is activated in this process by which the newly received environmental signals are compared to preexisting, genetically determined patterns of symbols. In other words, the genetically encoded supply of symbols that was handed down as a whole is adapted to the properties of the environment. It has been established that, with increasing developmental complexity in animals and the attendant differentiated organization of the nervous system, the amount of information acquired through learning processes plays an ever-greater role when compared to the amount acquired through innate stimulus-response patterns. One learns all one's life to distinguish new patterns, to recognize variations in situations, and to hone one's own responses.

In the course of these processes, new neuronal assemblies are being formed constantly through the activity-dependent stabilization of hitherto unstable synaptic contacts. One result of this is that extant information pathways are reinforced through frequent use. Secondly, new pathways are established continuously between functionally related neuronal assemblies. Individual information pathways are connected successively upon the simultaneous emergence of two or more stimulus inputs–in much the same way as conditioned reflexive processes are established in the process of conditioning. In this way, neuronal networks first emerge that, upon further histogenic development, become organized into wiring units (*columns*) that react to identical stimulus input as integrated neuronal assemblies by generating identical pulse volleys. Thus, over time, a wiring network of synaptically linked neuronal pathways emerges in the brain in which the genetically determined neuronal assemblies increasingly are integrated with newly acquired assemblies, i.e., those established by individual experience. By virtue of a certain degree of correspondence among innate and acquired neuronal information patterns (from elements of the environment), there emerge *logical* and *causal arrangements of circuitry* by way of which events can elapse in the brain on a trial, or exploratory basis, i.e., without direct verification by actual environmental stimuli. Processes of this sort are referred to as *thought* or *thinking*. Thinking proves to be an extremely wasteful means of information processing: an immense amount of individually acquired experience can be linked to new insights, only to be lost toward the end of one's life through processes of degeneration–unless they are maintained as "extracerebral chains of association," i.e., in the form of symbolic information carriers such as writing, sound and visual recordings, electronic data carriers, etc. These devices allow thoughts to be passed from generation to genaration.

Models such as these, which suggest the formation of extremely complex neuronal networks in the differentiated brain from the early developmental formation of the first simple neuronal assemblies, recently have inspired data scientists and computer experts to consider whether and to what extent the functional nature of the human brain can be explained in physicotechnical terms. Presently, it is possible to relate this complexity to relatively simple, yet entirely plausible, models (Gerke, 1987; Palm, 1988).

11.4 Localizing Memory

Indeed, the field of neurobiology has made enormous strides toward clarifying the myriad neuronal processes. Methodologies borrowed from the areas of electrophysiology, histo-autoradiography, and computer tomography have been invaluable aids in this pursuit; however, we are still not in a position to understand all the details of the immensely complex issue of information processing and storage, i.e., memory formation, in large part, because so many factors still stand in the way of researchers in their search for the engram (Table 11.2).

The question as to the exact *location* of memory within the nervous system still cannot be answered today. The only view that is widely accepted is the one presented above in detail, i.e., that memory ultimately is stored in the form of molecular changes in the synapses of those neuronal structures involved in perception, analysis, and further processing of acquired (learned) information. There is also unanimity of opinion in regard to the suggestion that memory content is stored neither in the sensory nor motor neuronal pathways, but rather only in the CNS (the brain and spinal cord).

Currently, various research methods are used to determine which brain structures are involved in storing memory content:

TABLE 11.2. Limiting factors in memory research.

Organization of the human brain consists of
—hundreds of billions (10^{11}) of neurons
—several trillion (10^{12}) glial cells
—several hundred trillion (10^{14}) synapses
Total length of all neuronal fibers in the brain =
$2 \times 384,000$ km, or similar in length to the distance from the Earth to the moon and back again
Impulse conductions speeds of 100 to 120 m/sec (360 to 400 km/h)
4×10^9 impulses/sec are exchanged between the two hemispheres of the brain via the corpus callosum
> 40 various chemical transmitter substances differentiate stimulatory synapses from inhibitory synapses
15,000 protein molecules are transformed each second in an active neuron
Between 10^7 and 10^8 neurons are believed to be activated during each memory event

- In animal models, lesions (injuries) are caused in the CNS; then, the degree to which behavior is altered and the reduction in the ability to perform memory-oriented tasks are noted.
- Upon the administering of marker substances (radioactively labeled *D*-oxyglucose, for example), neuronal activity is measured in the various brain regions prior to and after adequate stimulation of the sensory organs. Measurement can be either by means of electrophysiological techniques (using lead electrodes) or by measurement of metabolic change.
- Case studies of certain patients help to round out the present picture of memory. Patients who have suffered damage to specific areas of the brain through illness and injury have lost some of their ability to learn and recall. Thus, it has been known for more than 100 years that memory capability is severely impaired upon bilateral trauma to the middle region of the temporal lobes of the forebrain. Injuries to these regions result in extreme difficulty in forming new memory (*anterograde amnesia*) and recalling events that were stored prior to loss of memory (*retrograde amnesia*). However, with this type of injury, normal intelligence remains uncompromised, as do the abilities to learn and to form short-term memory. The ability to recall experiences of the distant past remains intact, as well. Possibly the best known case in this regard is that of a *patient known as H.M.* This individual underwent surgery in 1953 for the bilateral removal of the medial temporal lobe of the prosencephalic cortex (thus, the *hippocampus* region) in order to alleviate his epileptic seizures. In the wake of the operation, H.M. lost the ability to recall anything for more than just a few minutes. Curiously, his short-term memory remained intact. He was able to converse intelligently without any difficulty. His recollection of events that occurred long before the operation remained intact, as well. However, through the operation, he lost the ability to transfer events from short-term memory into long-term storage. One concludes that, in the human, the hippocampus must be involved significantly in the formation of long-term memory. This does not imply, however, that this region of

the brain is the actual repository of long-term memory. Apparently, in addition to the medial temporal lobe of the forebrain cortex, the diencephalon (interbrain), especially a group of nerves in the *thalamus* and *hypothalamus*, is involved extensively in the formation of memory, since injury to these structures leads to serious *amnesias*. Thus, in the case of *Korsakoff's syndrome,* a wide-ranging amnesia that accompanies chronic *alcoholism*, structures degenerated that lie near the median line of the diencephalon. Injury to the diencephalon brought on by stroke, infection, or tumor can induce similar manifestations of amnesia.

Based on extensive experimental research using rats whose cerebral cortices were destroyed in varying degree, Lashley (1950) concluded that it was not possible to identify one specific, isolated repository of memory content anywhere in the nervous system. Rather, Lashley suggested that memory is stored in a diffuse manner, i.e., that it is distributed among many structures. Moreover, he suggested that the same neurons that were responsible for producing memory events were also involved in their storage. Accordingly, memory then would involve a synergistic activity or a kind of resonance among a great number of neurons.

On the basis of proportionate decreases in the ability to perform memory-oriented tasks in relation to the mass of cortex removed, Lashley formulated the "*principle of mass action*" and that of "*equipotentiality,* " i.e., functional reciprocity among the various brain regions. The theory suggests that memory formation is sooner dependent upon the quantity of available brain substances than specific regions. Since Lashley studied the extent of lesion-induced memory loss in rats that had learned to run a maze, and since this kind of learning is associated with several types of information channels (visual, olfactory, and kinesthetic), a great variety of brain regions might be involved in forming the memory of such a complex learning exercise. A definitive statement regarding the possible location of the various components involved in the *learning of a maze* cannot be expected from this kind of learning.

Research has concentrated on less ambiguous systems to address the question as to whether memory is located in the brain in a strictly localized or in a diffuse manner. Since the greatest portion of our memory content results from visual sensory impressions, recent research has focused on the neuronal processing pathways responsible for sight, or *visual perception*. Chapter 3 contains a discussion about the information flow from the retina via the optic nerve (*nervus opticus*) to the lateral geniculate bodies (*corpora geniculata lateralia*) in the diencephalon. This flow is relevant to the processing of visual impressions in mammals and, thus, in humans. The majority of neuronal pathways are relayed to the occipital region of the cerebral cortex, the *area striata*, where the information relative to visual impressions undergoes its initial processing. In the area striata, individual neurons respond to simple, spatially defined elements within the field of view such as sharply defined borders or colored dots. Within this section of the visual cortex, other neuronal pathways project to other neurons that analyze less subtle properties of the object such as its overall structure or its color. At the end of this object-oriented pathway, i.e., in the inferior temporal gyrus (*gyrus temporalis inferior*), other neurons communicate information about a great number of properties as well as about a broad section of the field of view. It is reasonable to conclude that information about the object ultimately collects here.

Presumably, there is one other projection pathway in the anterior parietal lobe that is responsible for processing spatial relationships within a visual scene. This other pathway for processing visual information, which possibly is involved in memory formation of visual impressions (i.e., from the cerebral cortex into lower lying brain regions and, thus, into the underlying structures), can be identified very clearly as a result of analyses of various surgical procedures that were performed on monkeys to test visual *recall*. According to these analyses, a visual perception that materialized in the last section of the projection pathway, the sub-parietal cortex, stimulates two parallel neuronal assemblies, one of which originates in the amygdaloid nucleus (*nucleus amygdalae*) and the

other derives from the *hippocampus*. Apparently, both regions are responsible for many kinds of *cognitive learning*, i.e., for the ability to recognize objects, to recall properties from memory that are not being perceived at that moment and to assign emotional significance to those properties. Apparently, however, neither of the regions represents the final location of those structures involved in perceiving and storing visual information, since projection pathways lead from here to the striate body (*corpus striatum*) and to the diencephalon (*thalamus* and *hypothalamus*) which, in turn, closes the circle of information processing insofar as it sends neuronal pathways back to the cerebral cortex (Fig.11.4).

It is presumed now that the most likely *repositories of memory* are those locations in which the sensory impressions develop. The neuronal assemblies that exist beneath the cerebral cortex might serve as feedback pathways to the cortex as memory is being formed: the neuronal assemblies involved in memory must interact with the sensory areas of the cortex after a sensory impulse, having been processed already in the cerebral cortex, has stimulated the hippocampus and the amygdala.

The structural correlate to a memory storage event of visual impressions may well be represented by the reciprocal feedback mechanisms between the neuronal representation centers in the *cortex* (see Chap. 3) and the neuronal assemblies of the *hippocampus* and *amygdala*. Furthermore, lesion experimentation in apes suggests that the amygdala also could be responsible for linking together items of the memory inventory that are conveyed through the various senses. It might be activated as *intermediate recall*, for example, when we picture the face of a caller upon hearing a familiar voice over the telephone. Moreover, it is believed that the amygdala is partially responsible for the *feelings* that accompany the processing of sensory stimuli. Reciprocity between the amygdala and the cerebral cortex could explain why events that are charged with emotion produce particularly long-lasting memories.

Indeed, the previous statements all clearly suggests that the storage of specific memory events occurs in the region of particular

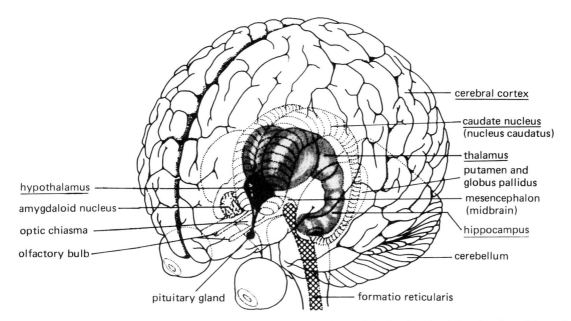

Fig. 11.4. Diagram of the human brain. Those structures that might be involved in the depositing of memory content are underscored.

representation centers in the cortex in conjuction with subcortical regions such as the hippocampus and the amygdala. It must be persumed that, within areas as expansive as these, memory storage is *not highly localized*, but rather *dispersed*, i.e., distributed diffusely over wide regions of these neuronal networks. Thus, individual memory inventories are stored not so much as the individual reference points of a photograph, but as of a *hologram*, whereby it is possible to store and recall three-dimensional spacial scenes.

Ultimately, the phenomenon of storing individually acquired information, i.e., the formation of memory, might be viewed as being bound to an unwrought, genetically predetermined representational model of the various sensory modalities that manifests itself in certain arrangements and combinations of great numbers of neurons in larger brain regions. As these neuronal arrangements begin to function, neuronal memory assemblies emerge within them by means of molecular facilitation through feedback in the participating synapses (see Sect. 11.2.5). This occurs in such a way that the developing combinations of patterns are preserved, thus, preserving the individual perceptions. Subsequent recognition and even recall can be explained in terms of restimulating the same grouping of neurons with the same kind of sensory stimuli that established it initially.

Thereafter, in the origination of new memory content, each newly formed synapse stores a certain portion of the newly incorporated information. Moreover, each new perception and, thus, each new bit of information, is preserved within a broadly diffused network of new synaptic connections. Thus, every individually acquired bit of information is stored within the nervous system in a diffuse manner, i.e., over wide areas while, at the same time, vast amounts of information can be layered in storage throughout the entire system.

Bibliography

Chapter 1: The Cellular Basis of Memory

Akert K (1971) Struktur und Ultrastruktur von Nerven und Synapsen. Klin Wochenschr 49: 509–519

Barondes SH (1969) Cellular dynamics of the neuron. Academic Press, New York

Eccles JC (1964) The physiology of synapses. Springer, Berlin Göttingen Heidelberg New York

Greenough WT (1984) Structural correlates of information storage in the mammalian brain: a review and hypothesis. Trends in Neurosciences 7 (7):229–233

Hydén H (ed) (1967) The neuron. Elsevier, Amsterdam

Kandel ER (1976) Cellular basis of behavior. Freeman, San Francisco

Kandel ER, Schwartz JH (eds) (1985) Principles of neural science, 2nd edn. Elsevier Science, Amsterdam

Lockwood APM, Lee AG (1984) The membranes of animal cells, 3rd edn. Arnold, Baltimore (Studies in Biology, vol 27)

Mill JP (1982) Comparative neurobiology: Contemporary biology. Arnold, London

Möllendorf W v (1943) Handbuch der mikroskopischen Anatomie des Menschen. Berlin

Morell P, Norton WT (1984) Myelin. In: Gehirn und Nervensystem. Spektrum der Wissenschaft: 65–74

Palay SL, Palade GE (1955) The fine structure of neuron. J Biophys Biochem Cytol 1:69–88

Popper KR, Eccles JC (1982) Das Ich und sein Gehirn. Piper, München Zürich

Rahmann H (1976) Neurobiologie. Ulmer, Stuttgart (UTB)

Ramon y Cajal S (1909) Histologie du système nerveux de l'homme et des vertèbres (2). Maloine, Paris

Ramon y Cajal S (1933) Histology, 10th edn. Wood, Baltimore

Rasmussen AT (1952) The principal nervous pathways. MacMillan, London

Rohen JW (1971) Funktionelle Anatomie des Nervensystems. Schattauer, Stuttgart New York

Schadé J (1973) Die Funktion des Nervensystems. Fischer, Stuttgart

Sester U, Probst W, Rahmann H (1984) Einfluß unterschiedlicher Akklimationstemperaturen auf die Ultrastruktur neuronaler Synapsen von Buntbarschen (Tilapia mariae; Cichlidae, Teleostei). Z Hirnforsch 25 (6):701–711

Shepherd GM (1983) Neurobiology. Oxford University Press, New York Oxford

Sherrington CS (1897) The central nervous system. In: A textbook of physiology. MacMillan, London

Sherrington CS (1947) The integrative action of the nervous system, 2nd edn. Yale University Press, New Haven

Snyder SH (1985) Signalübertragung zwischen Zellen. Spektrum der Wissenschaft: 126–135

Stevens CF (1984) Die Nervenzelle. In: Gehirn und Nervensystem. Spektrum der Wissenschaft: 3–14

Stöhr P, Möllendorf W v, Görttler K (1963) Lehrbuch der Histologie und der mikroskopischen Anatomie des Menschen. Fischer, Stuttgart

Vrenzen G, Nunes Cardozo J, Mueller L, Van der Went J (1980) The presynaptic grid: A new approach. Brain Res 184:23–40

Waldeyer W (1891) Über einige neuere Forschungen im Gebiete der Anatomie des Zentralnervensystems. Dtsch Med Wochenschr: 1352–1356

Weinberg CB, Sanes JR, Hall ZW (1981) Formation of neuromuscular junction in adult rats: accumulation of acetylcholine receptors, acetylcholinesterase, and compounds of synaptic basal lamina. Dev Biol 84:255–266

Wolburg H, Neuhaus J, Mack A (1986) The glio-axonal-interaction and the problem of regeneration of axons in the central nervous system – concept and perspectives. Z Naturforsch [C] 41:1147–1155

Gehirn und Nervensystem, 4. Aufl. (1984) Spektrum der Wissenschaft, Heidelberg

The Neuroscience I, II, III. The MIT Press, London Cambridge, Mass

Chapter 2: Development of the Nervous System in Vertebrates

Blinkov SM, Glezer II (1968) Das Zentralnervensystem in Zahlen und Tabellen. Fischer, Jena

Brauer K, Schober W (1970) Katalog der Säugetiergehirne. Fischer, Jena

Cowan WM (1984) Die Entwicklung des Gehirns. In: Gehirn und Nervensystem. Spektrum der Wissenschaft: 102–110

Eccles JC (1964) The physiology of synapses. Springer, Berlin Göttingen Heidelberg New York

Giersberg H, Rietschel P (1967) Vergleichende Anatomie der Wirbeltiere. Fischer, Jena

Hilbig R, Lauke G, Rahmann H (1983/84) Brain gangliosides during the life span (embryogenesis to senescence) of the rat. Dev Neurosci 6: 260–270

Hilbig R, Rösner H, Merz G, Segler-Stahl K, Rahmann H (1982) Developmental profile of gangliosides in mouse and rat cerebral cortex. Roux's Arch Dev Biol 191:281–284

Jacobson M (1978) Developmental neurobiology. Plenum Press, New York

Kandel ER, Schwartz JH (eds) (1985) Principles of neural science, 2nd edn. Elsevier Science, Amsterdam

Mill JP (1982) Comparative neurobiology: Contemporary biology. Arnold, London

Möllendorf W v (1943) Handbuch der mikroskopischen Anatomie des Menschen. Berlin

Morell P, Norton WT (1984) Myelin. In: Gehirn und Nervensystem. Spektrum der Wissenschaft: 65–74

Nauta WJH, Freitag M (1984) Die Architektur des Gehirns. In: Gehirn und Nervensystem. Spektrum der Wissenschaft: 89–99

Patterson PH, Potter DD, Furshpan EJ (1984) Chemische Differenzierung von Nervenzellen. In: Gehirn und Nervensystem. Spektrum der Wissenschaft: 45–62

Popper KR, Eccles JC (1982) Das Ich und sein Gehirn. Piper, München Zürich

Portmann A (1948) Einführung in die vergleichende Morphologie der Wirbeltiere. Schwabe, Basel

Rahmann H (1976) Neurobiologie. Ulmer, Stuttgart (UTB)

Ramon y Cajal S (1909) Histologie du système nerveux de l'homme et des vertèbres (2). Maloine, Paris

Ramon y Cajal S (1933) Histology, 10th edn. Wood, Baltimore

Ribchester RR (1986) Molecule, nerve and embryo. Blackie, Glasgow

Rösner H, Rahmann H (1987) Ontogeny of vertebrate brain gangliosides. In: Rahmann H (ed) Gangliosides and modulation of neuronal functions. Springer, Berlin Heidelberg New York London Paris Tokyo (NATO ASI Series H: Cell Biology, vol 7, pp 373–390)

Rösner H, Willibald CJ, Schwarzmann B, Rahmann H (1987) Uptake of exogenous gangliosides by the CNS? In: Rahmann H (ed) Gangliosides and modulation of neuronal functions. Springer, Berlin Heidelberg New York London Paris Tokyo (NATO ASI Series H: Cell Biology, vol 7, pp 581–592)

Rohen JW (1971) Funktionelle Anatomie des Nervensystems. Schattauer, Stuttgart New York

Romer AS (1966) Vergleichende Anatomie der Wirbeltiere. Parey, Hamburg Berlin

Schäfer C (1987) Gehirnzellen sterben nicht ab.

Bild der Wissenschaft 9:60–69

Seybold U, Rahmann H (1985) Brain gangliosides in birds with different types of postnatal development (nidifugous and nidicolous type). Dev Brain Res 17:201–208

Seybold V, Rahmann H (1985) Changes in developmental profiles of brain gangliosides during ontogeny of a teleost fish (Sarotherodon mossambisus, Cichlidae). Roux's Arch Dev Biol 194:166–172

Shepherd GM (1983) Neurobiology. Oxford University Press, New York Oxford

Sherrington CS (1947) The integrative action of the nervous system, 2nd edn. Yale University Press, New Haven

Singer W (1985) Hirnentwicklung und Umwelt. Spektrum der Wissenschaft: 48–61

Sperry RW (1951) Mechanisms of neural maturation. In: Stevens SS (ed) Handbook of experimental psychology. Wiley, New York, pp 236–280

Sperry RW (1963) Chemoaffinity in the orderly growth of nerve fibre patterns and connections. Proc Natl Acad Sci USA 50:703–710

Sperry RW, Preilowski B (1972) Die beiden Gehirne des Menschen. Bild der Wissenschaft 9: 921–928

Squire LR, Zola-Morgan S (1988) Memory: brain system and behavior. Trends in Neurosciences 11 (4):170–175

Stevens CF (1984) Die Nervenzelle. In: Gehirn und Nervensystem. Spektrum der Wissenschaft: 3–14

Stöhr P, Möllendorf W v, Görttler K (1963) Lehrbuch der Histologie und der mikroskopischen Anatomie des Menschen. Fischer, Stuttgart

Weiss PA (1934) In vitro experiments on the factors determining the course of the outgrowing nerve fibre. J Exp Zool 68:393–448

Wiesel TN (1982) Postnatal development of the visual cortex and the influence of the environment. Nature 299:583–591

Zeutzius I, Rahmann H (1980) Quantitative ultrastructural investigations on synaptogenesis in the cerebellum and the optic tectum of light-reared and dark-reared rainbow trout (Salmo gairdneri Rich.). Differentiation 17:181–186

Zeutzius I, Rahmann H (1980) Synaptogenesis in cerebellum and optic tectum of dark- and light-reared rainbow trout. IRCS Medical Science 8:47–48

Gehirn und Nervensystem, 4. Aufl. (1984) Spektrum der Wissenschaft, Heidelberg

The Neuroscience I, II, III. The MIT Press, London Cambridge, Mass

Chapter 3: Functional Morphology of the Nervous System in Vertebrates

Blinkov SM, Glezer II (1968) Das Zentralnervensystem in Zahlen und Tabellen. Fischer, Jena

Brauer K, Schober W (1970) Katalog der Säugetiergehirne. Fischer, Jena

Cowan WM (1984) Die Entwicklung des Gehirns. In: Gehirn und Nervensystem. Spektrum der Wissenschaft: 102—110

Creutzfeldt O, Innocenti GH, Brooks D (1974) Vertical organization in the visual cortex (Area 17) in the cat. Exp Brain Res 21:315—336

Douglas RJ, Pribram KH (1966) Learning and limbic lesions. Neurophysiologica 4:197—220

Eccles JC (1979) Das Gehirn des Menschen. Piper, München Zürich

Eccles JC, Ito M, Szentagothai J (1967) The cerebellum as a neuronal machine. Springer, New York Berlin Heidelberg

Forssmann WG, Heym C (1974) Grundriß der Neuroanatomie. Springer, Berlin Heidelberg New York

McGeer PL, Eccles JC, McGeer EG (1987) Molecular neurobiology of the mammalian brain. Plenum Press, New York London

Giersberg H, Rietschel P (1967) Vergleichende Anatomie der Wirbeltiere. Fischer, Jena

Ihle JEW, van Kampen PN, Nierstrasz HF, Versluys J (1927) Vergleichende Anatomie der Wirbeltiere. Springer, Berlin

Ito M (1984) The cerebellum and neural control. Raven Press, New York

Kandel ER, Schwartz JH (eds) (1985) Principles of neural science, 2nd edn. Elsevier Science, Amsterdam

Kleist K (1934) Gehirnpathologie. In: Schjerning OV (Hrsg) Handbuch der ärztlichen Erfahrungen im Weltkriege, Bd 4. Barth, Leipzig

Kuhlenbeck H (1967) The central nervous system of vertebrates, vol 2: Invertebrates and origin of vertebrates. Academic Press, New York

Lashley KS (1950) In search of the engram. Symp Soc Exp Biol 4:454—481

Mill JP (1982) Comparative neurobiology: Contemporary biology. Arnold, London

Mishkin M, Appenzeller T (1987) Die Anatomie des Gedächtnisses. Spektrum der Wissenschaft (8):94—104

Nauta WJH, Freitag M (1984) Die Architektur des Gehirns. In: Gehirn und Nervensystem. Spektrum der Wissenschaft: 89—99

Penfield W, Rasmussen T (1950) The cerebral cortex of man: Additional study on localisation of function. MacMillan, New York

Penfield W, Roberts L (1959) Speech and brain mechanisms. Princeton University Press, Princeton, NJ

Popper KR, Eccles JC (1982) Das Ich und sein Gehirn. Piper, München Zürich

Portmann A (1948) Einführung in die vergleichende Morphologie der Wirbeltiere. Schwabe, Basel

Rahmann H (1976) Neurobiologie. Ulmer, Stuttgart (UTB)

Rohen JW (1971) Funktionelle Anatomie des Nervensystems. Schattauer, Stuttgart New York

Romer AS (1966) Vergleichende Anatomie der Wirbeltiere. Parey, Hamburg Berlin

Schadé J (1973) Die Funktion des Nervensystems. Fischer, Stuttgart

Shepherd GM (1983) Neurobiology. Oxford University Press, New York Oxford

Sherrington CS (1947) The integrative action of the nervous system, 2nd edn. Yale University Press, New Haven

Sinz R (1978) Gehirn und Gedächtnis. Fischer, Stuttgart New York (UTB 852)

Sperry RW (1951) Mechanisms of neural maturation. In: Stevens SS (ed) Handbook of experimental psychology. Wiley, New York, pp 236—280

Sperry RW (1963) Chemoaffinity in the orderly growth of nerve fibre patterns and connections. Proc Natl Acad Sci USA 50:703—710

Sperry RW, Preilowski B (1972) Die beiden Gehirne des Menschen. Bild der Wissenschaft 9: 921—928

Squire LR (1986) Mechanisms of memory. Science 232:1612—1619

Stöhr P, Möllendorf W v, Görttler K (1963) Lehrbuch der Histologie und der mikroskopischen Anatomie des Menschen. Fischer, Stuttgart

Wiesel TN, Hubel DH (1963) Single-cell responses in striate cortex of kittens deprived of vision in one eye. J Neurophysiol 26:1003—1017

Wiesel TN, Hubel DH (1971) Long-term changes in the cortex after visual deprivation. Proceedings of the 25th International Congress of Psychological Science

Gehirn und Nervensystem, 4. Aufl. (1984) Spektrum der Wissenschaft, Heidelberg

The Neuroscience I, II, III. The MIT Press, London Cambridge, Mass

Chapter 4: Evolution and Architecture of the Nervous System in Invertebrates

Bullock TH, Horridge A (1965) The structure and function of the nervous system in invertebrates. Freeman, San Francisco

Crow T (1988) Cellular and molecular analysis of associative learning and memory in Hermissenda. Trends in Neuroscience 11 (4):136—141

Hanstroem B (1928) Vergleichende Anatomie des Nervensystems der wirbellosen Tiere. Springer, Berlin

Kandel ER, Schwartz JH (eds) (1985) Principles of neural science, 2nd edn. Elsevier Science, Amsterdam
Mill JP (1982) Comparative neurobiology: Contemporary biology. Arnold, London
Popper KR, Eccles JC (1982) Das Ich und sein Gehirn. Piper, München Zürich
Rahmann H (1976) Neurobiologie. Ulmer, Stuttgart (UTB)
Shepherd GM (1983) Neurobiology. Oxford University Press, New York Oxford
Wells M (1968) Lower animals. McGraw-Hill, New York

Gehirn und Nervensystem, 4. Aufl. (1984) Spektrum der Wissenschaft, Heidelberg

Chapter 5: Principles of Circuitry in Neurobiological Information Processing

Bruggencate G ten (1972) Experimentelle Neurophysiologie − Funktionsprinzipien der Motorik. Goldmann, München
Eccles JC (1964) The physiology of synapses. Springer, Berlin Göttingen Heidelberg New York
Eccles JC, Ito M, Szentagothai J (1967) The cerebellum as a neuronal machine. Springer, New York Berlin Heidelberg
Gerke PR (1987) Wie denkt der Mensch? Bergmann, München
Hassenstein B (1973) Biologische Kybernetik. Springer, Berlin Heidelberg New York
Ito M (1984) The cerebellum and neural control. Raven Press, New York
Kandel ER, Schwartz JH (eds) (1985) Principles of neural science, 2nd edn. Elsevier Science, Amsterdam
Katz B (1971) Nerv, Muskel und Synapse. Thieme, Stuttgart
Palm G (1982) Neural assemblies. Springer, Berlin Heidelberg New York
Palm G (1988) Assoziatives Gedächtnis und Gehirntheorie. Spektrum der Wissenschaft 6: 54−64
Popper KR, Eccles JC (1982) Das Ich und sein Gehirn. Piper, München Zürich
Rahmann H (1976) Neurobiologie. Ulmer, Stuttgart (UTB)
Schmidt RF (Hrsg) (1987) Grundriß der Neurophysiologie, 6. Aufl. Springer, Berlin Heidelberg New York London Paris Tokyo (HTB)
Schmidt RF, Thews G (Hrsg) (1980) Physiologie des Menschen, 20. Aufl. Springer, Berlin Heidelberg New York
Shepherd GM (1978) Microcircuits in the nervous system. Sci Am 238 (2):92−103
Shepherd GM (1983) Neurobiology. Oxford University Press, New York Oxford

Sherrington CS (1947) The integrative action of the nervous system, 2nd edn. Yale University Press, New Haven
Sinz R (1978) Gehirn und Gedächtnis. Fischer, Stuttgart New York (UTB 852)
Wiesel TN (1982) Postnatal development of the visual cortex and the influence of the environment. Nature 299:583−591

Gehirn und Nervensystem, 4. Aufl. (1984) Spektrum der Wissenschaft, Heidelberg

Chapter 6: Electrophysiological Aspects of Information Processing

Berger H (1935) Über das Elektroencephalogramm des Menschen. Arch Psychiatr 102: 538−557
Bingmann D (1984) Lernen und Gedächtnis: Neurophysiologische Grundlagen. Therapiewoche 34:7155−7162
Bruggencate G ten (1972) Experimentelle Neurophysiologie − Funktionsprinzipien der Motorik. Goldmann, München
Creutzfeld OD (1971) Neurophysiologische Grundlagen des Elektroenzephalogramms. In: Haider M (Hrsg) Neuropsychologie. Huber, Bern
Creutzfeldt O, Innocenti GH, Brooks D (1974) Vertical organization in the visual cortex (Area 17) in the cat. Exp Brain Res 21:315−336
Eccles JC (1964) The physiology of synapses. Springer, Berlin Göttingen Heidelberg New York
Eccles JC (1979) Das Gehirn des Menschen. Piper, München Zürich
Eccles JC, Ito M, Szentagothai J (1967) The cerebellum as a neuronal machine. Springer, New York Berlin Heidelberg
Eckert R, Randall D (1986) Tierphysiologie. Thieme, Stuttgart New York
McGeer PL, Eccles JC, McGeer EG (1987) Molecular neurobiology of the mammalian brain. Plenum Press, New York London
Goldman DE (1943) Potential, impedance and rectification in membranes. J Gen Physiol 27: 36−60
Gustafsson B, Wigstroem H (1988) Physiological mechanisms underlying long-term potentiation. Trends in Neurosciences 11 (4):156−162
Hodgkin AL (1964) The conduction of the nerve impulse. Liverpool University Press, Liverpool
Ito M (1984) The cerebellum and neural control. Raven Press, New York
Kandel ER (1984) Kleine Verbände von Nervenzellen. In: Gehirn und Nervensystem. Spektrum der Wissenschaft: 77−85
Kandel ER, Schwartz JH (eds) (1985) Principles

of neural science, 2nd edn. Elsevier Science, Amsterdam

Katz B (1971) Nerv, Muskel und Synapse. Thieme, Stuttgart

Krnjevic K (1974) Chemical nature of synaptic transmission in vertebrates. Physiol Rev 54: 418–505

Lajtha A (ed) (1979–1986) Handbook of neurochemistry, vol I–VII. Plenum Press, New York

Morell P, Norton WT (1984) Myelin. In: Gehirn und Nervensystem. Spektrum der Wissenschaft: 65–74

Popper KR, Eccles JC (1982) Das Ich und sein Gehirn. Piper, München Zürich

Rahmann H (1976) Neurobiologie. Ulmer, Stuttgart (UTB)

Reckhaus W, Rahmann H (1983) Longterm thermal adaptation of evoked potentials in fish brain. J Therm Biol 8:456–457

Regan D (1984) Reaktionspotentiale im menschlichen Hirn. In: Gehirn und Nervensystem. Spektrum der Wissenschaft: 144–151

Schadé J (1973) Die Funktion des Nervensystems. Fischer, Stuttgart

Schmidt RF (Hrsg) (1987) Grundriß der Neurophysiologie, 6. Aufl. Springer, Berlin Heidelberg New York London Paris Tokyo (HTB)

Schmidt RF, Thews G (Hrsg) (1980) Physiologie des Menschen, 20. Aufl. Springer, Berlin Heidelberg New York

Segler K, Rahmann H, Rösner H (1978) Chemotaxonomical investigations on the occurrence of sialic acids in Protostomia and Deuterostomia. Biochem System Ecol 6:87–93

Shepherd GM (1978) Microcircuits in the nervous system. Sci Am 238 (2):92–103

Shepherd GM (1983) Neurobiology. Oxford University Press, New York Oxford

Sherrington CS (1947) The integrative action of the nervous system, 2nd edn. Yale University Press, New Haven

Singer W (1985) Hirnentwicklung und Umwelt. Spektrum der Wissenschaft: 48–61

Sinz R (1978) Gehirn und Gedächtnis. Fischer, Stuttgart New York (UTB 852)

Smith SJ, Augustine GJ, Charlton MP (1985) Transmission at voltage-clamped giant synapse of the squid: Evidence for cooperativity of presynaptic calcium action. Proc Natl Acad Sci USA 82:622–625

Snyder SH (1985) Signalübertragung zwischen Zellen. Spektrum der Wissenschaft: 126–135

Stevens CF (1984) Die Nervenzelle. In: Gehirn und Nervensystem. Spektrum der Wissenschaft: 3–14

Wiesel TN, Hubel DH (1963) Single-cell responses in striate cortex of kittens deprived of vision in one eye. J Neurophysiol 26:1003–1017

Wiesel TN, Hubel DH (1971) Long-term changes in the cortex after visual deprivation. Proceedings of the 25th International Congress of Psychological Science

Gehirn und Nervensystem, 4. Aufl. (1984) Spektrum der Wissenschaft, Heidelberg

The Neuroscience I, II, III. The MIT Press, London Cambridge, Mass

Trends in Neurosciences, vol 11 (4) (1988) Special issue: Learning, Memory. Elsevier, Amsterdam Cambridge

Chapter 7: Chemical Aspects of Neuronal Information Transmission in Synapses

Abrams TW, Kandel ER (1985) Roles of calcium and adenylate cyclase in activity-dependent facilitation, a cellular mechanism for classical conditioning in Aplysia. Neurosci Abstr.

McGeer PL, Eccles JC, McGeer EG (1987) Molecular neurobiology of the mammalian brain. Plenum Press, New York London

Gibson GE, Peterson C (1985) Calcium and the aging nervous system. Neurobiol Aging 8: 329–343

Hayashi K, Mühleisen M, Probst W, Rahmann H (1984) Binding of (Ca^{2+}) to phosphoinositols, phosphatidyl-serines and gangliosides. Chem Phys Lipids 34:317–322

Hucho F (1986) Neurochemistry, fundamentals and concepts. VCH, Weinheim

Iversen LL (1984) Die Chemie der Signalübertragung im Gehirn. In: Gehirn und Nervensystem. Spektrum der Wissenschaft: 21–31

Kandel ER (1981) Calcium and the control of synaptic strength by learning. Nature 293: 697–700

Kandel ER (1984) Kleine Verbände von Nervenzellen. In: Gehirn und Nervensystem. Spektrum der Wissenschaft: 77–85

Kandel ER, Schwartz JH (eds) (1985) Principles of neural science, 2nd edn. Elsevier Science, Amsterdam

Katz JJ, Halstead WC (1950) Protein organisation and mental function. Comp Psychol Monogr 20:1–38

Keynes RD (1984) Ionenkanäle in Nervenmembranen. In: Gehirn und Nervensystem. Spektrum der Wissenschaft: 15–19

Krnjevic K (1974) Chemical nature of synaptic transmission in vertebrates. Physiol Rev 54: 418–505

Mühleisen M, Probst W, Hayashi K, Rahmann H (1983) Calcium binding to liposomes composed of negatively charged lipid moieties. Jpn J Exp Med 53:103–107

Nishizuka Y (1984) Turnover of phospholipids and signal transduction. Science 225: 1365–1370

Patterson PH, Potter DD, Furshpan EJ (1984) Chemische Differenzierung von Nervenzellen. In: Gehirn und Nervensystem. Spektrum der Wissenschaft: 45−62

Pfenninger KH (1973) Synaptic morphology and cytochemistry. Fischer, Stuttgart

Popper KR, Eccles JC (1982) Das Ich und sein Gehirn. Piper, München Zürich

Rahmann H (1976) Neurobiologie. Ulmer, Stuttgart (UTB)

Rahmann H (1983) Functional implication of gangliosides in synaptic transmission (Critique). Neurochemistry International 5: 539−547

Rahmann H (1983) Lernen und Gedächtnis sowie Aspekte der Gedächtnissteigerung vom Standpunkt der Neurobiologie. In: Fischer B, Lehrl S (Hrsg) Gehirn-Jogging (biologische und informationspsychologische Grundlagen des zerebralen Jogging). Narr, Tübingen, S 28−44

Rahmann H (1985) Hirnganglioside der Wirbeltiere und ihre funktionelle Bedeutung bei der synaptischen Informationsübertragung. In: Evolution, Festschrift f. Bernhard Rensch. Aschendorff, Münster (Schriftenreihe d. Westf. Wilhelmsuniversität Münster, Bd 4, S 8−50)

Rahmann H (1985) Gedächtnisbildung durch molekulare Bahnung in Synapsen mit Gangliosiden. Funkt Biol Med 4:249−261

Rahmann H, Probst W (1986) Ultrastructural localization of calcium at synapses and modulatory interactions with gangliosides. In: Tettamanti G, Ledeen RW, Sandhoff K, Nagai Y, Toffano G (eds) Gangliosides and neuronal plasticity. Liviana Press, Padova (Fidia Research Series, pp 125−135)

Routtenberg A (1984) Das Belohnungssystem des Gehirns. In: Gehirn und Nervensystem. Spektrum der Wissenschaft: 160−167

Schadé J (1973) Die Funktion des Nervensystems. Fischer, Stuttgart

Schmidt RF (Hrsg) (1987) Grundriß der Neurophysiologie, 6. Aufl. Springer, Berlin Heidelberg New York London Paris Tokyo (HTB)

Schmidt RF, Thews G (Hrsg) (1980) Physiologie des Menschen, 20. Aufl. Springer, Berlin Heidelberg New York

Shapiro E, Castellucci VF, Kandel ER (1980) Presynaptic inhibition in Aplysia involves a decrease in the Ca^{2+}-current of the presynaptic neuron. Proc Natl Acad Sci USA 77: 1185−1189

Shepherd GM (1983) Neurobiology. Oxford University Press, New York Oxford

Singer W (1985) Hirnentwicklung und Umwelt. Spektrum der Wissenschaft: 48−61

Sinz R (1978) Gehirn und Gedächtnis. Fischer, Stuttgart New York (UTB 852)

Smith SJ, Augustine GJ, Charlton MP (1985) Transmission at voltage-clamped giant synapse of the squid: Evidence for cooperativity of presynaptic calcium action. Proc Natl Acad Sci USA 82:622−625

Snyder SH (1985) Signalübertragung zwischen Zellen. Spektrum der Wissenschaft: 126−135

Stevens CF (1984) Die Nervenzelle. In: Gehirn und Nervensystem. Spektrum der Wissenschaft: 3−14

Weinberg CB, Sanes JR, Hall ZW (1981) Formation of neuromuscular junction in adult rats: accumulation of acetylcholine receptors, acetylcholinesterase, and compounds of synaptic basal lamina. Dev Biol 84:255−266

Whittaker VP, Gray EG (1962) The synapse: Biology and morphology. Br Med Bull 18: 223−228

Whittaker VP, Michaelson IA, Kirkland RJA (1964) The separation of synaptic vesicles from nerve ending particles (synaptosomes). Biochem J 90:293−303

Gehirn und Nervensystem, 4. Aufl. (1984) Spektrum der Wissenschaft, Heidelberg

The Neuroscience I, II, III. The MIT Press, London Cambridge, Mass

Trends in Neurosciences, vol 11 (4) (1988) Special issue: Learning, Memory. Elsevier, Amsterdam Cambridge

Chapter 8: Modulation of Neuronal Information Transmission

Berridge M (1986) Second messenger dualism in neuromodulation and memory. Nature 323: 294−295

Hayashi K, Mühleisen M, Probst W, Rahmann H (1984) Binding of (Ca^{2+}) to phosphoinositols, phosphatidyl-serines and gangliosides. Chem Phys Lipids 34:317−322

Hilbig R, Lauke G, Rahmann H (1983/84) Brain gangliosides during the life span (embryogenesis to senescence) of the rat. Dev Neurosci 6: 260−270

Hilbig R, Rahmann H (1980) Variability in brain gangliosides of fishes. J Neurochem 34: 236−240

Hilbig R, Rahmann H (1987) Phylogeny of vertebrate brain gangliosides. In: Rahmann H (ed) Gangliosides and Modulation of Neuronal Functions. Springer, Berlin Heidelberg New York London Paris Tokyo (NATO ASI Series H: Cell Biology, vol 7, pp 373−390)

Hilbig R, Rösner H, Merz G, Segler-Stahl K, Rahmann H (1982) Developmental profile of gangliosides in mouse and rat cerebral cortex. Roux's Arch Dev Biol 191:281−284

Kandel ER, Schwartz JH (eds) (1985) Principles of neural science, 2nd edn. Elsevier Science, Amsterdam

Mühleisen M, Probst W, Wiegandt H, Rahmann H (1979) In-vitro-studies on the influence of cations, neurotransmitters and tubocurarine on calcium-ganglioside-interactions. Life Sci 25: 791–796

Mühleisen M, Probst W, Hayashi K, Rahmann H (1983) Calcium binding to liposomes composed of negatively charged lipid moieties. Jpn J Exp Med 53:103–107

Popper KR, Eccles JC (1982) Das Ich und sein Gehirn. Piper, München Zürich

Probst W, Möbius D, Rahmann H (1984) Modulatory effects of different temperatures and Ca^{2+} concentrations on gangliosides and phospholipids in monolayers at air/water interfaces and their possible functional role. Cell Mol Neurobiol 4 (2):157–176

Probst W, Rahmann H, Rösner H (1977) In vitro studies of neuronal Ca^{2+}-ganglioside complexes. IRCS Medical Science 5:124

Probst W, Rahmann H (1987) Peculiarities of ganglioside-Ca^{2+}-interactions. In: Rahmann H (ed) Gangliosides and modulation of neuronal functions. Springer, Berlin Heidelberg New York London Paris Tokyo (NATO ASI Series H: Cell Biology, vol 7, pp 139–154)

Rahmann H (1976) Neurobiologie. Ulmer, Stuttgart (UTB)

Rahmann H (1978) Gangliosides and thermal adaptation in vertebrates (review). Jpn J Exp Med 48 (2):85–96

Rahmann H (1979) The possible functional role of gangliosides for synaptic transmission and memory formation. In: Matthies H, Krug M, Popov N (eds) Biological aspects of learning, memory formation and ontogeny of the CNS. Abhdlg. Akad. Wiss. DDR, Akademie Verlag, Berlin, pp 83–110

Rahmann H (1981) Die Bedeutung der Hirnganglioside bei der Temperaturadaptation. Zool Jb Physiol 85:209–248

Rahmann H (1982) Correlations among neuronal ganglioside metabolism, bioelectrical activity and memory formation in teleost fishes. In: Marsen CA, Matthies H (eds) Neuronal plasticity and memory formation. Raven Press, New York, pp 203–211

Rahmann H (1983) Functional implication of gangliosides in synaptic transmission (Critique). Neurochemistry International 5: 539–547

Rahmann H (1983) Lernen und Gedächtnis sowie Aspekte der Gedächtnissteigerung vom Standpunkt der Neurobiologie. In: Fischer B, Lehrl S (Hrsg) Gehirn-Jogging (biologische und informationspsychologische Grundlagen des zerebralen Jogging). Narr, Tübingen, S 28–44

Rahmann H (1984) Memory formation by means of molecular facilitation in synapses with Ca^{2+}-ganglioside modulation. Jpn J Neuropsycho-

pharmacol 6:383–391

Rahmann H (1985) Hirnganglioside der Wirbeltiere und ihre funktionelle Bedeutung bei der synaptischen Informationsübertragung. In: Evolution, Festschrift f. Bernhard Rensch. Aschendorff, Münster (Schriftenreihe d. Westf. Wilhelmsuniversität Münster, Bd 4, S 8–50)

Rahmann H (1985) Gedächtnisbildung durch molekulare Bahnung in Synapsen mit Gangliosiden. Funkt Bio Med 4:249–261

Rahmann H (1986) Brain gangliosides: Neuromodulator for synaptic transmission and memory formation. In: Matthies H (ed) Learning and memory. Mechanisms of information storage in the nervous system. Pergamon Press, New York, pp 235–245

Rahmann H (ed) (1987) Gangliosides and modulation of neuronal functions. Springer, Berlin Heidelberg New York London Paris Tokyo (NATO ASI Series H: Cell Biology, vol 7)

Rahmann H (1987) Brain gangliosides, bio-electrical activity and poststimulation effects. In: Rahmann H (ed) Gangliosides and modulation of neuronal functions. Springer, Berlin Heidelberg New York London Paris Tokyo (NATO ASI Series H: Cell Biology, vol 7, pp 501–521)

Rahmann H, Hilbig R (1983) Phylogenetical aspects of brain gangliosides in vertebrates. J Comp Physiol 151:215–224

Rahmann H, Probst W, Mühleisen M (1982) Gangliosides and synaptic transmission (review). Jpn J Exp Med 52:275–286

Rahmann H, Rösner H, Breer H (1975) Sialomacromolecules in synaptic transmission and memory formation. IRCS Medical Science Forum 3:110–112

Rahmann H, Rösner H, Breer H (1976) A functional model of sialo-glycomacromolecules in synaptic transmission and memory formation. J Theor Biol 57:231–237

Rahmann H, Schneppenheim R, Hilbig R, Lauke G (1984) Variability in brain ganglioside composition: A further molecular mechanism beside serum antifreeze-glycoprotein for adaptation to cold in antarctic and arctic-boreal fishes. Polar Biol 3:119–125

Rösner H, Rahmann H (1987) Ontogeny of vertebrate brain gangliosides. In: Rahmann H (ed) Gangliosides and modulation of neuronal functions. Springer, Berlin Heidelberg New York London Paris Tokyo (NATO ASI Series H: Cell Biology, vol 7, pp 373–390)

Rösner H, Willibald CJ, Schwarzmann B, Rahmann H (1987) Uptake of exogenous gangliosides by the CNS? In: Rahmann H (ed) Gangliosides and modulation of neuronal functions. Springer, Berlin Heidelberg New York London Paris Tokyo (NATO ASI Series H: Cell Biology, vol 7, pp 581–592)

Seybold U, Rahmann H (1985) Brain gangliosides in birds with different types of postnatal
development (nidifugous and nidicolous type).
Dev Brain Res 17:201—208

Seybold V, Rahmann H (1985) Changes in developmental profiles of brain gangliosides during
ontogeny of a teleost fish (Sarotherodon mossambisus, Cichlidae). Roux's Arch Dev Biol
194:166—172

Smith SJ, Augustine GJ, Charlton MP (1985)
Transmission at voltage-clamped giant synapse
of the squid: Evidence for cooperativity of presynaptic calcium action. Proc Natl Acad Sci
USA 82:622—625

Gehirn und Nervensystem, 4. Aufl. (1984) Spektrum der Wissenschaft, Heidelberg

Chapter 9: Neuronal Plasticity

Allen RD, Weiss DG (1987) Mikrotubuli als in-
trazelluläres Transportsystem. Spektrum der
Wissenschaft: 76—85

Barondes SH (1969) Cellular dynamics of the
neuron. Academic Press, New York

Breer H, Rahmann H (1974) Axonal transport of
(3H) glucose radioactivity in the optic system
of Scardinius erythophthalmus. J Neurochem
22:245—250

Changeux J, Danchin A (1976) Selective stabili-
sation of the developing synapses as a mechanism for the specification of neuronal networks
(review). Nature 264:705—711

Eckert R, Randall D (1986) Tierpysiologie.
Thieme, Stuttgart New York

Greenough WT (1984) Structural correlates of information storage in the mammalian brain: a
review and hypothesis. Trends in Neurosciences 7 (7):229—233

Greenough WT, Bailey CH (1988) The anatomy
of the memory: convergence of results across a
diversity of tests. Trends in Neurosciences 11
(4):142—146

Jeserich G, Rahmann H (1979) Effect of light deprivation on fine structural changes in the optic
tectum of the rainbow trout (Salmo gairdneri
Rich.) during ontogeny. Dev Neurosci 2:
19—24

Kandel ER (1976) Cellular basis of behavior.
Freeman, San Francisco

Kandel ER (1981) Calcium and the control of
synaptic strength by learning. Nature 293:
697—700

Kandel ER (1984) Kleine Verbände von Nervenzellen. In: Gehirn und Nervensystem. Spektrum der Wissenschaft: 77—85

Kandel ER, Castelluci VF, Goelet P, Schacher S
(1987) Cell-biological interrelationships between short-term and long-term memory. In:

Kandel ER (ed) Molecular neurobiology in
neurology and psychiatry. Raven Press, New
York, pp 111—132

Kandel ER, Schwartz JH (eds) (1985) Principles
of neural science, 2nd edn. Elsevier Science,
Amsterdam

Nottebohm F (1975) Vocal behavior in birds. In:
Farner DS, King JR (eds) Avian biology, vol
V. Academic Press, New York, pp 287—332

Nottebohm F (1980) Brain pathways for vocal
learning in birds: A review of the first 10 years.
In: Sprague JM, Epstein AN (eds) Progress in
psychobiology and psychology. Academic
Press, New York

Popper KR, Eccles JC (1982) Das Ich und sein
Gehirn. Piper, München Zürich

Rahmann H (1965) Zum Stofftransport im Zentralnervensystem der Vertebraten. Autoradiographische Untersuchungen mit P-32-Ortho-
phosphat, H-3-Histidin, H-3-Cytidin und H-3-
Uridin an Mäusen und Fischen. Z Zellforsch
66:878—890

Rahmann H (1967) Darstellung des intraneuro-
nalen Proteintransports vom Auge in das Tectum opticum und die Cerebrospinalflüssigkeit
von Teleosteern nach intraocularer Injektion
von 3H-Histidin. Naturwissenschaften 54:
174—175

Rahmann H (1968) Syntheseort und Ferntrans-
port von Proteinen im Fischhirn. Z Zellforsch
86:214—237

Rahmann H (1970) Entstehungsorte und Ver-
bleib von Syntheseprodukten im Zentralnervensystem von Vertebraten (Übersichtsrefe-
rat). Zool Anz [Suppl] 33:430—460 (Verh
Dtsch Zool Ges, Würzburg 1969)

Rahmann H (1970) Transport von 3H-Palmitin-
säure im ZNS von Teleosteern. Z Zellforsch
110:444—456

Rahmann H (1971) Different modes of substance
flow in the optic tract. Acta Neuropathol
[Suppl V]:1962—1970

Rahmann H (1973) Rolltreppe Nervenzelle. Bild
der Wissenschaft 10:1130—1136

Rahmann H (1976) Neurobiologie. Ulmer, Stuttgart (UTB)

Rahmann H (1979) The possible functional role
of gangliosides for synaptic transmission and
memory formation. In: Matthies H, Krug M,
Popov N (eds) Biological aspects of learning,
memory formation and ontogeny of the CNS.
Abhdlg. Akad. Wiss. DDR. Akademie Verlag, Berlin, pp 83—110

Rahmann H (1981) Die Bedeutung der Hirnganglioside bei der Temperaturadaptation. Zool
Jb Physiol 85:209—248

Rahmann H (1982) Correlations among neuronal
ganglioside metabolism, bioelectrical activity
and memory formation in teleost fishes. In:
Marsen CA, Matthies H (eds) Neuronal plasti-

city and memory formation. Raven Press, New York, pp 203−211

Rahmann H (1983) Functional implication of gangliosides in synaptic transmission (Critique). Neurochemistry International 5: 539−547

Rahmann H (1983) Lernen und Gedächtnis sowie Aspekte der Gedächtnissteigerung vom Standpunkt der Neurobiologie. In: Fischer B, Lehrl S (Hrsg) Gehirn-Jogging (biologische und informationspsychologische Grundlagen des zerebralen Jogging). Narr, Tübingen, S 28−44

Rahmann H (1985) Hirnganglioside der Wirbeltiere und ihre funktionelle Bedeutung bei der synaptischen Informationsübertragung. In: Evolution, Festschrift f. Bernhard Rensch. Aschendorff, Münster (Schriftenreihe d. Westf. Wilhelmsuniversität Münster, Bd 4, S 8−50)

Rahmann H (1985) Gedächtnisbildung durch molekulare Bahnung in Synapsen mit Gangliosiden. Funkt Biol Med 4:249−261

Rahmann H (ed) (1987) Gangliosides and modulation of neuronal functions. Springer, Berlin Heidelberg New York London Paris Tokyo (NATO ASI Series H: Cell Biology, vol 7)

Rahmann H (1987) Brain gangliosides, bio-electrical activity and poststimulation effects. In: Rahmann H (ed) Gangliosides and modulation of neuronal functions. Springer, Berlin Heidelberg New York London Paris Tokyo (NATO ASI Series H: Cell Biology, vol 7, pp 501−521)

Rahmann H, Probst W, Mühleisen M (1982) Gangliosides and synaptic transmission (review). Jpn J Exp Med 52:275−286

Rahmann H, Probst W (1986) Ultrastructural localization of calcium at synapses and modulatory interactions with gangliosides. In: Tettamanti G, Ledeen RW, Sandhoff K, Nagai Y, Toffano G (eds) Gangliosides and neuronal plasticity. Liviana Press, Padova (Fidia Research Series, pp 125−135)

Reckhaus W, Rahmann H (1983) Longterm thermal adaptation of evoked potentials in fish brain. J Therm Biol 8:456−457

Schaefer C (1987) Gehirnzellen sterben nicht ab. Bild der Wissenschaft 9:60−69

Schauer R (ed) (1982) Sialic acids. Chemistry, metabolism and function. Springer, Wien New York

Schönharting M, Breer H, Rahmann H, Siebert G, Rösner H (1977) Colchiceine, a novel inhibitor of the fast axonal transport without tubulin binding properties. Cytobiologie 16: 106−117

Sester U, Probst W, Rahmann H (1984) Einfluß unterschiedlicher Akklimationstemperaturen auf die Ultrastruktur neuronaler Synapsen von Buntbarschen (Tilapia mariae; Cichlidae, Teleostei). Z Hirnforsch 25 (6):701−711

Simon H (1981) Geht es auch ohne Gehirn? Naturwiss Rdsch 34 (3):126

Singer W (1985) Hirnentwicklung und Umwelt. Spektrum der Wissenschaft: 48−61

Teyler TJ, Discenna P (1985) The role of hippocampus in memory: A hypothesis. Neurosci Behav Rev 9:377−389

Voronin LL (1983) Longterm potentiation in the hippocampus. Neuroscience 10 (4):1021−1069

Vrenzen G, Nunes Cardozo J, Mueller L, Van der Went J (1980) The presynaptic grid: A new approach. Brain Res 184:23−40

Weiss PA (1934) In vitro experiments on the factors determining the course of the outgrowing nerve fibre. J Exp Zool 68:393−448

Weiss PA (1969) Neuronal dynamics and axonal flow. In: Barondes SH (ed) Symposium of the International Society of Cell Biology, New York

Weiss PA, Hiscoe HB (1948) Experiments on the mechanism of nerve growth. J Exp Zool 107: 315−396

Wiegandt H (ed) (1985) Glycolipids. Elsevier, Amsterdam New York Oxford

Wiesel TN, Hubel DH (1963) Single-cell responses in striate cortex of kittens deprived of vision in one eye. J Neurophysiol 26:1003−1017

Wiesel TN, Hubel DH (1971) Long-term changes in the cortex after visual deprivation. Proceedings of the 25th International Congress of Psychological Science

Yamakawa T, Nagai Y (1978) Glycolipids at the cell surface and their biological functions. Trends in Biochem Sci 3:127−132

Zeutzius I, Probst W, Rahmann H (1984) Influence of dark-rearing on the ontogenetic development of Sarotherodon mossambicus (Cichlidae, Teleostei): II Effects on allometrical growth relations and differentiation of the optic tectum. Exp Biol 43:87−96

Zeutzius I, Rahmann H (1980) Quantitative ultrastructural investigations on synaptogenesis in the cerebellum and the optic tectum of light-reared and dark-reared rainbow trout (Salmo gairdneri Rich.). Differentiation 17:181−186

Zeutzius I, Rahmann H (1980) Synaptogenesis in cerebellum and optic tectum of dark- and light-reared rainbow trout. IRCS Medical Science 8:47−48

Gehirn und Nervensystem, 4. Aufl. (1984) Spektrum der Wissenschaft, Heidelberg

The Neuroscience I, II, III. The MIT Press, London Cambridge, Mass

Chapter 10: Behavioral-Physiological Basis of Memory

Alkon DL (1987) Memory traces in the brain. Cambridge University Press, Cambridge, p 261

Barnett SA (1971) Instinkt und Intelligenz. Fischer, Frankfurt

Basar E (1988) Dynamics of sensory and cognitive processing by the brain. Springer, Berlin Heidelberg New York London Paris Tokyo

Bouer GH, Hawkins L (eds) (1988) The psychology of learning and motivation. Wiley, Sussex

Buchholtz Ch (1973) Das Lernen bei Tieren. Fischer, Stuttgart

Byrne J, Berry W (eds) (1988) Neural models of plasticity. Academic Press, New York

Changeux JP, Konishi M (1982) Animal mind − human mind (Dahlem Konferenzen). Springer, Berlin Heidelberg New York

Changeux JP, Konishi M (1986) Neural and molecular basis of learning (Dahlem Konferenzen). Springer, Berlin Heidelberg New York

Eibl-Eibesfeldt I (1969) Grundriß der vergleichenden Verhaltensforschung. Piper, München

Esser M (1963) Vermögen zum Erlernen von Handlungsrhythmen bei bin- und monokular sehenden Hühnern. Naturwissenschaften 50: 602−603

Ewert JP (1976) Neuro-Ethologie. Einführung in die neurophysiologischen Grundlagen des Verhaltens. Springer, Berlin Heidelberg New York

Franck D (1985) Verhaltensbiologie, 2. Aufl. Thieme, Stuttgart New York

Frank HG (1969) Kybernetische Grundlagen der Pädagogik, Bd 2. Agis, Baden-Baden

Guthrie DM (1980) Neuroethology (an introduction). Blackwell Scientific, Oxford London Edinburgh Boston Melbourne

Immelmann K (1976) Einführung in die Verhaltensforschung. Parey, Berlin Hamburg (Pareys Studientexte 13)

Kandel ER (1976) Cellular basis of behavior. An introduction to behavioral neurobiology. Freeman, San Francisco

Kandel ER, Schwartz JH (eds) (1985) Principles of neural science, 2nd edn. Elsevier Science, Amsterdam

Landfield PW, Readwyter SA (eds) (1988) Longterm potentiation: From biophysics to behavior. Liss, New York

Lashley KS (1919) Brain mechanisms and intelligence. A quantitative study of injuries to the brain. Chicago University Press, Chicago

Lashley KS (1950) In search of the engram. Symp Soc Exp Biol 4:454−481

Lorenz K (1950) The comparative method in studying innate behavior patterns. Symp Soc Exp Biol 4:221−268

Lorenz K (1965) Evolution and modification of behavior. Chicago University Press, Chicago

Nottebohm F (1975) Vocal behavior in birds. In: Farner DS, King JR (eds) Avian biology, vol V. Academic Press, New York, pp 287−332

Pawlov IP (1926) Die höchste Nerventätigkeit (das Verhalten) von Tieren. Bergmann, München

Pawlov IP (1927) Conditioned reflexes: An investigation of the physiological activity of the cerebral cortex. Oxford University Press, London

Peeke HVS, Merz MJ (1973) Habituation. Academic Press, New York

Ploog D, Gottwald P (1974) Verhaltensforschung. Instinkt − Lernen − Hirnfunktionen. Urban & Schwarzenberg, München Berlin Wien

Popper KR, Eccles JC (1982) Das Ich und sein Gehirn. Piper, München Zürich

Rahmann H (1961) Einfluß des Pervitins auf Gedächtnisleistungen, Verhaltensweisen und einige physiologische Funktionen von Goldhamstern. Pflügers Arch Ges Physiol 273:247−263

Rahmann H (1970) The influence of methamphetamine on learning, longterm memory and transposition ability in golden hamsters. In: Costa E, Garattini S (eds) Amphetamines and related compounds. Raven Press, New York, pp 813−817

Rahmann H (1976) Neurobiologie. Ulmer, Stuttgart (UTB)

Rahmann H (1983) Lernen und Gedächtnis sowie Aspekte der Gedächtnissteigerung vom Standpunkt der Neurobiologie. In: Fischer B, Lehrl S (Hrsg) Gehirn-Jogging (biologische und informationspsychologische Grundlagen des zerebralen Jogging). Narr, Tübingen, S 28−44

Rahmann H, Rahmann M (1972) Visual acuity in animals. Informa 31. Boehringer, Ingelheim

Rahmann H, Schmidt W, Schmidt B (1980) Influence of longterm thermal acclimation on the conditionability of fish. J Therm Biol 5:11−16

Rahmann-Esser M (1964) Erlernen rhythmischer Handlungsfolgen bei Hühnern. Z Tierpsychol 21 (7):837−853

Rahmann M (1983) Zur Bedeutung der Motivationslage für die höhere assoziative Hirntätigkeit. In: Fischer B, Lehrl S (Hrsg) Gehirn-Jogging (biologische und informationspsychologische Grundlagen des zerebralen Jogging). Narr, Tübingen, S 218−220

Rauschecker JP (1987) Imprinting and cortical plasticity. Wiley, Sussex, p 392

Rensch B (1954) Das Problem der Residuen bei Lernvorgängen. Westdeutscher Verlag, Köln Opladen

Rensch B (1962) Gedächtnis, Abstraktion und Generalisation. Westdeutscher Verlag, Köln

Opladen

Rensch B (1973) Gedächtnis, Begriffsbildung und Planhandlungen bei Tieren. Parey, Berlin Hamburg

Rensch B, Rahmann H (1960) Einfluß des Pervitins auf das Gedächtnis von Goldhamstern. Pflügers Arch Ges Physiol 271:693−704

Rensch B, Rahmann H (1966) Autoradiographische Untersuchung über visuelle „Engramm"-Bildung bei Zahnkarpfen. Pflügers Arch Ges Physiol 290:158−166

Rensch B, Rahmann H, Skrzipek KH (1968) Autoradiographische Untersuchungen über visuelle „Engramm"-Bildung bei Fischen (II). Pflügers Arch Ges Physiol 304:242−252

Sinz R (1978) Gehirn und Gedächtnis. Fischer, Stuttgart New York (UTB 852)

Squire LR (1987) Memory and brain. Oxford University Press, London

Squire LR, Zola-Morgan S (1988) Memory: brain system and behavior. Trends in Neurosciences 11 (4):170−175

Tembrock G (1974) Grundlagen der Tierpsychologie. Vieweg, Braunschweig

Terrace A, Marter P (1984) The biology of learning (Dahlem Konferenzen). Springer, Berlin Heidelberg New York

Thorndike EL (1911) Animal intelligence: Experimental studies. MacMillan, New York London

Thorpe WH (1964) Learning and instinct in animals. Methuen, London

Tinbergen N (1966) Instinktlehre. Parey, Berlin Hamburg

Vester F (1981) Denken, Lernen, Vergessen. Deutscher Taschenbuch Verlag, München

Weinberger N, Lynch G, McGaugh J (eds) (1985) Memory systems of the brain. Gilford Press, New York

Whitaker HA (ed) (1988) Phonological processes and brain mechanisms. Springer, Berlin Heidelberg New York London Paris Tokyo

Woody C, Alkon D, McGaugh JL (eds) (1988) Cellular mechanisms of conditioning and behavioral plasticity. Plenum Press, New York

Zimmermann H, Zimmermann E (1985) Der gelbe Pfeilgiftfrosch Phyllobates terribilis (Werbung und Eiablage). Aquarienmagazin 19:460−463

Gehirn und Nervensystem, 4. Aufl. (1984) Spektrum der Wissenschaft, Heidelberg

Trends in Neurosciences, vol 11 (4) (1988) Special issue: Learning, Memory. Elsevier, Amsterdam Cambridge

Chapter 11: Neurobiological Models of Memory

Abrams TW, Kandel ER (1988) Is contiguity detection in classical conditioning a system or a cellular property? Trends in Neurosciences 11: 128−135

Alkon DL (1984) Calcium-mediated reduction of ionic currents: A biophysical memory trace. Science 226:1037−1045

Berridge M (1986) Second messenger dualism in neuromodulation and memory. Nature 323: 294−295

Bingmann D (1984) Lernen und Gedächtnis. Neurophysiologische Grundlagen. Therapiewoche 34:7155−7162

Crow T (1988) Cellular and molecular analysis of associative learning and memory in Hermissenda. Trends in Neurosciences 11 (4): 136−141

Douglas RJ, Pribram KH (1966) Learning and limbic lesions. Neurophysiologica 4:197−220

McGeer PL, Eccles JC, McGeer EG (1987) Molecular neurobiology of the mammalian brain. Plenum Press, New York London

Gerke PR (1987) Wie denkt der Mensch? Bergmann, München, S 150

Greenough WT, Bailey CH (1988) The anatomy of the memory: convergence of results across a diversity of tests. Trends in Neurosciences 11 (4):142−146

Hydén H (1967) Biochemical change accompanying learning. In: Quarton GE, Melnechuk T, Schmitt FO (eds) The neurosciences. Rockefeller University Press, New York, pp 765−771

Hydén H (ed) (1967) The neuron. Elsevier, Amsterdam

Kandel ER (1976) Cellular basis of behavior. Freeman, San Francisco

Kandel ER (1981) Calcium and the control of synaptic strength by learning. Nature 293: 697−700

Kandel ER (1984) Kleine Verbände von Nervenzellen. In: Gehirn und Nervensystem. Spektrum der Wissenschaft: 77−85

Kandel ER, Schwartz JH (eds) (1985) Principles of neural science, 2nd edn. Elsevier Science, Amsterdam

Kandel ER, Castelluci VF, Goelet P, Schacher S (1987) Cell-biological interrelationships between short-term and long-term memory. In: Kandel ER (ed) Molecular neurobiology in neurology and psychiatry. Raven Press, New York, pp 111−132

Katz JJ, Halstead WC (1950) Protein organisation and mental function. Comp Psychol Monogr 20:1−38

Lashley KS (1950) In search of the engram. Symp Soc Exp Biol 4:454−481

Lynch G, Baudry M (1984) The biochemistry of

memory: A new and specific hypothesis. Science 224:1057–1063

Lynch G, Baudry M (1987) Brain spectrin, calpain and long-term changes in synaptic efficacy. Brain Res Bull 18:801–815

Menzel R (1983) Neurobiology of learning and memory: The honeybee as model system. Naturwissenschaften 70:504–511

Mishkin M, Appenzeller T (1987) Die Anatomie des Gedächtnisses. Spektrum der Wissenschaft (8):94–104

Ott T, Matthies H (1978) Lernen und Gedächtnis. In: Die Psychologie des 20. Jahrhunderts. Kindler, Zürich, S 988–1018

Palm G (1988) Assoziatives Gedächtnis und Gehirntheorie. Spektrum der Wissenschaft 6:54–64

Penfield W, Rasmussen T (1950) The cerebral cortex of man: Additional study on localisation of function. MacMillan, New York

Penfield W, Roberts L (1959) Speech and brain mechanisms. Princeton University Press, Princeton NJ

Popper KR, Eccles JC (1982) Das Ich und sein Gehirn. Piper, München Zürich

Rahmann H (1976) Neurobiologie. Ulmer, Stuttgart (UTB)

Rahmann H (1979) The possible functional role of gangliosides for synaptic transmission and memory formation. In: Matthies H, Krug M, Popov N (eds) Biological aspects of learning, memory formation and ontogeny of the CNS. Abhdlg. Akad. Wiss. DDR. Akademie Verlag, Berlin, pp 83–110

Rahmann H (1982) Die Bausteine der Erinnerung. Bild der Wissenschaft 19:74–86

Rahmann H (1982) Correlations among neuronal ganglioside metabolism, bioelectrical activity and memory formation in teleost fishes. In: Marsen CA, Matthies H (eds) Neuronal plasticity and memory formation. Raven Press, New York, pp 203–211

Rahmann H (1983) Functional implication of gangliosides in synaptic transmission (Critique). Neurochemistry International 5:539–547

Rahmann H (1983) Lernen und Gedächtnis sowie Aspekte der Gedächtnissteigerung vom Standpunkt der Neurobiologie. In: Fischer B, Lehrl S (Hrsg) Gehirn-Jogging (biologische und informationspsychologische Grundlagen des zerebralen Jogging). Narr, Tübingen, S 28–44

Rahmann H (1984) Lernen und Gedächtnis vom Standpunkt der Neurobiologie. Therapiewoche 34:7139–7154

Rahmann H (1984) Memory formation by means of molecular facilitation in synapses with Ca^{2+}-ganglioside modulation. Jpn J Neuropsychopharmacol 6:383–391

Rahmann H (1985) Hirnganglioside der Wirbel-

tiere und ihre funktionelle Bedeutung bei der synaptischen Informationsübertragung. In: Evolution, Festschrift f. Bernhard Rensch. Aschendorff, Münster (Schriftenreihe d. Westf. Wilhelmsuniversität Münster, Bd 4, S 8–50)

Rahmann H (1985) Gedächtnisbildung durch molekulare Bahnung in Synapsen mit Gangliosiden. Funkt Biol Med 4:249–261

Rahmann H (1986) Brain gangliosides: Neuromodulator for synaptic transmission and memory formation. In: Matthies H (ed) Learning and memory. Mechanisms of information storage in the nervous system. Pergamon Press, New York, pp 235–245

Rahmann H (ed) (1987) Gangliosides and modulation of neuronal functions. Springer, Berlin Heidelberg New York London Paris Tokyo (NATO ASI Series H: Cell Biology, vol 7)

Rahmann H (1987) Brain gangliosides, bio-electrical activity and poststimulation effects. In: Rahmann H (ed) Gangliosides and modulation of neuronal functions. Springer, Berlin Heidelberg New York London Paris Tokyo (NATO ASI Series H: Cell Biology, vol 7, pp 501–521)

Rahmann H, Rösner H, Breer H (1975) Sialomacromolecules in synaptic transmission and memory formation. IRCS Medical Science Forum 3:110–112

Rahmann H, Rösner H, Breer H (1976) A functional model of sialo-glycomacromolecules in synaptic transmission and memory formation. J Theor Biol 57:231–237

Rahmann H, Probst W, Mühleisen M (1982) Gangliosides and synaptic transmission (Review). Jpn J Exp Med 52:275–286

Rawlins JN (1985) Associations across time: The hippocampus as a temporary memory store. Behav Brain Sci 8:479–496

Rensch B (1954) Das Problem der Residuen bei Lernvorgängen. Westdeutscher Verlag, Köln Opladen

Rensch B (1973) Gedächtnis, Begriffsbildung und Planhandlungen bei Tieren. Parey, Berlin Hamburg

Rensch B, Rahmann H (1966) Autoradiographische Untersuchung über visuelle „Engramm"-Bildung bei Zahnkarpfen. Pflügers Arch Ges Physiol 290:158–166

Rensch B, Rahmann H, Skrzipek KH (1968) Autoradiographische Untersuchungen über visuelle „Engramm"-Bildung bei Fischen (II). Pflügers Arch Ges Physiol 304:242–252

Routtenberg A (1987) Phospholipid and fatty acid regulation of signal transduction at synapses: potential role for protein kinase C in information storage. J Neurol Transm 24:239–245

Shapiro E, Castellucci VF, Kandel ER (1980) Presynaptic inhibition in Aplysia involves a decrease in the Ca^{2+}-current of the presynaptic neuron. Proc Natl Acad Sci USA 77: 1185−1189

Shashoua VE (1982) Molecular and cell biological aspects of learning: toward a theory of memory. Adv Cell Neurobiol 3:97−141

Shashoua VE (1985) The role of brain extracellular proteins in learning and memory. In: Alkon DL, Woody CD (eds) Neural mechanisms of conditioning. Plenum Press, New York, pp 459−490

Sinz R (1978) Gehirn und Gedächtnis. Fischer, Stuttgart New York (UTB 852)

Smith SJ, Hughes H (1987) Progress on LTP at hippocampal synapses: a post-synaptic Ca^{2+} trigger for memory storage? Trends in Neurosciences 10 (4):142−144

Squire LR (1986) Mechanisms of memory. Science 232: 1612−1619

Squire LR, Zola-Morgan S (1988) Memory: brain system and behavior. Trends in Neurosciences 11 (4):170−175

Teyler TJ, Discenna P (1985) The role of hippocampus in memory: A hypothesis. Neurosci Behav Rev 9:377−389

Vester F (1981) Denken, Lernen, Vergessen. Deutscher Taschenbuch Verlag, München

Trends in Neurosciences, vol 11 (4) (1988) Special issue: Learning, Memory. Elsevier, Amsterdam Cambridge

Neuroscience Research (1986) Special issue 3: Synaptic Plasticity, Memory, and Learning (In Memory of the late Dr. Nakaakira Tsukahara), pp 469−698

Index